DISCRETE MATHEMATICS FOR COMPUTER SCIENCE

Kenneth Bogart
Dartmouth College

Clifford Stein
Columbia University

Robert L. Drysdale
Dartmouth College

Key College Publishing
Innovators in Higher Education

www.keycollege.com

Kenneth Bogart
Dartmouth College
Department of Mathematics
Hanover, NH 03755

Clifford Stein
Columbia University
Department of Industrial Engineering and
Operations Research
New York, NY 10027

Robert L. Drysdale
Dartmouth College
Department of Computer
Science
Hanover, NH 03755

Key College Publishing was founded in 1999 as a division of Key Curriculum Press® in cooperation with Springer Science and Business Media. We publish innovative texts and courseware for the undergraduate curriculum in mathematics and statistics as well as mathematics and statistics education. For more information, visit us at www.keycollege.com.

Key College Publishing
1150 65th Street
Emeryville, CA 94608
(510) 595-7000
info@keycollege.com
www.keycollege.com

This material is based upon work supported by the National Science Foundation initiative, Mathematical Sciences and Their Applications Throughout the Curriculum, under grant DUE-9552462 from the Division of Undergraduate Education. Any opinions, findings, and conclusions or recommendations expressed in this publication are those of the authors and do not necessarily reflect the views of the National Science Foundation.

Development Editor: Allyndreth Cassidy
Production Director: McKinley Williams
Production Coordinator: Ken Wischmeyer
Editorial Production Project Manager: Laura Ryan
Freelance Project Manager: Eric Houts
Copyeditor: Tara Joffe
Proofreader: Andrea Fox
Indexer: James Minkin
Text Designer: Charles Oey
Compositor: Interactive Composition Corporation
Illustrators: Interactive Composition Corporation
Art and Design Coordinator: Kavitha Becker
Cover Designer: Gary Harman
Cover Photo Credit: R.G.K. Photography/Gettyimages
Printer: RR Donnelley

Editorial Director: Richard Bonacci
General Manager: Mike Simpson
Publisher: Steven Rasmussen

Library of Congress Cataloging-in-Publication Data

Bogart, Kenneth P.
    Discrete mathematics for computer science / Kenneth Bogart, Clifford Stein, Robert L. Drysdale
        p. cm.
    Includes bibliographical references and index.
    ISBN 1-930190-86-7
        1. Computer science—Mathematics. I. Stein, Clifford, 1965- II. Drysdale, Robert L., 1951- III. Title.

QA76.9.M35B64 2005
004′.01′51—dc22   2005043351

Printed in the United States of America
10 9 8 7 6 5 4 3 2 1   09 08 07 06 05

This book is dedicated to our friend and co-author, Ken Bogart, whose untimely death on March 30, 2005, prevented his seeing it in final published form. Ken was the driving force behind this book, having guided it from conception through page proofs. We miss him.

Brief Contents

Detailed Contents

Preface

Our Motivation and Vision

Many colleges and universities offer a course in discrete mathematics. Students taking these courses are from many disciplines, one of the largest being computer science. As a part of the Mathematics Across the Curriculum project at Dartmouth[1] we proposed to create a discrete mathematics course that directly addresses the needs of computer science students. In analyzing what topics in discrete mathematics we want our computer science students to know and why we want them to know these topics, we made two observations. First, there are a few topics we consider important to computer science that are not always covered thoroughly, if at all, in traditional discrete mathematics courses. Among these are recursion trees, the relationship between induction and recursive problem decomposition, and randomized algorithms. Second, for each important discrete mathematics topic, there is a motivating computer science topic that can be understood at the level of a first or second computer science course. The second observation makes it possible to answer the age-old question students ask in applied mathematics courses: "Why do we have to learn this?" We therefore chose to write a textbook with computer science students in mind, with the objective of providing these students the foundation they need in discrete mathematics, motivated by computer science problems that students can understand early in their studies.

In many computer science departments, discrete mathematics is one of the first courses taken by majors. It may even be a prerequisite to the first computer science course. In this case, instructors are faced with a dilemma—teach the concepts purely mathematically with little or no visible application to computer science, or teach computer science examples to create a context relevant to computer science students. The first leaves students complaining that they are being forced to study too much "irrelevant" mathematics before they can take their first computer science course. The second leaves professors (who are often mathematicians) trying to explain fairly advanced computer science topics, such as hashing, binary trees, and recursive programs, to students who may have never written a program. Even under the best circumstances, this latter approach significantly reduces the depth of the mathematics that can be taught.

---

[1]Supported by the National Science Foundation grant DUE-9552462.

Our analysis led to a different approach, creating a course that appears slightly later in students' studies. This course should be taken after an introductory computer science course and either concurrently with or after a second computer science course. (This text is designed to enhance a course in data structures, but we discuss crucial issues in data structures that we feel are necessary to make it compatible with almost any first and second computer science course.) We feel that there are several advantages to this placement, including

- Students have already had serious experience writing code and thinking about problem solving and algorithms.

- Students have learned or are ready to learn several important computer science concepts, such as hashing, recursion, sorting and searching, and basic data structures.

- Students know enough computer science to comprehend the motivating examples, or the examples are straightforward enough for students to understand. For example

  - We use finding the minimum element of a list to motivate harmonic numbers.

  - We use hashing to motivate the study of probability.

  - We use the analysis of recursive programs, such as mergesort and quicksort, to motivate recurrence relations and their solutions.

  - We use binary trees to teach proof by induction on data structures, as well as to motivate the study of rooted trees as examples of graphs.

- In our own teaching experiences, in which this class is a prerequisite to an algorithms class and to several other computer science classes, students often take the algorithms course immediately after the discrete mathematics course. In doing so, they find themselves immediately using the mathematics they have just learned.

Our Educational Philosophy

This text is driven by activities presented as exercises. The material is then fleshed out through explanations and extensions of the exercises. The most effective way for students to use the book is to attempt the exercises seriously before reading the explanation that follows. Most of the exercises are designed to be done in groups; thus, even for exercises done out of class, we recommend that students form groups to work together. This style of course design helps students develop their own habits of mathematical thought. Our reading of the research in how undergraduate students learn mathematics leads us to several conclusions:

- Students who actively discover what they are learning (engaging in what is often called "active learning") remember concepts longer than those who don't. These students are also more likely to use those concepts outside the context in which they were learned.

- Students are more likely to ask questions until they understand a subject when they are working in a small group of peers rather than in a larger class with an instructor. (However, this isn't always the case. Many students need to feel comfortable with their group before they ask questions. We try to develop this comfort level in class by allowing students to choose their groups and to change from group to group on different days as attendance patterns allow or require. One can also use fixed heterogeneous grouping and have one group member responsible for checking to see if everyone understands.)

- Explaining ideas to someone else helps students organize ideas in their minds and familiarizes students with the language of mathematics.

In this text, though we do not explicitly assume students have taken calculus, we assume they have familiarity with, and we make significant use of, summation notation, unions and intersections of sets, logarithms, and exponential functions. Thus, a strong understanding of precalculus material is very helpful.[2]

The Role of Proofs

One of our purposes in writing this book is to give students a background for the kinds of proofs they will need to understand and do in subsequent courses. Our view is that students learn how to do proofs by hearing, seeing, discussing, and trying to do proofs. To discuss proofs, we need a common language that classifies the ingredients of a proof and provides a framework for discussion. For this reason, we have included Chapter 3 on logic, designed both to give students this language and to assist them in the process of reflecting on the proofs they have already seen. So that students have something significant to talk about in Chapter 3, we have introduced it after students have learned some combinatorial and number theoretic proofs. We also placed this chapter before Chapter 4 on mathematical induction so that we can use the language of logic and proofs to discuss and reflect on mathematical induction.

The Use of Pseudocode

We describe algorithms both in prose and by using pseudocode. The pseudocode should be easily readable by anyone who has programmed in a language such as Pascal, C, C++, or Java. We do not strive to give syntactically correct code in any language; rather, we strive for clarity. For example to say, "Swap the values held in variables x and y," we write, "exchange x and y," rather than writing three lines of code. Similarly, we write, "if points i, j, and k are not collinear" without concern for how a more detailed computation proceeds. We also do not worry about type consistency—that is, a function may return both a value and a message. Following are some particular conventions we use in the text.

- Blocks of code are denoted by indentation. There is no begin, end, or { }, as in many languages.
- Loops are written as "for i = 1 to n" to denote that the variable i ranges from 1 to n.

---

[2]Most of our students have had calculus. In isolated places, we make use of elementary derivatives and in optional subsections in probability, we use natural logarithm and exponential functions and elementary power series. By ignoring the few proofs or problems using derivatives, integrals, or power series and the optional subsections in probability, the instructor can avoid calculus.

- Arrays are subscripted using [ ].
- Assignment is represented with =, and comparison for equality with ==.
- Shorthand for incrementing and decrementing x is "x++" and "x- -."
- The logical operator "not" is symbolized with !, so "!true" is `false`, and "!(x < y)" is true when x is not less than y.

Acknowledgments

We would like to thank Eddie Cheng, Oakland University; Alice Dean, Skidmore College; Ruth Hass, Smith College; and Italo Dejter, University of Puerto Rico for their thoughtful review comments on an early version of the manuscript. As the book was being developed, preliminary versions were used to teach discrete mathematics at Dartmouth by the authors and by Neal Young, Prasad Jayanti, Tom Shemanske, Rosa Orellana, April Rasala, Amit Chakrabarti, Carl Pomerance, and Ramgopal Mettu. Each of them had an impact on the final product—some very substantial. We thank them for their advice. We offer a special thanks to Carl Pomerance for his thorough and insightful commentary as he taught the course. Qun Li was a graduate assistant to us as we were initially preparing the manuscript, and he had the job of making sure that the problems we created really did have solutions! His work still forms the core of the solutions available to the instructor. While we and others taught from the manuscript, the graduate teaching assistants from the Computer Science and Mathematics Departments gave us valuable insights into what students were and were not learning and provided further help with solutions to problems. In order of their service, they were S. Agrawal, Elishiva Werner-Reiss, Robert Savell, Virgiliu Pavlu, Libo Song, Geeta Chaudhry, King Tan, Yurong Xu, Gabriella Dumitrascu, Florin Constantin, Alin Popescu, Wei Zhang, and Paritosh Kavathekar. Our students and homework graders over the years have provided us with valuable feedback. In particular, Eric Robinson carefully read a near-final version, looking explicitly for passages that were hard to understand.

Each of the authors would like to thank the other two for the time they have taken from other professional activities to work on this project. Because of the time required to meld the points of view of our disciplines, it was only with the support of the National Science Foundation (Grant DUE-9552462) to the Dartmouth College Mathematics Across the Curriculum project that we were able to undertake this project. We believe that the staff of the Division of Undergraduate Education showed excellent insight into the needs of undergraduates and the difficulties of interdisciplinary curricular development when they conceived their program of Mathematical Sciences and Their Applications Throughout the Curriculum. We would like to acknowledge the positive impact this program had on undergraduate mathematics education and on the development of interdisciplinary collaborations in curriculum development. The authors would also like to thank their families for their patience, understanding, and support during the writing of this book.

To the Student

This book is organized in a nontraditional way, interleaving exercises with more traditional explanations. This organization was chosen to enhance student learning. Your instructor may ask you to work on the numbered exercises (not the end-of-section problems) in class. Your instructor might also ask you to work on the exercises before coming to class and to be prepared to discuss them with other students when you get to class. The reason for working on these exercises in class, and perhaps before class, is so that when the instructor starts the discussion of an idea in class, you will already have thought about it. When you have already tried to do something before your instructor starts talking about it, your mind is better prepared to engage the idea. Thus, instead of simply sitting in class and listening, you can be actively thinking about the issues the instructor is discussing. Our students have found this helps them stay engaged in the class, even if they don't solve every exercise. Our students have told us that although you may cover less material in a given class in this manner, you will understand it better and find it easier to learn related topics on your own. Most of our students have found they use class time more effectively when they are actually learning in class rather than simply taking notes and "digging it out for themselves" afterward.

To focus our students' attention on these exercises, we normally ask them to work in groups; your instructor may ask you to do so as well. Not only does working in groups help keep you "on topic," but it also makes it easier for you to ask questions than in a whole-class discussion. In addition, it lets the instructor understand how well the groups are understanding concepts by listening to your discussions. This process results in a significant improvement in student-instructor communication. Following group work, the instructor can tailor a discussion to the areas where students seem to have the most difficulty. Thus, although it may seem that your instructor is not "teaching" while you are working with others in class, in fact he or she is carrying out one of the most important teaching functions, namely, finding out what you know and need to know. If your class does not complete everything your instructor wants to cover on a given day, you may be asked to complete the work out of class, just as in any other course. We recommend that whenever possible, you get together in groups to work through these assigned topics, just as you would in class.

You will notice a difference between the homework problems in this book and those in many other texts. Some homework problems are quite similar to the class exercises and allow you to practice the basic ideas. Usually, however, the problems are designed to help you apply and extend what is covered in the section. Although you can do the problems using methods introduced in the section, or in earlier sections, you often will not be able to solve them simply by trying out formulas found in the section or using the exact same technique used in an example. This feature is intentional—the problems you see in applying discrete mathematics don't normally fit into specific patterns; instead, they require you to integrate your knowledge and think about how various solution techniques you previously learned might be applied. The homework problems take you one step in this direction. Some homework problems are marked by an a/h, indicating that there is an answer or hint in the back of the book. In the appendix of answers and hints, the problem number is given in black if the entry is an answer and in blue if it is a hint. Each section also ends with a list of important concepts that are intended to help you put the section into perspective.

Our students have told us that they use all the concepts they learn in our course in computer science courses that rely on discrete mathematics. Perhaps as importantly, they also find themselves using the learning habits and the problem-solving skills they develop in our course. We hope you will also find that your experience with this book helps you make the transition from a student of mathematics to a user of mathematics.

To the Instructor

There is ample material in the book for a four semester–hour course. At Dartmouth, we use the book for a fast-paced course that meets three days a week for just over nine weeks and that covers all but the last few sections of the book and leaves out some material marked with an asterisk.

Instructor CD

A CD that contains teaching suggestions is available to adopting instructors. On this CD, we discuss how to use the book in courses ranging from our quarter-long three-day-a-week course to a four semester–hour course. In addition to containing solutions to all homework problems, the CD also has a PDF version of the book, with links from the problems to the solutions for instructors who prefer to use it electronically. There is also a detailed discussion of how we have students work on exercises in groups in order to stimulate discussion. In particular, we come to class each day with a handout containing the exercises from a section of the book so that students can work on them in groups without needing to open their books. These handouts are included on the CD. We strongly recommend that instructors who have adopted the book or who are considering adopting the book request this CD.

For more information, call your Key College Publishing sales representative toll free at **888-877-7240**.

The Vision of Mathematics Across the Curriculum

Dear Reader,

In 1994, Dartmouth College received a generous grant from the National Science Foundation to integrate mathematics throughout the undergraduate college curriculum in a five-year project, Mathematics Across the Curriculum (MATC). The project has involved more than forty faculty members from Dartmouth and various other colleges and universities, representing departments of biology, chemistry, music, drama, English, art history, computer science, physics, earth science, economics, engineering, medicine, mathematics, and Spanish, producing lesson plans, short books, videotapes, and a web site with images and text. The series of volumes published by Key College Publishing represents some of the best of the MATC collection.

These materials will make it easier for students to become more quantitatively literate as they tackle complex, real-world problems that must be approached through the door of mathematics. We hope that you, the reader, will appreciate our efforts to place the mathematics in this book completely in the context of your field of interest. Our goal is to help you see that applied mathematics is a powerful form of inquiry, and ever so much richer than mere "word problems." We trust that you will like this approach and want to explore some of the other volumes in the series.

Sincerely,
Dorothy Wallace
Professor of Mathematics
Principal Investigator: Mathematics Across the Curriculum Project
Dartmouth College

1 COUNTING

1.1 BASIC COUNTING

The Sum Principle

In this book, we introduce ideas through exercises. Trying to figure out how to do each exercise will help you understand the explanation that follows. Our first exercise illustrates the sum principle.

Exercise 1.1-1

The following loop is part of an implementation of selection sort, which sorts a list of items chosen from an ordered set (numbers, alphabet characters, words, etc.) into nondecreasing order.

```
(1)    for i = 1 to n−1
(2)        for j = i+1 to n
(3)            if (A[i] > A[j])
(4)                exchange A[i] and A[j]
```

How many times is the comparison $A[i] > A[j]$ made in Line 3?

In Exercise 1.1-1, the segment of code from Lines 2–4 is executed $n − 1$ times, once for each value of i between 1 and $n − 1$, inclusive. The first time, it makes $n − 1$ comparisons. The second time, it makes $n − 2$ comparisons. The ith time, it makes $n − i$ comparisons. Thus, the total number of comparisons is

$$(n − 1) + (n − 2) + \cdots + 1 . \tag{1.1}$$

This formula is not as important as the reasoning that led to it. To put the reasoning into a broadly applicable format, we use the language of sets to describe what we are doing. Think about the set S containing all comparisons made by the algorithm in

Exercise 1.1-1, in which we divided set S into $n-1$ pieces (i.e., smaller sets): the set S_1 of comparisons made when $i = 1$, the set S_2 of comparisons made when $i = 2$, and so on through the set S_{n-1} of comparisons made when $i = n - 1$. We figured out the number of comparisons in each piece and then added the sizes of all the pieces to get the size of the set of all comparisons.

To describe a general version of this process, we now introduce some set-theoretic terminology. Two sets are called **disjoint** when they have no elements in common. Each set S_i described above is disjoint from each of the others because the comparisons made for one value of i are different from those made for another value of i. We say that the set of sets $\{S_1, \ldots, S_m\}$ (above, m was $n - 1$) is a family of **mutually disjoint sets**, to mean that it is a family (set) of sets, any two of which are disjoint. With this language, we can state a general principle that explains what we did without making any specific reference to the problem we solved.

Principle 1.1	**(Sum Principle)** The size of a union of a family of mutually disjoint finite sets is the sum of the sizes of the sets.

Thus, in effect, we used the sum principle to solve Exercise 1.1-1. We can also describe the sum principle using an algebraic notation. Let $|S|$ denote the size of the set S. For example, $|\{a, b, c\}| = 3$ and $|\{a, b, a\}| = 2$.[1] Using this notation, we can state the sum principle as follows. If $S_1, S_2, \ldots, S_m$ are disjoint sets, then

$$|S_1 \cup S_2 \cup \cdots \cup S_m| = |S_1| + |S_2| + \cdots + |S_m| . \tag{1.2}$$

We can also use the standard notation for union, as follows, to avoid writing the dots that indicate left-out material (as was done in Equation 1.2). The union notation is used exactly as summation notation and is read as "the union from i equals 1 to m of S sub i."

$$\left| \bigcup_{i=1}^{m} S_i \right| = \sum_{i=1}^{m} |S_i| .$$

When we can write a set S as a union of disjoint sets $S_1, S_2, \ldots, S_k$, we say that we have **partitioned** S into the sets $S_1, S_2, \ldots, S_k$ and that the sets $S_1, S_2, \ldots, S_k$ form a **partition** of S. Thus, $\{\{1\}, \{3, 5\}, \{2, 4\}\}$ is a partition of the set $\{1, 2, 3, 4, 5\}$, and the set $\{1, 2, 3, 4, 5\}$ can be partitioned into the sets $\{1\}, \{3, 5\}, \{2, 4\}$. However, it is clumsy to say we are partitioning a set into sets; instead, we call the sets S_i, into which we partition a set S, the **blocks** of the partition. Thus, the sets $\{1\}, \{3, 5\}, \{2, 4\}$ are the blocks of a partition of $\{1, 2, 3, 4, 5\}$. In this language, we can restate the sum principle as follows.

---

[1]It may look strange to have $|\{a, b, a\}| = 2$, but an element either is or is not in a set. An element cannot be in a set multiple times. (This situation leads to the idea of multisets, which are introduced in Section 1.4.) This example emphasizes that the notation $\{a, b, a\}$ means the same thing as $\{a, b\}$. Why would someone even contemplate the notation $\{a, b, a\}$? Suppose we wrote $S = \{x \mid x$ is the first letter of Ann, Bob, or Alice$\}$. Explicitly following this description of S would lead us to first write down $\{a, b, a\}$ and then realize it equals $\{a, b\}$.

(Sum Principle) If a finite set S has been partitioned into blocks, then the size of S is the sum of the sizes of the blocks.

Abstraction

The process of figuring out a general principle that explains why a certain computation makes sense is an example of the mathematical process of **abstraction**. In this book, we don't give a precise definition of abstraction; instead, we provide examples of the process as we proceed. In a course in set theory, we would further abstract our work and derive the sum principle from the axioms of set theory. In a course in discrete mathematics, however, this level of abstraction is unnecessary. We simply use the sum principle as the basis of computations when it is convenient to do so. If our goal were to solve only Exercise 1.1-1, then our abstraction would have been almost a mindless exercise that complicated what was an "obvious" solution. However, the sum principle will prove to be useful in a variety of problems. Thus, the value of abstraction is that recognizing the abstract elements of a problem often helps us solve subsequent problems.

Summing Consecutive Integers

Returning to the problem in Exercise 1.1-1, it would be nice to find a simpler form for the sum given in Equation 1.1. We may write this sum as

$$\sum_{i=1}^{n-1}(n-i) \; .$$

To avoid summing the values of $n - i$, we observe that the values we are summing are $n - 1, n - 2, \ldots, 1$; so, we may write

$$\sum_{i=1}^{n-1}(n-i) = \sum_{i=1}^{n-1} i \; .$$

A clever trick, usually attributed to Carl Friedrich Gauss, gives us a shorter formula for this sum:

$$
\begin{array}{ccccccc}
1 & + & 2 & + \cdots + & (n-2) & + & (n-1) \\
+ \, (n-1) & + & (n-2) & + \cdots + & 2 & + & 1 \\
\hline
n & + & n & + \cdots + & n & + & n
\end{array}
$$

The sum below the horizontal line has $n - 1$ terms, each equal to n. Thus, the sum is $n(n - 1)$, or the sum of the two sums above the line. Because these sums are equal (identical except for being in reverse order), the sum below the line must be twice either sum above. Therefore, either of the sums above the line must be $n(n - 1)/2$. In other words, we may write

$$\sum_{i=1}^{n-1}(n-i) = \sum_{i=1}^{n-1} i = \frac{n(n-1)}{2} \; .$$

This lovely trick is quite helpful in other similar situations involving a sum of variables. There are other ways to get the formula that don't use a trick. At the end of this section, after we analyze Exercise 1.1-2 and abstract the process used there, we will come back to this problem to see how we could have discovered this formula for ourselves without any tricks.

The Product Principle

Exercise 1.1-2

The following loop is part of a program that computes the product of two matrices. (You don't need to know how to find the product of two matrices to do this exercise.)

```
(1)   for i = 1 to r
(2)       for j = 1 to m
(3)           S = 0
(4)               for k = 1 to n
(5)                   S = S + A[i,k] * B[k,j]
(6)           C[i,j] = S
```

How many multiplications (expressed in terms of r, m, and n) does this pseudocode carry out in total among all the iterations of Line 5?

Exercise 1.1-3

Consider the following longer piece of pseudocode that sorts a list of numbers and then counts big gaps in the list. (For this exercise, a "big gap" is a place where a number in the list is more than twice the previous number.)

```
(1)   for i = 1 to n−1
(2)       minval = A[i]
(3)       minindex = i
(4)       for j = i to n
(5)           if (A[j] < minval)
(6)               minval = A[j]
(7)               minindex = j
(8)       exchange A[i] and A[minindex]
(9)   bigjump = 0
(10)  for i = 2 to n
(11)      if (A[i] > 2 * A[i−1])
(12)          bigjump = bigjump + 1
```

How many comparisons does the pseudocode make in Lines 5 and 11?

In Exercise 1.1-2, the program segment in Lines 4–5, which we call the "inner loop," takes exactly n steps. Thus, it makes n multiplications, regardless of what the variables i and j are. The program segment in Lines 2–5 repeats the inner loop exactly m times, regardless of what i is. Therefore, this program segment makes n multiplications m times, or nm multiplications.

Why did we add in Exercise 1.1-1 but multiply here? We can answer this question using the abstract point of view we adopted in discussing Exercise 1.1-1. The algorithm in Exercise 1.1-2 performs a certain set of multiplications. For any given i, the set of multiplications performed in Lines 2–5 can be divided into the set S_1 of multiplications performed when $j = 1$, the set S_2 of multiplications performed when $j = 2$, and, in general, the set S_j of multiplications performed for any given value of j. Each set S_j consists of those multiplications that the inner loop carries out for a particular value of j, and there are exactly n multiplications in this set. Let T_i be the set of multiplications that our program segment carries out for a certain value of i. The set T_i is the union of the sets S_j. We use the standard notation for unions to write

$$T_i = \bigcup_{j=1}^{m} S_j .$$

By the sum principle, the size of the set T_i is the sum of the sizes of the sets S_j. A sum of m numbers, each equal to n, is mn. Stated as an equation,

$$|T_i| = \left| \bigcup_{j=1}^{m} S_j \right| = \sum_{j=1}^{m} |S_j| = \sum_{j=1}^{m} n = mn . \tag{1.3}$$

Thus, we multiplied because multiplication is repeated addition.

From our solution, we extract a second principle that simply shortcuts the use of the sum principle.

Principle 1.3	**(Product Principle)** The size of a union of m disjoint sets, each of size n, is mn.

We now complete our discussion of Exercise 1.1-2. Lines 2–5 are executed once for each value of i from 1 to r. A different i value is used each time those lines are executed; so, the set of multiplications in one execution is disjoint from the set of multiplications in any other. Thus, the set of all multiplications that the program carries out is a union of r disjoint sets T_i, each of which consists of mn multiplications. By the product principle, the set of all multiplications has size rmn. Therefore, the program carries out rmn multiplications.

Exercise 1.1-3 shows us that thinking about whether the sum or product principle is appropriate for a problem can help decompose the problem into easily solvable pieces. If we can decompose the problem into smaller pieces and solve the smaller pieces, then we may be able to either add or multiply solutions to smaller problems in order to solve the larger problem. In this exercise, the number of comparisons in the program fragment is the sum of the number of comparisons in the first loop (Lines 1–8) and the number of comparisons in the second loop (Lines 10–12)—(What two disjoint sets are we talking about here?). Furthermore, the first loop makes $n(n + 1)/2 - 1$ comparisons,[2]

---

[2]To see why this is true, ask yourself first where the $n(n + 1)/2$ comes from and then why we subtracted 1.

and the second loop has $n - 1$ comparisons. The number of comparisons made by the fragment is $n(n + 1)/2 - 1 + n - 1 = n(n + 1)/2 + n - 2$ comparisons.

Two-Element Subsets

There are often several ways to solve a problem. We originally solved Exercise 1.1-1 using the sum principle, but it is also possible to solve it using the product principle. Solving a problem two ways not only increases our confidence that we have found the correct solution, but it can also allow us to make new connections and yield valuable insight.

Consider the set of comparisons made by the entire execution of the code in Exercise 1.1-1. When $i = 1$, variable j takes on every value from 2 to n. When $i = 2$, variable j takes on every value from 3 to n. Thus, for each two numbers i and j, we compare $A[i]$ and $A[j]$ exactly once in our loop. (The order in which we compare them depends on whether i or j is smaller.) Thus, the number of comparisons we make is the same as the number of two-element subsets of the set $\{1, 2, \ldots, n\}$.[3] In how many ways can we choose two elements from this set? If we choose a first and second element, there are n ways to choose a first element, and for each choice of the first element, there are $n - 1$ ways to choose a second element. Thus, the set of all such choices is the union of n sets of size $n - 1$, one set for each first element. It might appear that, by the product principle, there are $n(n - 1)$ ways to choose two elements from our set. However, what we have chosen is an **ordered pair**, or a pair of elements in which one comes first and the other comes second. For example, we could choose 2 first and 5 second to get the ordered pair $(2, 5)$, or we could choose 5 first and 2 second to get the ordered pair $(5, 2)$. Because each pair of distinct elements of $\{1, 2, \ldots, n\}$ can be ordered in two ways, we get twice as many ordered pairs as two-element sets. Thus, because the number of ordered pairs is $n(n - 1)$, the number of two-element subsets of $\{1, 2, \ldots, n\}$ is $n(n - 1)/2$. Therefore, the answer to Exercise 1.1-1 is $n(n - 1)/2$. This number comes up so often, it has its own name and notation: we call this number "n choose 2" and denote it by $\binom{n}{2}$. To summarize, $\binom{n}{2}$ stands for the number of two-element subsets of an n-element set and equals $n(n - 1)/2$. Because one answer to Exercise 1.1-1 is $1 + 2 + \cdots + (n - 1)$ and a second answer to Exercise 1.1-1 is $\binom{n}{2}$, we see that

$$1 + 2 + \cdots + (n - 1) = \binom{n}{2} = \frac{n(n - 1)}{2}.$$

Important Concepts, Formulas, and Theorems

1. *Set.* A *set* is a collection of objects. In a set, order is not important. Thus, the set $\{A, B, C\}$ is the same as the set $\{A, C, B\}$. An element either is or is not in a set; it cannot be in a set more than once, even if a description of a set names that element more than once.

---

[3]The relationship between the set of comparisons and the set of two-element subsets of $\{1, 2, \ldots, n\}$ is an example of a bijection, an idea that we examine more in Section 1.2.

2. *Disjoint.* Two sets are *disjoint* if they have no elements in common.

3. *Mutually disjoint sets.* A set of sets $\{S_1, \ldots, S_n\}$ is a family of *mutually disjoint sets* if each two of the sets S_i are disjoint.

4. *Size of a set.* Given a set S, the size of S, denoted $|S|$, is the number of distinct elements in S.

5. *Sum principle.* The size of a union of a family of mutually disjoint sets is the sum of the sizes of the sets. In other words, if $S_1, S_2, \ldots, S_n$ are disjoint sets, then

$$|S_1 \cup S_2 \cup \cdots \cup S_n| = |S_1| + |S_2| + \cdots + |S_n| .$$

To avoid the "dots" that indicate left-out material, we write

$$\left| \bigcup_{i=1}^{n} S_i \right| = \sum_{i=1}^{n} |S_i| .$$

6. *Partition of a set.* A *partition of a set* S is a set of mutually disjoint subsets (sometimes called *blocks*) of S whose union is S.

7. *Sum of first $n - 1$ numbers.*

$$\sum_{i=1}^{n-1} n - i = \sum_{i=1}^{n-1} i = \frac{n(n-1)}{2} .$$

8. *Product principle.* The size of a union of m disjoint sets, each of size n, is mn.

9. *Two-element subsets.* The number of *two-element subsets* of an n-element set, denoted $\binom{n}{2}$, equals $n(n-1)/2$. $\binom{n}{2}$ is read as "n choose 2."

Problems

1. The following segment of code is part of a program that uses insertion sort to sort a list A.

```
for i = 2 to n
    j = i
    while (j ≥ 2) and (A[j] < A[j−1])
        exchange A[j] and A[j−1]
        j = j−1
```

What is the maximum number of times (considering all lists of n items that you could be asked to sort) the program makes the comparison $A[j] < A[j - 1]$? Describe as succinctly as you can those lists that require this number of comparisons. a/h

2. Five schools are going to send their baseball teams to a tournament in which each team must play each other team exactly once. How many games are required?

3. In how many ways can you draw a first card and then a second card from a deck of 52 cards? a/h

4. In how many ways can you draw two cards from a deck of 52 cards? a/h

5. In how many ways can you draw a first, second, and third card from a deck of 52 cards? a/h

6. In how many ways can a ten-person club select a president and a secretary-treasurer from among its members? a/h

7. In how many ways can a ten-person club select a two-person executive committee from among its members? a/h

8. In how many ways can a ten-person club select a president and a two-person executive advisory board from among its members (assuming that the president is not on the advisory board)? a/h

9. Using the formula for $\binom{n}{2}$, it is straightforward to show that

$$n \binom{n-1}{2} = \binom{n}{2} (n-2).$$

However, this proof simply uses blind substitution and simplification. Find a more conceptual explanation of why this formula is true. (*Hint:* Think in terms of officers and committees in a club.) a/h

10. If M is an m-element set and N is an n-element set, how many ordered pairs are there with the first member in M and the second member in N?

11. The local ice cream shop sells ten different flavors of ice cream. How many different two-scoop cones are there? (Following your mother's rule that it all goes to the same stomach, a cone with a vanilla scoop on top of a chocolate scoop is considered the same as a cone with chocolate on top of vanilla.)

12. Suppose you decide to disagree with your mother in Problem 11—the order of the scoops does matter. How many different possible two-scoop cones are there? a/h

13. Suppose on Day 1 you receive one penny, and, for $i > 1$, on Day i you receive twice as many pennies as you did on Day $i - 1$. How many pennies will you have on Day 20? How many will you have on Day n? Can you justify your answer by using the sum or product principle?

14. The Pile High Deli offers a simple sandwich, consisting of your choice of one of five different kinds of bread; either butter, mayonnaise, or no spread; one of three different kinds of meat; and one of three different kinds of cheese, with the meat and cheese piled high on the bread. In how many ways can you choose a simple sandwich? a/h

15. Do you see any unnecessary steps in the pseudocode of Exercise 1.1-3? Explain.

1.2 COUNTING LISTS, PERMUTATIONS, AND SUBSETS

Using the Sum and Product Principles

Exercise 1.2-1

A password for a certain computer system is supposed to be between 4 and 8 characters long and composed of lowercase and/or uppercase letters. How many passwords are possible? What counting principles did you use? Estimate the percentage of the possible passwords that have exactly 4 letters.

A good way to attack a counting problem is to ask if we can use either the sum principle or the product principle to simplify or completely solve it. For this exercise, that question might lead us to think about the fact that a password can have 4, 5, 6, 7, or 8 characters. Because the set of all passwords is the union of those with 4, 5, 6, 7, and 8 letters, the sum principle might help us. To write the problem algebraically, let P_i be the set of i-letter passwords and P be the set of all possible passwords. Clearly,

$$P = P_4 \cup P_5 \cup P_6 \cup P_7 \cup P_8 \,.$$

The P_i are mutually disjoint; thus, we can apply the sum principle to obtain

$$|P| = \sum_{i=4}^{8} |P_i| \,.$$

We still need to compute $|P_i|$. For an i-letter password, there are 52 choices for the first letter, 52 choices for the second, and so on. By the product principle, $|P_i|$—the number of passwords with i letters—is 52^i. Therefore, the total number of passwords is

$$52^4 + 52^5 + 52^6 + 52^7 + 52^8 \,.$$

Of these, 52^4 have 4 letters, so the percentage with 4 letters is

$$\frac{100 \cdot 52^4}{52^4 + 52^5 + 52^6 + 52^7 + 52^8} \,.$$

Although this is a nasty formula to evaluate by hand, we can get quite a good estimate as follows: Notice that 52^8 is 52 times as big as 52^7 and even more dramatically larger than any other term in the sum in the denominator. The ratio is thus a bit less than

$$\frac{100 \cdot 52^4}{52^8} \,,$$

which is $100/52^4$, or approximately 0.000014. Thus, to five decimal places, only 0.00001% of the passwords have 4 letters. It is therefore much easier to guess a password that we know has 4 letters than it is to guess one that has between 4 and 8 letters—roughly 7 million times easier!

Our solution to Exercise 1.2-1 casually refers to the use of the product principle in computing the number of passwords with i letters. We didn't write any set as a union of sets of equal size. We could have, but it would have been clumsy and repetitive. For this reason, we now state a second version of the product principle, which we can derive from the version for unions of sets by using the idea of mathematical induction (see Chapter 4).

<table>
<tr><td>Principle 1.4</td><td>

(Product Principle, Version 2) If a set S of lists of length m has the properties that

1. there are i_1 different first elements of lists in S, and

2. for each $j > 1$ and each choice of the first $j - 1$ elements of a list in S, there are i_j choices of elements in position j of those lists,

then there are $i_1 i_2 \cdots i_m = \prod_{k=1}^{m} i_k$ lists in S.

</td></tr>
</table>

Version 2 of the product principle introduces a new notation: the use of Π to stand for product. This is called the **product notation**, and it is used just like summation notation. In particular, $\prod_{k=1}^{m} i_k$ is read, "The product from $k = 1$ to m of i_k." Thus, $\prod_{k=1}^{m} i_k$ means the same thing as $i_1 i_2 \cdots i_m$.

Let's apply this version of the product principle to compute the number of m-letter passwords. Because an m-letter password is simply a list of m letters and because there are 52 different first elements of the password and 52 choices for each other position of the password, we have that $i_1 = 52$, $i_2 = 52$, ..., $i_m = 52$. Thus, this version of the product principle tells us immediately that the number of passwords of length m is $i_1 i_2 \cdots i_m = 52^m$.

Lists and Functions

Our discussion of version 2 of the product principle left the term "list" undefined. A **list** of three things chosen from a set T consists of a first member t_1 of T, a second member t_2 of T, and a third member t_3 of T, not necessarily all different. If we rewrite the list in a different order, we get a different list. A list of k things chosen from T consists of a first member of T through a kth member of T. To give a more precise definition of a list, we can use the word "function," which you probably recall from algebra or calculus.

Recall that a **function** from a set S (called the **domain** of the function) to a set T (called the **range** of the function) is a relationship between the elements of S and the elements of T that relates exactly one element of T to each element of S. We use a letter, such as f, to stand for a function and $f(x)$ to stand for the element of T related to the element x of S. You are probably used to thinking of functions in terms of formulas like $f(x) = x^2$. We use formulas like this in algebra and calculus because the functions studied in those classes have infinite sets of numbers as their domains and ranges. In discrete mathematics, however, functions often have finite sets as their domains and ranges, so it is possible to describe a function by saying exactly what it is. For example,

$$f(1) = \text{Sam}, \quad f(2) = \text{Mary}, \quad f(3) = \text{Sarah}$$

is a function that describes a list of three names. This suggests a precise definition of a list of k elements from a set T: a **list of k elements from a set T** is a function from $\{1, 2, \ldots, k\}$ to T.

Exercise 1.2-2 Write down all the functions from the two-element set $\{1, 2\}$ to the two-element set $\{a, b\}$.

Exercise 1.2-3 How many functions are there from a two-element set to a three-element set?

Exercise 1.2-4 How many functions are there from a three-element set to a two-element set?

In Exercise 1.2-2, it is difficult to choose a notation for writing the functions. We use f_1, f_2, and so on to stand for the various functions we find. To describe a function f_i from $\{1, 2\}$ to $\{a, b\}$, we have to specify $f_i(1)$ and $f_i(2)$. We can write

$$f_1(1) = a \qquad f_1(2) = b$$
$$f_2(1) = b \qquad f_2(2) = a$$
$$f_3(1) = a \qquad f_3(2) = a$$
$$f_4(1) = b \qquad f_4(2) = b \, .$$

In this case, we simply wrote the functions as they occurred to us; but how do we know we have all of them? The set of all functions from $\{1, 2\}$ to $\{a, b\}$ is the union of the functions f_i with $f_i(1) = a$ and those with $f_i(1) = b$. The set of functions with $f_i(1) = a$ has two elements, one for each choice of $f_i(2)$. Therefore, by the product principle, the set of all functions from $\{1, 2\}$ to $\{a, b\}$ has size $2 \cdot 2 = 4$.

To compute the number of functions from a two-element set (say $\{1, 2\}$) to a three-element set, we can again think of using f_i to stand for a typical function. The set of all functions is the union of three sets, one for each choice of $f_i(1)$. Each of these sets has three elements, one for each choice of $f_i(2)$. Thus, by the product principle, we have $3 \cdot 3 = 9$ functions from a two-element set to a three-element set.

To compute the number of functions from a three-element set (say $\{1, 2, 3\}$) to a two-element set, we observe that the set of functions is a union of four sets, one for each choice of $f_i(1)$ and $f_i(2)$ (as we saw in our solution to Exercise 1.2-2). However, each of these sets has two functions in it, one for each choice of $f_i(3)$. Thus, by the product principle, we have $4 \cdot 2 = 8$ functions from a three-element set to a two-element set.

A function f is called **one-to-one**, or an **injection**, if $f(x) \neq f(y)$ whenever $x \neq y$.[4] Notice that the two functions f_1 and f_2 in our solution to Exercise 1.2-2 are one-to-one, but f_3 and f_4 are not.

---

[4]To understand the concept of one-to-one, it may help to contrast "one-to-one" with "many-to-one."

A function f is called **onto**, or a **surjection**, if every element y in the range is $f(x)$ for some x in the domain. Notice that the functions f_1 and f_2 in our solution to Exercise 1.2-2 are onto, but f_3 and f_4 are not.

Exercise 1.2-5   Using two- or three-element sets as domains and ranges, find an example of a one-to-one function that is not onto.

Exercise 1.2-6   Using two- or three-element sets as domains and ranges, find an example of an onto function that is not one-to-one.

The function given by $f(1) = c$, $f(2) = a$ is an example of a function from $\{1, 2\}$ to $\{a, b, c\}$ that is one-to-one but not onto. Also, the function given by $f(1) = a$, $f(2) = b$, $f(3) = a$ is an example of a function from $\{1, 2, 3\}$ to $\{a, b\}$ that is onto but not one-to-one.

The Bijection Principle

Exercise 1.2-7   The following loop is part of a program to determine the number of triangles formed by n points in the plane.

```
(1)  trianglecount = 0
(2)  for i = 1 to n
(3)        for j = i+1 to n
(4)              for k = j+1 to n
(5)                    if points i, j, and k are not collinear
(6)                          trianglecount = trianglecount + 1
```

Among all iterations of line 5 of the pseudocode, what is the total number of times this line checks three points to see if they are collinear?

Exercise 1.2-7 has a loop embedded in a loop embedded in another loop. Because the second loop, starting in Line 3, begins with $j = i + 1$ and j increases up to n and because the third loop, starting in Line 4, begins with $k = j + 1$ and k increases up to n, the code examines each triple of values i, j, k, with $i < j < k$, exactly once. For example, if n is 4, then the triples (i, j, k) used by the algorithm, in order, are $(1, 2, 3), (1, 2, 4), (1, 3, 4)$, and $(2, 3, 4)$. Thus, one way to solve Exercise 1.2-7 would be to compute the number of such triples, which we call increasing triples. As with the earlier case of two-element subsets, the number of such triples is the number of three-element subsets of an n-element set. This is the second time we have proposed counting the elements of one set (in this case, the set of increasing triples chosen from the set $\{1, 2, \ldots, n\}$) by saying that the number of elements of the set is equal to the number of elements of some other set (in this case, the set of three-element subsets of the set $\{1, 2, \ldots, n\}$).

When are we justified in making the assertion that two sets have the same size? There is a fundamental principle that abstracts our concept of what it means for two sets to have the same size. Intuitively, two sets have the same size if we can match their elements in such a way that each element of one set corresponds to exactly one element of the other set. This description carries with it some of the same words that appeared in the definitions of one-to-one and onto functions. Thus, it should be no surprise that one-to-one and onto functions are part of our abstract principle.

Principle 1.5	**(Bijection Principle)** Two sets have the same size if and only if there is a one-to-one function from one set onto the other.

This principle is called the **bijection principle** because a one-to-one and onto function is called a **bijection**. Another name for a bijection is a **one-to-one correspondence**. A bijection from a set to itself is called a **permutation** of that set.

What bijection is behind our assertion that the number of increasing triples equals the number of three-element subsets? We define the function f as the function that takes the increasing triple (i, j, k) to the subset $\{i, j, k\}$. Because the three elements of an increasing triple are different, the subset is a three-element set; so, we have a function from increasing triples to three-element sets. Because two different triples can't be the same set in two different orders, they must be associated with different sets. Thus, f is one-to-one. Because each set of three integers can be listed in increasing order, it is thus the image of an increasing triple under f. Therefore f is onto. Thus, we have a one-to-one correspondence, or bijection, between the set of increasing triples and the set of three-element sets.

k-Element Permutations of a Set

Because counting increasing triples is equivalent to counting three-element subsets, we can count increasing triples by counting three-element subsets instead. We use a method similar to the one used to compute the number of two-element subsets of a set. Recall that the first step of that method was to compute the number of ordered pairs of distinct elements that we can choose from the set $\{1, 2, \ldots, n\}$. So we now ask, in how many ways can we choose an ordered triple of distinct elements from $\{1, 2, \ldots, n\}$? Or more generally, in how many ways can we choose a list of k distinct elements from $\{1, 2, \ldots, n\}$? A list of k distinct elements chosen from a set N is called a **k-element permutation**[5] of N.

How many three-element permutations of $\{1, 2, \ldots, n\}$ can we make? Recall that a k-element permutation is a list of k distinct elements. There are n choices for the first number in the list. For each way of choosing the first element, there are $n - 1$

---

[5]In particular, a k-element permutation of $\{1, 2, \ldots, k\}$ is a list of k distinct elements of $\{1, 2, \ldots, k\}$, which, by our definition of a list, is a function from $\{1, 2, \ldots, k\}$ to $\{1, 2, \ldots, k\}$. This function must be one-to-one because the elements of the list are distinct. Because there are k distinct elements of the list, every element of $\{1, 2, \ldots, k\}$ appears in the list, so the function is onto. This means our function is a bijection. Thus, our definition of a permutation of a set is consistent with our definition of a k-element permutation in the case where the set is $\{1, 2, \ldots, k\}$.

choices for the second. For each choice of the first two elements, there are $n - 2$ ways to choose a third (distinct) number. So, by version 2 of the product principle, there are $n(n-1)(n-2)$ ways to choose the list of numbers. For example, if $n = 4$, the three-element permutations of $\{1, 2, 3, 4\}$ are

$$L = \{123, 124, 132, 134, 142, 143, 213, 214, 231, 234, 241, 243,$$
$$312, 314, 321, 324, 341, 342, 412, 413, 421, 423, 431, 432\} . \quad (1.4)$$

There are indeed $4 \cdot 3 \cdot 2 = 24$ lists in this set. Notice that this list is in the order that it would appear in a dictionary (assuming we treated numbers as we treat letters). This ordering of lists is called **lexicographic ordering**.

A general pattern is emerging. To compute the number of k-element permutations of the set $\{1, 2, \ldots, n\}$, we first recall that those permutations are lists. Then we note the following:

- We have n choices for the first element of the list.

- Regardless of which choice we make, we have $n - 1$ choices for the second element of the list.

- More generally, given the first $i - 1$ elements of a list, we have $n - (i - 1) = n - i + 1$ choices for the ith element of the list.

Thus, by version 2 of the product principle, we have $n(n-1) \cdots (n-k+1)$ (which is the first k terms of $n!$) ways to choose a k-element permutation of $\{1, 2, \ldots, n\}$. A very handy notation for this product, first suggested by Donald E. Knuth, is $n^{\underline{k}}$, which stands for

$$n(n-1) \cdots (n-k+1) = \prod_{i=0}^{k-1} (n-i) .$$

We call this the **kth falling factorial power of n**. We summarize our observations in a theorem.

Theorem 1.1 The number of k-element permutations of an n-element set is

$$n^{\underline{k}} = \prod_{i=0}^{k-1} (n-i) = n(n-1) \cdots (n-k+1) = \frac{n!}{(n-k)!} .$$

Counting Subsets of a Set

We now return to the question of counting the number of three-element subsets of $\{1, 2, \ldots, n\}$. We use $\binom{n}{3}$, which we read as "n choose 3," to stand for the number of three-element subsets of $\{1, 2, \ldots, n\}$, or, more generally, of any n-element set. We just carried out the first step of computing $\binom{n}{3}$ by counting the number of three-element permutations of $\{1, 2, \ldots, n\}$.

Exercise 1.2-8	Let L be the set of all three-element permutations of $\{1, 2, 3, 4\}$, as in Equation 1.4. How many of the lists (permutations) in L are lists of the three-element set $\{1, 3, 4\}$? What are these lists?

We see that this set appears in L as six different lists: 134, 143, 314, 341, 413, and 431. In general, when given three different numbers with which to create a list, there are three ways to choose the first number in the list; given the first, there are two ways to choose the second; and given the first two, there is only one way to choose the third element. Thus, by version 2 of the product principle, there are $3 \cdot 2 \cdot 1 = 6$ ways to make the list.

Because there are $n(n-1)(n-2)$ three-element permutations of an n-element set and each three-element subset appears in exactly six of these lists, the number of three-element permutations is six times the number of three-element subsets—that is, $n(n-1)(n-2) = \binom{n}{3} \cdot 6$. Whenever we see that one number that counts something is the product of two other numbers that count something, we should expect there to be an argument using the product principle explaining why. Thus, we should be able to see how to break the set of all three-element permutations of $\{1, 2, \ldots, n\}$ either into six disjoint sets of size $\binom{n}{3}$ or into $\binom{n}{3}$ subsets of size six. Because we argued that each three-element subset corresponds to six lists, we have described how to get a set of six lists from one three-element set. Two different subsets could never give us the same lists, so our sets of three-element lists are disjoint. In other words, we have divided the set of all three-element permutations into $\binom{n}{3}$ mutually disjoint sets of size six. Thus, the product principle explains why $n(n-1)(n-2) = \binom{n}{3} \cdot 6$. By division, we find that

$$\binom{n}{3} = \frac{n(n-1)(n-2)}{6}$$

is the number of three-element subsets of $\{1, 2, \ldots, n\}$. For $n = 4$, the number is $4(3)(2)/6 = 4$, and the sets are $\{1, 2, 3\}, \{1, 2, 4\}, \{1, 3, 4\}$, and $\{2, 3, 4\}$. It is straightforward to verify that each of these sets appears six times in L as six different lists.

Essentially the same argument gives us the number of k-element subsets of $\{1, 2, \ldots, n\}$. We denote this number by $\binom{n}{k}$, which is read as "n choose k." Here is the argument: The set of all k-element permutations of $\{1, 2, \ldots, n\}$ can be partitioned into $\binom{n}{k}$ disjoint blocks,[6] with each block comprising all k-element permutations of a k-element subset of $\{1, 2, \ldots, n\}$. However, the number of k-element permutations of a k-element set is $k!$, either by version 2 of the product principle or by Theorem 1.1. Thus, by version 1 of the product principle, we get

$$n^{\underline{k}} = \binom{n}{k} k! .$$

Division by $k!$ gives us our next theorem.

---

[6] Here we are using the language introduced for partitions of sets in Section 1.1.

Theorem 1.2	For integers n and k with $0 \le k \le n$, the number of k-element subsets of an n-element set is

$$\frac{n^{\underline{k}}}{k!} = \frac{n!}{k!\,(n-k)!}.$$

Proof: The proof is given above, except for when $k = 0$. However, the only subset of our n-element set of size zero is the empty set, so we have exactly one such subset. This is exactly what the formula gives us as well. (Note that the cases $k = 0$ and $k = n$ both use the fact[7] that $0! = 1$.) The equality in the theorem comes from the definition of $n^{\underline{k}}$. ∎

Another notation for the numbers $\binom{n}{k}$ is $C(n, k)$. Thus, we have that

$$C(n, k) = \binom{n}{k} = \frac{n!}{k!\,(n-k)!}. \tag{1.5}$$

These numbers are called **binomial coefficients** for reasons that will become clear later.

Important Concepts, Formulas, and Theorems

1. *List.* A list of k items chosen from a set X is a function from $\{1, 2, \ldots, k\}$ to X.

2. *Lists versus sets.* In a *list,* the order in which elements appear matters, and an element may appear more than once. In a *set,* the order of the elements does not matter, and an element may appear at most once.

3. *Product principle, version 2.* If a set S of lists of length m has the properties that

   a. there are i_1 different first elements of lists in S, and

   b. for each $j > 1$ and each choice of the first $j - 1$ elements of a list in S, there are i_j choices of elements in position j of those lists,

   then there are $i_1 i_2 \cdots i_m$ lists in S.

4. *Product notation.* We use the Greek letter Π to stand for "product," just as we use the Greek letter Σ to stand for "sum." This notation, called the *product notation,* is used just like summation notation. In particular, $\prod_{k=1}^{m} i_k$ is read as "the product from $k = 1$ to m of i_k." Thus, $\prod_{k=1}^{m} i_k$ means the same thing as $i_1 i_2 \cdots i_m$.

5. *Function.* A *function* f from a set S to a set T is a relationship between S and T that relates exactly one element of T to each element of S. We write $f(x)$ for the element of T related to the element x of S. The same element of T may be related to different members of S.

---

[7]There are many reasons why $0!$ is defined to be 1. Making the formula for $\binom{n}{k}$ work out is one of those reasons.

6. *One-to-one, Injection.* A function f from a set S to a set T is *one-to-one* if, for each $x \in S$ and $y \in S$ with $x \neq y$, $f(x) \neq f(y)$. A one-to-one function is also called an *injection.*

7. *Onto, Surjection.* A function f from a set S to a set T is *onto* if, for each element $y \in T$, there is at least one $x \in S$ such that $f(x) = y$. An onto function is also called a *surjection.*

8. *Bijection, One-to-one correspondence.* A function from a set S to a set T is a *bijection* if it is both one-to-one and onto. A bijection is sometimes called a *one-to-one correspondence.*

9. *Permutation.* A one-to-one function from a set S to S is called a *permutation* of S.

10. *k-element permutation.* A *k-element permutation* of a set S is an ordered list of k distinct elements of S.

11. *k-element subsets, n choose k, Binomial coefficients.* For integers n and k with $0 \leq k \leq n$, the number of k-element subsets of an n-element set is $\frac{n!}{k!\,(n-k)!}$. The number of k-element subsets of an n-element set is usually denoted by $\binom{n}{k}$ or $C(n, k)$, both of which are read as "n choose k." These numbers are called *binomial coefficients.*

12. *The number of k-element permutations.* The number of k-element permutations of an n-element set is

$$n^{\underline{k}} = n(n-1)\cdots(n-k+1) = \frac{n!}{(n-k)!} \, .$$

13. *Interpreting a product combinatorially.* When we have a formula to count something and the formula expresses the result as a product, it is useful to try to understand whether and how we could use the product principle to prove the formula.

Problems

1. In how many ways can we pass out k distinct pieces of fruit to n children (with no restriction on how many pieces of fruit a child may get)?

2. List all the functions from the three-element set $\{1, 2, 3\}$ to the set $\{a, b\}$. Which functions, if any, are one-to-one? Which functions, if any, are onto? `a/h`

3. List all the functions from the two-element set $\{1, 2\}$ to the three-element set $\{a, b, c\}$. Which functions, if any, are one-to-one? Which functions, if any, are onto?

4. There are more functions from the real numbers to the real numbers than most of us can imagine. In discrete mathematics, however, we often work with functions from a finite set S with s elements to a finite set T with t elements. Thus, there are only a finite number of functions from S to T. How many functions are there from S to T in this case? `a/h`

5. Assuming $k \le n$, in how many ways can we pass out k distinct pieces of fruit to n children if each child may get at most one piece? What if $k > n$? Assume for both questions that we pass out all the fruit.

6. Assuming $k \le n$, in how many ways can we pass out k identical pieces of fruit to n children if each child may get at most one? What if $k > n$? Assume for both questions that we pass out all the fruit. a/h

7. How many base ten numbers have five digits? How many five-digit numbers have no two consecutive digits equal? How many have at least one pair of consecutive digits equal?

8. Suppose you are organizing a panel discussion on allowing alcohol on campus. You need to arrange a list of participants—four administrators and four students—who will sit behind a table in the order listed. In how many ways can you list them if the administrators must sit together in a group and the students must sit together in a group? In how many ways can you list them if you must alternate students and administrators? a/h

9. (This problem is for students who are working on the relationship between k-element permutations and k-element subsets.) List in lexicographic order all three-element permutations of the five-element set $\{1, 2, 3, 4, 5\}$. Underline those elements that correspond to the set $\{1, 3, 5\}$. Draw a rectangle around those that correspond to the set $\{2, 4, 5\}$. How many three-element permutations of $\{1, 2, 3, 4, 5\}$ correspond to a given three-element set? How many three-element subsets does the set $\{1, 2, 3, 4, 5\}$ have?

10. In how many ways can a class of 20 students choose a group of 3 students from among themselves to go to the professor to explain that the 3-hour labs actually take 10 hours? a/h

11. Suppose you are choosing participants for a panel discussion on allowing alcohol on campus. You must choose 4 administrators from a group of 10 and 4 students from a group of 20. In how many ways can this be done?

12. Suppose you are organizing a panel discussion on allowing alcohol on campus. Participants will sit behind a table in the order in which you list them. You must choose 4 administrators from a group of 10 and 4 students from a group of 20. If the administrators must sit together in a group and the students must sit together in a group, in how many ways can you choose and list the 8 people? If you must alternate students and administrators, in how many ways can you choose and list them? a/h

13. At the local ice cream shop, you may get a sundae with two scoops of ice cream chosen from ten flavors; any one of three flavors of topping; and any (or all, some, or none) of whipped cream, nuts, and a cherry. How many different sundaes are possible? (In accordance with your mother's rule from Problem 11 in Section 1.1, the way the scoops sit in the dish does not matter.)

14. At the local ice cream shop, you may get a three-way sundae with up to three of the ten flavors of ice cream; any one of three flavors of topping; and any (or all, some, or none) of whipped cream, nuts, and a cherry. How many different

sundaes are possible? (In accordance with your mother's rule from Problem 11 in Section 1.1, the way the scoops sit in the dish does not matter.) a/h

15. A tennis club has $2n$ members. We want to pair up the members by twos for singles matches. In how many ways can we pair up all the members of the club? Suppose that in addition to specifying who plays whom, we also determine who serves first for each pairing. Now in how many ways can we specify our pairs?

16. A basketball team has 12 players. However, only 5 players play at any given time during a game. In how may ways can the coach choose the 5 players? To be more realistic, the 5 players playing a game normally consist of 2 guards, 2 forwards, and 1 center. If there are 5 guards, 4 forwards, and 3 centers on the team, in how many ways can the coach choose 2 guards, 2 forwards, and 1 center? What if one of the centers is equally skilled at playing forward? a/h

17. Explain why a function from an n-element set to an n-element set is one-to-one if and only if it is onto.

18. The function g is called an *inverse* to the function f if the domain of g is the range of f, if $g(f(x)) = x$ for every x in the domain of f, and if $f(g(y)) = y$ for each y in the range of f.

    a. Explain why a function is a bijection if and only if it has an inverse function. a/h

    b. Explain why a function that has an inverse function has only one inverse function. a/h

1.3 BINOMIAL COEFFICIENTS

In this section, we explore various properties of binomial coefficients. Remember that we defined the quantity $\binom{n}{k}$ to be the number of k-element subsets of an n-element set.

Pascal's Triangle

Table 1.1 contains the values of the binomial coefficients $\binom{n}{k}$ for $n = 0$ to $n = 6$ and all relevant values of k. The table begins with a 1 for $n = 0$ and $k = 0$ because the empty set, which is the set with no elements, has exactly one zero-element subset: itself. We have not put any value into the table for a value of k larger than n, because we haven't directly said what we mean by the binomial coefficient $\binom{n}{k}$ in that case. However, because there are no subsets of an n-element set that have size larger than n, it is natural to say that $\binom{n}{k}$ is zero when $k > n$. Therefore, when $k > n$, we define $\binom{n}{k}$ to be zero.[8] Thus, although we could fill in the empty places in the table with zeros, we leave them out to make the table easier to read.

---

[8]If you are thinking "But we did define $\binom{n}{k}$ to be zero when $k > n$ by saying that it is the number of k-element subsets of an n-element set, so of course it is zero," then good for you.

Table 1.1: *Binomial coefficients*

$n \backslash k$	0	1	2	3	4	5	6
0	1						
1	1	1					
2	1	2	1				
3	1	3	3	1			
4	1	4	6	4	1		
5	1	5	10	10	5	1	
6	1	6	15	20	15	6	1

Several properties of binomial coefficients are apparent in Table 1.1. Each row begins with a 1 because $\binom{n}{0}$ is always 1. This is because there is just one subset of an n-element set with zero elements, namely, the empty set. Similarly, each row ends with a 1 because an n-element set S has just one n-element subset, namely, S itself. Each row increases at first and then decreases. Further, the second half of each row is the reverse of the first half. **Pascal's triangle** is an array of numbers that emphasizes this symmetry by rearranging the rows of the table so that they line up at their centers (see Table 1.2). When we write Pascal's triangle, we leave out the values of n and k.

Table 1.2: *Pascal's triangle*

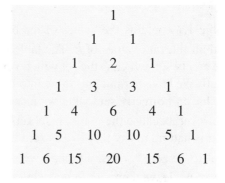

You may know a method for creating Pascal's triangle that creates each row from the row above, rather than computing binomial coefficients. Each entry in Table 1.2, except for the 1s, is the sum of the entry directly above it to the left and the entry directly above it to the right. We call this the **Pascal relationship**. This relationship gives another way to compute binomial coefficients without multiplying and dividing, as was done in Equation 1.5. If we wish to compute many binomial coefficients, the Pascal relationship often yields a more efficient way to do so. Once the coefficients in a row have been computed, the coefficients in the next row can be computed using only one addition per entry.

We now verify that the two methods for computing Pascal's triangle always yield the same result. To do so, we need an algebraic statement of the Pascal relationship. In Table 1.1, each entry is the sum of the one above it and the one above it and to the left. In algebraic terms, then, the Pascal relationship says

$$\binom{n}{k} = \binom{n-1}{k-1} + \binom{n-1}{k} , \tag{1.6}$$

whenever $n > 0$ and $0 < k < n$. It is possible to give a purely algebraic (and rather dreary) proof of this formula by plugging our earlier formula for binomial coefficients into all three terms and verifying that we get an equality. A guiding principle of discrete mathematics is that when we have a formula relating the numbers of elements of several sets, we should find an explanation that involves a relationship among the sets.

A Proof Using the Sum Principle

From Theorem 1.2 and Equation 1.5, we know that the expression $\binom{n}{k}$ stands for the number of k-element subsets of an n-element set. Each of the three terms in Equation 1.6, therefore, represents the number of subsets of a particular size chosen from an appropriately sized set. In particular, the three terms are the number of k-element subsets of an n-element set, the number of $(k-1)$-element subsets of an $(n-1)$-element set, and the number of k-element subsets of an $(n-1)$-element set. Thus, we should be able to use the sum principle to explain the relationship among these three quantities. This explanation will provide a proof that is just as valid as an algebraic derivation. Often, a proof using the sum principle will be less tedious and will yield more insight into the problem at hand.

Before giving such a proof in Theorem 1.3, we work out a special case. Suppose $n = 5$ and $k = 2$. Equation 1.6 says that

$$\binom{5}{2} = \binom{4}{1} + \binom{4}{2} . \tag{1.7}$$

Because the numbers are small, we could verify this simply by using the formula for binomial coefficients. However, let us instead consider subsets of a five-element set. Equation 1.7 says that the number of two-element subsets of a five-element set equals the number of one-element subsets of a four-element set plus the number of two-element subsets of a four-element set. But to apply the sum principle, we need to say something stronger. Namely, we should be able to partition the set of two-element subsets of a five-element set into two disjoint sets, one of which has size equal to the number of one-element subsets of a four-element set and one of which has size equal to the number of two-element subsets of a four-element set. Such a partition provides a proof of Equation 1.7. Consider now the set $S = \{A, B, C, D, E\}$. The set of two-element subsets is

$$S_1 = \{\{A, B\}, \{A, C\}, \{A, D\}, \{A, E\}, \{B, C\}, \{B, D\}, \{B, E\}, \\ \{C, D\}, \{C, E\}, \{D, E\}\} \,.$$

We now partition S_1 into two blocks, S_2 and S_3. S_2 consists of all sets in S_1 that do contain the element E, while S_3 consists of all sets in S_1 that do not contain the element E. Thus,

$$S_2 = \{\{A, E\}, \{B, E\}, \{C, E\}, \{D, E\}\}$$

and

$$S_3 = \{\{A, B\}, \{A, C\}, \{A, D\}, \{B, C\}, \{B, D\}, \{C, D\}\} \,.$$

Each set in S_2 must contain E; thus, each set contains one other element from S. Because there are four other elements in S that we can choose along with E, we have $|S_2| = \binom{4}{1}$. Each set in S_3 contains two elements from the set $\{A, B, C, D\}$. There are $\binom{4}{2}$ ways to choose such a two-element subset of $\{A, B, C, D\}$, but $S_1 = S_2 \cup S_3$ and S_2 and S_3 are disjoint. By the sum principle, Equation 1.7 must hold.

We now give a proof for general n and k.

Theorem 1.3	If n and k are integers with $n > 0$ and $0 < k < n$, then

$$\binom{n}{k} = \binom{n-1}{k-1} + \binom{n-1}{k} \,.$$

Proof: The formula says that the number of k-element subsets of an n-element set is the sum of two numbers. As in our example, we apply the sum principle. To do so, we need to represent the set of k-element subsets of an n-element set as a union of two other disjoint sets. Suppose our n-element set is $S = \{x_1, x_2, \ldots, x_n\}$. Let us take S_1 to be the $\binom{n}{k}$-element set of all k-element subsets of S. To apply the sum principle, we partition S_1 into two disjoint blocks of k-element subsets, S_2 and S_3. The sizes of S_2 and S_3 are $\binom{n-1}{k-1}$ and $\binom{n-1}{k}$, respectively. We do this as follows: Note that $\binom{n-1}{k}$ stands for the number of k-element subsets of the first $n-1$ elements $x_1, x_2, \ldots, x_{n-1}$ of S. Thus, we let S_3 be the set of k-element subsets of S that don't contain x_n. The only possibility for S_2 is the set of k-element subsets of S that *do* contain x_n. We see that the number of elements of this set S_2 is $\binom{n-1}{k-1}$ by observing that removing x_n from one of the elements of S_2 gives a $(k-1)$-element subset of $S' = \{x_1, x_2, \ldots, x_{n-1}\}$. Further, each $(k-1)$-element subset of S' arises in this way from one, and only one, k-element subset of S containing x_n. Thus, the number of elements of S_2 is the number of $(k-1)$-element subsets of S', which is $\binom{n-1}{k-1}$. Because S_2 and S_3 are two disjoint sets whose union is S, the sum principle shows that the number of elements of S is $\binom{n-1}{k-1} + \binom{n-1}{k}$. ∎

Notice that in this proof, we used a bijection that we did not explicitly describe. Namely, there is a bijection f between S_3 (the k-element sets of S that contain x_n) and the $(k-1)$-element subsets of S'. For any subset K in S_3, we let $f(K)$ be the set we obtain by removing x_n from K. It is immediate that this is a bijection. The bijection principle tells us that the size of S_3 is the size of the set of all $(k-1)$-element subsets of S'.

The Binomial Theorem

Exercise 1.3-3 What is $(x+y)^3$? What is $(x+1)^4$? What is $(2+y)^4$? What is $(x+y)^4$?

The number of k-element subsets of an n-element set is called a **binomial coefficient** because of the role that these numbers play in the algebraic expansion of a binomial $x+y$.

Theorem 1.4 **(Binomial Theorem)** For any integer $n \geq 0$,

$$(x+y)^n = \binom{n}{0}x^n + \binom{n}{1}x^{n-1}y + \binom{n}{2}x^{n-2}y^2 + \cdots + \binom{n}{n-1}xy^{n-1} + \binom{n}{n}y^n ,$$

(1.8)

or, in summation notation,

$$(x+y)^n = \sum_{i=0}^{n} \binom{n}{i} x^{n-i}y^i .$$

Unfortunately, when most people first see this theorem, they do not have the tools to see easily why it is true. However, armed with our new way of using relationships among sets to prove algebraic identities, we can prove this theorem.

Let us begin by considering the example $(x+y)^3$, which, by the binomial theorem, is

$$(x+y)^3 = \binom{3}{0}x^3 + \binom{3}{1}x^2y + \binom{3}{2}xy^2 + \binom{3}{3}y^3$$

(1.9)

$$= x^3 + 3x^2y + 3xy^2 + y^3 .$$

(1.10)

Suppose we did not know the binomial theorem but still wanted to compute $(x+y)^3$. We would write out $(x+y)(x+y)(x+y)$ and perform the multiplication. Probably we would multiply the first two terms, obtaining $x^2 + 2xy + y^2$, and then we would multiply this expression by $x+y$. Notice that by applying distributive laws to $(x+y)^2$, we would get

$$(x+y)(x+y) = (x+y)x + (x+y)y = xx + yx + xy + yy .$$

(1.11)

We could use the commutative law to put this into the usual form, but let us hold off for a moment so we can see a pattern evolve. To compute $(x+y)^3$, we can multiply the expression on the right side of Equation 1.11 by $x+y$, using the distributive laws to get

$$(xx + xy + yx + yy)(x + y)$$
$$= (xx + xy + yx + yy)x + (xx + xy + yx + yy)y \quad (1.12)$$
$$= xxx + xyx + yxx + yyx + xxy + xyy + yxy + yyy . \quad (1.13)$$

Now compare Equation 1.13 with $(x + y)(x + y)(x + y)$. Each of the eight terms that we get from the distributive law may be thought of as a product of terms—one from the first binomial, one from the second binomial, and one from the third binomial. Because multiplication is commutative, many of these products are the same. In fact, we have one xxx, or x^3, product; three products with two x's and one y, or x^2y; three products with one x and two y's, or xy^2; and one product that becomes y^3. Now look at Equation 1.9, which summarizes this process. There are $\binom{3}{0} = 1$ way to choose a product with three x's and zero y's, $\binom{3}{1} = 3$ ways to choose a product with two x's and one y, and so on. Thus, we can understand the binomial theorem in terms of counting subsets of binomial factors: the coefficient of $x^{n-k}y^k$ is the number of ways to select k of our n factors. From each of these k factors we choose a y term to get a product of variables in which k of the variables are y.

Essentially the same explanation gives us a proof of the binomial theorem. Note that when we multiply three factors of $x + y$ using the distributive law without collecting like terms, we get a sum of eight products. Each factor of $x + y$ doubles the number of summands. Thus, when we apply the distributive law as many times as possible (without applying the commutative law and collecting like terms) to a product of n binomials all equal to $x + y$, we get 2^n summands. Each summand is a product of a length-n list of x's and y's. In each list, the ith entry comes from the ith binomial factor. A list that becomes $x^{n-k}y^k$ when we use the commutative law will have a y in k of its places and an x in the remaining places. The number of lists that have a y in k places is thus the number of ways to select k binomial factors to contribute a y to our list. But the number of ways to select k binomial factors from n binomial factors is simply $\binom{n}{k}$. Therefore, $\binom{n}{k}$ is the coefficient of $x^{n-k}y^k$. This proves the binomial theorem.

Applying the binomial theorem to the remaining questions in Exercise 1.3-3 gives us

$$(x + 1)^4 = x^4 + 4x^3 + 6x^2 + 4x + 1 ,$$
$$(2 + y)^4 = 16 + 32y + 24y^2 + 8y^3 + y^4 , \text{ and}$$
$$(x + y)^4 = x^4 + 4x^3y + 6x^2y^2 + 4xy^3 + y^4 .$$

Labeling and Trinomial Coefficients

Exercise 1.3-4

Suppose we have k labels of one kind and $n - k$ labels of another. In how many different ways can we apply these labels to n objects?

Exercise 1.3-5

Show that if we have k_1 labels of one kind, k_2 labels of a second kind, and $k_3 = n - k_1 - k_2$ labels of a third kind, then there are $n!/(k_1!k_2!k_3!)$ ways to apply these labels to n objects.

What is the coefficient of $x^{k_1} y^{k_2} z^{k_3}$ in $(x + y + z)^n$?

We can think of Exercises 1.3-4 and 1.3-5 as immediate applications of binomial coefficients. For Exercise 1.3-4, there are $\binom{n}{k}$ ways to choose the k objects that get the first kind of label. The other objects get the second kind of label, so the answer is $\binom{n}{k}$. For Exercise 1.3-5, there are $\binom{n}{k_1}$ ways to choose the k_1 objects that get the first kind of label and $\binom{n-k_1}{k_2}$ ways to choose the objects that get the second kind of label. After that, the remaining $k_3 = n - k_1 - k_2$ objects get the third kind of label. Thus, by the product principle, the total number of labelings is the product of the two binomial coefficients, which simplifies as follows:

$$\binom{n}{k_1}\binom{n-k_1}{k_2} = \frac{n!}{k_1!\,(n-k_1)!}\frac{(n-k_1)!}{k_2!\,(n-k_1-k_2)!}$$
$$= \frac{n!}{k_1!\,k_2!\,(n-k_1-k_2)!}$$
$$= \frac{n!}{k_1!\,k_2!\,k_3!}\,.$$

A more elegant approach to Exercise 1.3-4, Exercise 1.3-5, and other related problems appears in Section 1.4.

Exercise 1.3-6 shows how Exercise 1.3-5 applies to computing powers of trinomials. In expanding $(x + y + z)^n$, we think of writing n copies of the trinomial $x + y + z$ side by side and applying the distributive laws until we have a sum of terms, each of which is a product of x's, y's, and z's. How many such terms do we have with k_1 x's, k_2 y's, and k_3 z's? Imagine we make our choice by choosing x from some number k_1 of the copies of the trinomial, choosing y from some number k_2, and z from the remaining k_3 copies; multiplying all the chosen terms together; and adding the results over all ways of picking the k_i's. If we choose x from a copy of the trinomial, that copy is "labeled" x. The same is true for y and z, so the number of choices that yield $x^{k_1} y^{k_2} z^{k_3}$ is the number of ways to label n objects with k_1 labels of one kind, k_2 labels of a second kind, and k_3 labels of a third. Notice that this requires that $k_3 = n - k_1 - k_2$. By analogy with our notation for a binomial coefficient, we define the **trinomial coefficient** $\binom{n}{k_1, k_2, k_3}$ to be $\frac{n!}{k_1!\,k_2!\,k_3!}$ if $k_1 + k_2 + k_3 = n$; otherwise we define it to be 0. Then $\binom{n}{k_1, k_2, k_3}$ is the coefficient of $x^{k_1} y^{k_2} z^{k_3}$ in $(x + y + z)^n$. This is sometimes called the **trinomial theorem**.

Important Concepts, Formulas, and Theorems

1. *Pascal relationship.* The *Pascal relationship* says

$$\binom{n}{k} = \binom{n-1}{k-1} + \binom{n-1}{k}\,,$$

whenever $n > 0$ and $0 < k < n$.

2. *Pascal's triangle.* *Pascal's triangle* is the triangular array of rows of numbers obtained by

- putting 1s in position 0 and position i of row i, and,
- for each positive integer n and each integer j between 1 and $n - 1$, inclusive, putting into row n and column j the sum of the numbers in row $n - 1$ and column $j - 1$ and row $n - 1$ and column j.

3. *Binomial theorem.* The *binomial theorem* states that for any integer $n \geq 0$,

$$(x + y)^n = \binom{n}{0}x^n + \binom{n}{1}x^{n-1}y + \binom{n}{2}x^{n-2}y^2 + \cdots + \binom{n}{n-1}xy^{n-1} + \binom{n}{n}y^n ,$$

$$= x^n + \binom{n}{1}x^{n-1}y + \binom{n}{2}x^{n-2}y^2 + \cdots + \binom{n}{n-1}xy^{n-1} + \binom{n}{n}y^n ,$$

or, in summation notation,

$$(x + y)^n = \sum_{i=0}^{n} \binom{n}{i} x^{n-i} y^i .$$

4. *Labeling.* The number of ways to apply k labels of one kind and $n - k$ labels of another kind to n objects is $\binom{n}{k}$.

5. *Trinomial coefficient.* The *trinomial coefficient* $\binom{n}{k_1, k_2, k_3}$ is $\frac{n!}{k_1! \, k_2! \, k_3!}$ if $k_1 + k_2 + k_3 = n$; otherwise it is 0.

6. *Trinomial theorem.* The coefficient of $x^i y^j z^k$ in $(x + y + z)^n$ is $\binom{n}{i, j, k}$.

Problems

1. Find $\binom{12}{3}$ and $\binom{12}{9}$. What can you say in general about $\binom{n}{k}$ and $\binom{n}{n-k}$? `a/h`

2. Find the row of the Pascal triangle that corresponds to $n = 8$.

3. Find the following

   a. $(x + 1)^5$ `a/h`

   b. $(x + y)^5$

   c. $(x + 2)^5$

   d. $(x - 1)^5$ `a/h`

4. Carefully explain the proof of the binomial theorem for $(x + y)^4$. That is, explain what each of the binomial coefficients in the theorem stands for and how it is related to the powers of x and y that follow it.

5. If you have ten distinct chairs to paint, in how many ways can you paint three of them green, three of them blue, and four of them red? What does this have to do with labelings? `a/h`

6. When $n_1, n_2, \ldots, n_k$ are nonnegative integers that add to n, the number $\frac{n!}{n_1!, n_2!, \ldots, n_k!}$ is called a **multinomial coefficient** and is denoted by $\binom{n}{n_1, n_2, \ldots, n_k}$. A polynomial of the form $x_1 + x_2 + \cdots + x_k$ is called a multinomial. Explain the relationship between powers of a multinomial and multinomial coefficients. This relationship is called the multinomial theorem.

7. Give a bijection that proves your statement about $\binom{n}{k}$ and $\binom{n}{n-k}$ in Problem 1 of this section. a/h

8. In a Cartesian coordinate system, how many paths are there from the origin to the point with integer coordinates (m, n) if the paths are built up of exactly $m + n$ horizontal and vertical line segments, each of length 1? a/h

9. What formula do you get for the binomial theorem if, instead of analyzing the number of ways to choose k distinct y's, you analyze the number of ways to choose k distinct x's?

10. Explain the difference between choosing four disjoint 3-element sets from a 12-element set and labeling a 12-element set with three labels of type 1, three labels of type 2, three labels of type 3, and three labels of type 4. In how many ways can you choose four disjoint 3-element subsets from a 12-element set? In how many ways can you choose three disjoint 4-element subsets from a 12-element set? a/h

11. A 20-member club must have a president, vice president, secretary, and treasurer, as well as a three-person nominating committee. If the officers must be different people, and if no officer may be on the nominating committee, in how many ways could the officers and nominating committee be chosen? Answer the same question if officers may be on the nominating committee. a/h

12. Prove Equation 1.6 by plugging in the formula for $\binom{n}{k}$.

13. Give two proofs that

$$\binom{n}{k} = \binom{n}{n-k}.$$ a/h

14. Give at least two proofs that

$$\binom{n}{k}\binom{k}{j} = \binom{n}{j}\binom{n-j}{k-j}.$$

15. Give at least two proofs that

$$\binom{n}{k}\binom{n-k}{j} = \binom{n}{j}\binom{n-j}{k}.$$ a/h

16. You need not compute all of rows 7, 8, and 9 of Pascal's triangle to use it to compute $\binom{9}{6}$. Figure out which entries of Pascal's triangle not given in Table 1.2 you actually need, and compute them to get $\binom{9}{6}$.

17. Explain why

$$\sum_{i=0}^{n}(-1)^i \binom{n}{i} = 0 .$$ a/h

18. Apply calculus and the binomial theorem to $(1+x)^n$ to show that

$$\binom{n}{1} + 2\binom{n}{2} + 3\binom{n}{3} + \cdots = n2^{n-1} .$$

19. True or False: $\binom{n}{k} = \binom{n-2}{k-2} + \binom{n-2}{k-1} + \binom{n-2}{k}$. If true, give a proof. If false, give values of n and k that show the statement is false, find an analogous true statement, and prove it. a/h

1.4 EQUIVALENCE RELATIONS AND COUNTING*

The Symmetry Principle

Consider again the example from Section 1.2 in which we wanted to count the number of three-element subsets of a four-element set. To do so, we first formed all possible lists of $k = 3$ distinct elements chosen from a set with $n = 4$ elements (see Equation 1.4). The number of lists of k distinct elements is $n^{\underline{k}} = n!/(n-k)!$. We then observed that two lists are equivalent as sets if one can be obtained by rearranging (or "permuting") the other. This divides the lists into classes, which we now call **equivalence classes**, each of size $k!$. In the discussion of Exercise 1.2-8, one such equivalence class was

$$\{134, 143, 314, 341, 413, 431\} .$$

The other three are

$$\{234, 243, 324, 342, 423, 432\} ,$$
$$\{123, 132, 213, 231, 312, 321\} , \quad \text{and}$$
$$\{124, 142, 214, 241, 412, 421\} .$$

The product principle told us that if q is the number of such equivalence classes, if each equivalence class has $k!$ elements, and if the entire set of lists has $n!/(n-k)!$ elements, then we must have that

$$q(k!) = \frac{n!}{(n-k)!} .$$

Dividing, we solve for q and get an expression for the number of k-element subsets of an n-element set. In fact, this is how we proved Theorem 1.2.

---

*This section, which extends and abstracts counting principles from Section 1.3, is not required for any other section of the book, except for Appendix A, which gives a more traditional approach to ideas of this section.

A principle that helps us learn and understand mathematics is that if we have a mathematical result showing a certain symmetry, it often helps our understanding to find a proof that reflects this symmetry. We call this the **symmetry principle**.

Principle 1.6
(Symmetry Principle) If a formula has a symmetry (i.e., interchanging two variables doesn't change the result), then a proof that explains this symmetry is likely to give us additional insight into the formula.

The proof of Theorem 1.2 does not account for the symmetry of the $k!$ term and the $(n-k)!$ term in the expression $\frac{n!}{k!\,(n-k)!}$. This symmetry arises because choosing a k-element subset is equivalent to choosing the $(n-k)$-element subset of elements we don't want. In Exercise 1.3-4, we saw that the binomial coefficient $\binom{n}{k}$ also counts the number of ways to label n objects, say with the labels "in" and "out," so that we have k "ins" and, therefore, $n-k$ "outs." For each labeling, the k objects that get the label "in" are in our subset. This explains the symmetry in our formula, but it doesn't prove the formula. Here is a new proof that the number of labelings is $\frac{n!}{k!\,(n-k)!}$, which explains the symmetry.

Suppose we have m ways to assign k blue and $n-k$ red labels to n elements. From each labeling, we can create a number of lists, using the convention of listing the k blue elements first and the remaining $n-k$ red elements last. For example, suppose we are considering the number of ways to label three elements blue (and two red) from a five-element set $\{A, B, C, D, E\}$. Consider the particular labeling in which A, B, and D are labeled blue and C and E are labeled red. Which lists correspond to this labeling? They are

$$
\begin{array}{cccccc}
ABDCE & ABDEC & ADBCE & ADBEC & BADCE & BADEC \\
BDACE & BDAEC & DABCE & DABEC & DBACE & DBAEC ,
\end{array}
$$

or all lists in which A, B, and D precede C and E. Because there are $3!$ ways to arrange A, B, and D and $2!$ ways to arrange C and E, there are, by the product principle, $3!\,2! = 12$ lists in which A, B, and D precede C and E. For each of the q ways to construct a labeling, we could find a similar set of 12 lists associated with that labeling. Because *every* possible list of five elements will appear exactly once via this process, and because there are $5! = 120$ five-element lists overall, we must have, by the product principle, that

$$q \cdot 12 = 120 ,$$

or $q = 10$. This agrees with our previous calculations of $\binom{5}{3} = 10$ for the number of ways to label five items so that three are blue and two are red.

Generalizing, we let q be the number of ways to label n objects with k blue labels and $n-k$ red labels. To create the lists associated with a labeling, we list the blue elements first and then the red elements. We can mix the k blue elements among themselves, and we can mix the $n-k$ red elements among themselves, giving us $k!\,(n-k)!$ lists, consisting of the elements with a blue label first followed by the elements with a red label. Because we can choose to label any k elements blue, each of our lists of n distinct elements arises from some labeling in this way. Each such list arises from only one labeling, because two different labelings will have different first k elements. (The first k elements listed have the blue label.) Therefore, by the product principle, $q(k!)(n-k)!$ is the number of lists we can form with n distinct objects. Therefore, $q(k!)(n-k)!$ must equal $n!$. This gives us

$$q(k!)(n-k)! = n! \, ,$$

and division gives us our original formula for q. Because the red and blue labels must be treated identically, our formula is symmetric. Recall that our proof of the formula in Exercise 1.3-5 did not explain why the product of three factorials appeared in the denominator, it simply proved the formula was correct. We can now explain *why* the product in the denominator of the formula in Exercise 1.3-5 for the number of labelings with three labels is what it is. The denominator counts the number of lists that come from a given labeling with k_1 labels of a first kind, k_2 labels of a second kind, and k_3 labels of a third kind. With this insight, we can generalize this formula to any number of labels.

Equivalence Relations

The preceding process divided the set of all $n!$ lists of n distinct elements into classes (another word for sets) of lists. In each class, all the lists are mutually equivalent with respect to labeling with two labels. More precisely, two lists of the n objects are equivalent for defining labelings if we get one from the other by mixing the first k elements among themselves and mixing the last $n-k$ elements among themselves. Relating objects that we want to count to sets of lists (so that each object corresponds to a set of equivalent lists) is a technique we can use to solve a variety of counting problems. (This is another example of abstraction.)

A relationship that divides a set into mutually exclusive classes is called an **equivalence relation**.[9] Thus, if

$$S = S_1 \cup S_2 \cup \cdots \cup S_m$$

and $S_i \cap S_j = \emptyset$ for all i and j with $i \neq j$, then the relationship that says any two elements $x \in S$ and $y \in S$ are equivalent if and only if they lie in the same set S_i is an equivalence relation. The sets S_i are called **equivalence classes**, and, as we noted in Section 1.1, the family $S_1, S_2, \ldots, S_m$ is called a partition of S. One partition of the set $S = \{a, b, c, d, e, f, g\}$ is $\{a, c\}, \{d, g\}, \{b, e, f\}$. This partition corresponds to the following (boring) equivalence relation: a and c are equivalent, d and g are

---

[9]The usual mathematical approach to equivalence relations, which is discussed in this section's Problems, is different from the one given here. Typically, one sees an equivalence relation defined as a reflexive (everything is related to itself), symmetric (if x is related to y, then y is related to x), and transitive (if x is related to y and y is related to z, then x is related to z) relationship on a set X. Examples of such relationships are equality (on any set), similarity (on a set of triangles), and having the same birthday (on a set of people). The two approaches are equivalent. The traditional approach, which takes a bit longer to study, is presented in Appendix A.1.

equivalent, and b, e, and f are equivalent. A slightly less boring equivalence relation is that two letters are equivalent if, typographically, their top and bottom are at the same height. This gives the partition $\{a, c, e\}, \{b, d\}, \{f\}, \{g\}$.

Exercise 1.4-1 On the set of integers 0–12 inclusive, define two integers to be related if they have the same remainder on division by 3. Which numbers are related to 0? To 1? To 2? To 3? To 4? Is this relationship an equivalence relation?

In Exercise 1.4-1, the set of numbers related to 0 is the set $\{0, 3, 6, 9, 12\}$, the set related to 1 is $\{1, 4, 7, 10\}$, the set related to 2 is $\{2, 5, 8, 11\}$, the set related to 3 is $\{0, 3, 6, 9, 12\}$, and the set related to 4 is $\{1, 4, 7, 10\}$. A little more precisely, a number is related to one of 0, 3, 6, 9, or 12 if and only if it is in the set $\{0, 3, 6, 9, 12\}$; a number is related to 1, 4, 7, or 10 if and only if it is in the set $\{1, 4, 7, 10\}$; and a number is related to 2, 5, 8, or 11 if and only if it is in the set $\{2, 5, 8, 11\}$. These are mutually disjoint sets whose union is $\{0, 1, 2, 3, 4, 5, 6, 7, 8, 9, 10, 11, 12\}$. Therefore, the relationship is an equivalence relation on the set $\{0, 1, 2, 3, 4, 5, 6, 7, 8, 9, 10, 11, 12\}$.

The Quotient Principle

In Exercise 1.4-1, the equivalence classes had two different sizes. In the examples of counting labelings and subsets that we have seen so far, all the equivalence classes had the same size. This was very important. The principle we have been using to count subsets and labelings is given in the following theorem, which we call the **quotient principle**.

Theorem 1.5 **(Quotient Principle)** If an equivalence relation on a p-element set S has q classes each of size r, then $q = p/r$.

Proof: By the product principle, $p = qr$; therefore, $q = p/r$. ∎

Another statement of the quotient principle that uses the idea of a partition is as follows.

Principle 1.7 **(Quotient Principle, Version 2)** If we can partition a set of size p into q blocks of size r, then $q = p/r$.

Returning to our example of three blue and two red labels, $p = 5! = 120$, $r = 12$, and, therefore, by Theorem 1.5,

$$q = \frac{p}{r} = \frac{120}{12} = 10.$$

Equivalence Class Counting

We now give several examples of the use of Theorem 1.5.

Exercise 1.4-2

When four people sit down at a round table to play cards, two lists of their four names are equivalent as seating charts if each person has the same person to the right in both lists.[10] (The person to the right of the person in Position 4 of the list is the person in Position 1.) We use Theorem 1.5 to count the number of possible ways to seat the players. We take our set S to be the set of all four-element permutations of the four people, that is, the set of all lists of the four people.

    a. How many lists are equivalent to a given one?

    b. What are the lists equivalent to ABCD?

    c. Is the relationship of equivalence an equivalence relation?

    d. Use the quotient principle to compute the number of equivalence classes and, hence, the number of possible ways to seat the players.

Exercise 1.4-3

We wish to count the number of ways to attach $n > 2$ distinct beads to the corners of a regular n-gon (or string them on a necklace). We say that two lists of the n beads are equivalent if each bead is adjacent to exactly the same beads in both lists. (The first bead in the list is considered to be adjacent to the last.)

    • How does this exercise differ from the previous exercise?

    • How many lists are in an equivalence class?

    • How many equivalence classes are there?

In Exercise 1.4-2, suppose we named the places at the table north, east, south, and west. Given a list, we get an equivalent one in two steps. First, we observe that we have four choices of people to sit in the north position. There is then one person who can sit to this person's right, one who can be next on the right, and one who can be the following on the right, all determined by the original list. Thus, there are exactly four lists equivalent to a given one, including that given one. The lists equivalent to ABCD are ABCD, BCDA, CDAB, and DABC. This shows that two lists are equivalent if and only if we can get one from the other by moving everyone the same number of places to the right around the table (or we can get one from the other moving everyone the same number of places to the left around the table). From this we see that we have an equivalence relation, because each list is in one, and only one, of these sets of four equivalent lists. This means our relationship divides the set of all lists of the four names into equivalence classes each of size four. There are a total of $4! = 24$

---

[10]Think of the four places at the table as being called north, east, south, and west or as numbered 1–4. You would get a list by starting with the person in the north position (Position 1), then the person in the east position (Position 2), and so on, clockwise around the table.

lists of four distinct names; so, by Theorem 1.5, we have $4!/4 = 3! = 6$ seating arrangements.

Exercise 1.4-3 is similar in many ways to Exercise 1.4-2. However, there is one significant difference. We can visualize Exercise 1.4-3 as one of dividing lists of n distinct beads into equivalence classes, but now two lists are equivalent if each bead is adjacent to exactly the same beads in both lists. Suppose we number the vertices of our polygon 1–n, clockwise. Given a list, we can count the equivalent lists as follows: we have n choices for which bead to put in Position 1. For Position 2, we can use either of the two beads adjacent to Position 1 in the given list.[11] But now, only one bead can go in Position 3, because the other bead adjacent to Position 2 is already in Position 1. We can continue in this way to fill in the rest of the list. For example, with $n = 4$, the lists ABCD, ADCB, BCDA, BADC, CDAB, CBAD, DABC, and DCBA are all equivalent. Notice the first, third, fifth, and seventh lists are obtained by shifting the beads around the polygon, as are the second, fourth, sixth, and eighth lists (though in the opposite direction). Also note that the eighth list is the reverse of the first, the third is the reverse of the second, and so on. Rotating a necklace in space corresponds to shifting the letters in the list. Flipping a necklace over in space corresponds to reversing the order of a list. We can always get $2n$ lists by shifting and reversing shifts of a list. The lists equivalent to a given one consist of everything we can get from the given list by rotations and reversals. Thus, the relationship of every bead being adjacent to the same beads divides the set of lists of beads into disjoint sets. These sets, which have size $2n$, are the equivalence classes of our equivalence relation. Because there are $n!$ lists, Theorem 1.5 says there are

$$\frac{n!}{2n} = \frac{(n-1)!}{2}$$

bead arrangements.

Multisets

Sometimes when we think about choosing elements from a set, we want to be able to choose an element more than once. For example, the set of letters of the word "roof" is $\{f, o, r\}$. However, it is often more useful to think of the multiset of letters, which, in this case, is $\langle\!\langle f, o, o, r \rangle\!\rangle$. We use the double angle brackets to distinguish a multiset from a set. We specify a **multiset** chosen from a set S by saying how many times each of its elements occurs. If S is the set of English letters, the "multiplicity" function for roof is given by $m(f) = 1$, $m(o) = 2$, $m(r) = 1$, and $m(\text{letter}) = 0$ for every other letter. In a multiset, order is not important (the multiset $\langle\!\langle r, o, f, o \rangle\!\rangle$ is the same as the multiset $\langle\!\langle f, o, o, r \rangle\!\rangle$). We know this is the case because the multisets have the same multiplicity function. We would like to say that the size of $\langle\!\langle f, o, o, r \rangle\!\rangle$ is 4, so we define the **size** of a multiset to be the sum of the multiplicities of its elements.

---

[11]Remember that the first and last beads are considered adjacent, so they each have two beads adjacent to them.

Explain how placing k identical books onto the n shelves of a bookcase can be thought of as giving us a k-element multiset of the shelves of the bookcase. Explain how distributing k identical apples to n children can be thought of as giving us a k-element multiset of the children.

In Exercise 1.4-4, we can think of the multiplicity of a bookshelf as the number of books it gets and the multiplicity of a child as the number of apples the child gets. In fact, this idea of distribution of identical objects to distinct recipients gives a great mental model for a multiset chosen from a set S. That is, to determine a k-element multiset chosen from S, we "distribute" k identical objects to the elements of S. The number of objects an element x gets is the multiplicity of x.

Notice that it makes no sense to ask for the number of multisets we may choose from a set with n elements, because $\langle\!\langle A \rangle\!\rangle$, $\langle\!\langle A, A \rangle\!\rangle$, $\langle\!\langle A, A, A \rangle\!\rangle$, and so on are infinitely many multisets chosen from the set $\{A\}$. However, it does make sense to ask for the number of k-element multisets we can choose from an n-element set. What strategy could we employ to figure out this number? To count k-element subsets, we first count k-element permutations and then divide by the number of different permutations of the same set. Here we need an analog of permutations that allows repeats. A natural idea is to consider lists with repeats. After all, one way to describe a multiset is to list it, and there could be many different orders for listing a multiset. However, the two-element multiset $\langle\!\langle A, A \rangle\!\rangle$ can be listed in just one way, while the two-element multiset $\langle\!\langle A, B \rangle\!\rangle$ can be listed in two ways. When we counted k-element subsets of an n-element set by using the quotient principle, it was essential that each k-element subset corresponded to the same number (namely, $k!$) of permutations (lists), because we were using the reasoning behind the quotient principle to do our counting. So, if we hope to use similar reasoning, we can't apply the quotient principle to lists with repeats because different k-element multisets can correspond to different numbers of lists.

Suppose, however, that we could count the number of ways to arrange k distinct books on the n shelves of a bookcase. We can still think of the multiplicity of a shelf as being the number of books on it. However, many different arrangements of distinct books will give us the same multiplicity function. In fact, any way of mixing the books among themselves that does not change the number of books on each shelf will give us the same multiplicities. But the number of ways to mix the books among themselves is the number of permutations of the books—namely, $k!$. Thus, it looks like we have an equivalence relation on the arrangements of distinct books on a bookshelf such that

1. each equivalence class has $k!$ elements, and

2. there is a bijection between the equivalence classes and k-element multisets of the n shelves.

Therefore, if we can compute the number of ways to arrange k *distinct* books on the n shelves of a bookcase, we should be able to apply the quotient principle to compute the number of k-element multisets of an n-element set.

The Bookcase Arrangement Problem

We have k books to arrange on the n shelves of a bookcase. The order in which the books appear on a shelf matters, and each shelf can hold all the books. We will assume that as the books are placed on the shelves, they are pushed as far to the left as they will go. Thus, all that matters is the order in which the books appear. When book i is placed on a shelf, it can go between two books already there or to the left or right of all the books on that shelf.

   a. Because the books are distinct, we may think of a first, second, third, ... book. In how many ways can we place the first book on the shelves?

   b. Once the first book has been placed, in how many ways can we place the second book?

   c. Once the first two books have been placed, in how many ways can we place the third book?

   d. Once we have placed $i - 1$ books, we may place book i on any of the shelves to the left of any of the books already there. But there are also some additional ways we may place it. In how many ways, in total, can we place book i?

   e. In how many ways can we place k distinct books on n shelves in accordance with the constraints above?

How many k-element multisets can we choose from an n-element set?

In Exercise 1.4-5, there are n places where the first book can go—namely, on the left side of any shelf. The next book can then go in any of the n places on the far left side of any shelf, or it can go to the right of Book 1. Thus, there are $n + 1$ places where Book 2 can go. At first, placing Book 3 appears to be more complicated, because we could create two different patterns by placing the first two books. However, Book 3 could go on the far left of any shelf or to the immediate right of any of the books already there. (Notice that if Book 2 and Book 1 are on Shelf 7 in that order, putting Book 3 to the immediate right of Book 2 means putting it between Book 2 and Book 1.) Thus, in any case, there are $n + 2$ ways to place Book 3. Similarly, once $i - 1$ books have been placed, there are $n + i - 1$ places to place Book i: it can go on the far left side of any of the n shelves or to the immediate right of any of the $i - 1$ books that we have already placed. Thus, the number of ways to place k distinct books is

$$n(n+1)(n+2)\cdots(n+k-1) = \prod_{i=1}^{k}(n+i-1)$$

$$= \prod_{j=0}^{k-1}(n+j) = \frac{(n+k-1)!}{(n-1)!}. \qquad (1.14)$$

The specific product that arose in Equation 1.14 is called a **rising factorial power**. It has a notation (also introduced by Donald E. Knuth) analogous to that for the falling factorial notation; namely, we write

$$n^{\overline{k}} = n(n+1)\cdots(n+k-1) = \prod_{i=1}^{k}(n+i-1)\,.$$

This is the product of k successive numbers beginning with n.

The Number of k-Element Multisets of an n-Element Set

We can apply the formula from Exercise 1.4-5 to solve Exercise 1.4-6. We define two bookcase arrangements of k books on n shelves to be equivalent if we get one from the other by permuting the books among themselves. Thus, if two arrangements put the same number of books on each shelf, then they are put into the same class by this relationship. On the other hand, if two arrangements put a different number of books on at least one shelf, then they are not equivalent, and therefore they are put into different classes by this relationship. Thus, the classes into which this relationship divides the arrangements are disjoint. Because every arrangement is in a class, the classes partition the set of all arrangements. Each class has $k!$ arrangements in it. The set of all arrangements has $n^{\overline{k}}$ arrangements in it. This leads to the following theorem.

Theorem 1.6	The number of k-element multisets chosen from an n-element set is

$$\frac{n^{\overline{k}}}{k!} = \binom{n+k-1}{k}\,.$$

Proof: In terms of bookcase arrangements, the relationship that two arrangements are equivalent if and only if we get one from the other by permuting the books is an equivalence relation. The set of all arrangements has $n^{\overline{k}}$ elements, and the number of elements in an equivalence class is $k!$. By the quotient principle, the number of equivalence classes is $n^{\overline{k}}/k!$. There is a bijection between equivalence classes of bookcase arrangements with k books and multisets with k elements. The second equality follows from the definition of binomial coefficients. ∎

The number of k-element multisets chosen from an n-element set is sometimes called "the number of **combinations with repetitions** of n elements taken k at a time."

The right side of the formula is a binomial coefficient, so it is natural to ask whether there is a way to interpret choosing a k-element *multi*set from an n-element set as choosing a k-element *sub*set of some different $(n+k-1)$-element set. This illustrates an important principle. When we have a quantity that turns out to be equal to a binomial coefficient, it helps our understanding to interpret it as counting the number of ways to choose a subset of an appropriate size from a set of an appropriate size. We explore this idea for multisets in Problem 8.

Using the Quotient Principle to Explain a Quotient

Because the last expression in Equation 1.14 is a quotient of two factorials, it is natural to ask whether it is counting equivalence classes of an equivalence relation. If so, the set on which the relation is defined has size $(n + k - 1)!$. Thus, it might be all lists or permutations of $n + k - 1$ distinct objects. The size of an equivalence class is $(n - 1)!$. Thus, what makes two lists equivalent might be permuting $n - 1$ of the objects among themselves. Said differently, the quotient principle suggests that we look for an explanation of the formula involving lists of $n + k - 1$ objects, of which $n - 1$ are identical, so that the remaining k elements are distinct. Can we find such an interpretation?

Exercise 1.4-7 In how many ways can we arrange k distinct books and $n - 1$ identical blocks of wood in a straight line?

Exercise 1.4-8 How does Exercise 1.4-7 relate to arranging books on the shelves of a bookcase?

In Exercise 1.4-7, if we tape numbers to the wood so that the pieces of wood are distinguishable, there are $(n + k - 1)!$ arrangements of the books and wood. But because the pieces of wood are actually indistinguishable, $(n - 1)!$ of these arrangements are equivalent. Thus, by the quotient principle, there are $(n + k - 1)!/(n - 1)!$ arrangements. Such an arrangement allows us to put the books on the shelves as follows: put all the books before the first piece of wood on Shelf 1, all the books between the first and second on Shelf 2, and so on, until you put all the books after the last piece of wood on Shelf n. This explains why there are $(n + k - 1)!/(n - 1)!$ arrangements of k distinct books on n shelves in a bookcase. Problem 8 explores a similar relationship for multisets.

Important Concepts, Formulas, and Theorems

1. *Symmetry principle.* If a mathematical result shows a certain symmetry, finding a proof that reflects this symmetry often helps our understanding.

2. *Partition.* Given a set S of items, a *partition* of S consists of m sets $S_1, S_2, \ldots, S_m$, sometimes called *blocks*, so that $S_1 \cup S_2 \cup \cdots \cup S_m = S$, and for each i and j with $i \neq j$, $S_i \cap S_j = \emptyset$.

3. *Equivalence relation/equivalence class.* A relationship that partitions a set into mutually exclusive classes is called an *equivalence relation*. Thus, if $S = S_1 \cup S_2 \cup \cdots \cup S_m$ is a partition of S, the relationship that says any two elements $x \in S$ and $y \in S$ are equivalent if and only if they lie in the same set S_i is an equivalence relation. The sets S_i are called *equivalence classes*.

4. *Quotient principle.* If a set of p objects can be partitioned into q classes of size r, then $q = p/r$. Equivalently, if an equivalence relation on a set of size p has q equivalence classes of size r, then $q = p/r$. The quotient principle is frequently used for counting the number of equivalence classes of an equivalence relation. If a quantity is a quotient of two others, it is often helpful

to our understanding to find a way to use the quotient principle to explain why we have this quotient.

5. *Multiset.* A multiset is similar to a set, except each item can appear multiple times. We can specify a *multiset* chosen from a set S by saying how many times each of its elements occurs.

6. *Choosing k-element multisets.* The number of k-element multisets that can be chosen from an n-element set is

$$\frac{(n + k - 1)!}{k! \, (n - 1)!} = \binom{n + k - 1}{k}.$$

This is sometimes called the formula for combinations with repetitions.

7. *Interpreting binomial coefficients.* When a quantity turns out to be a binomial coefficient (or some other formula we recognize), it is often useful to try to interpret the quantity as the result of choosing a subset of a set (or creating objects that the formula we recognize counts.)

Problems

1. In how many ways can n people be seated around a round table? (Remember that two seating arrangements around a round table are equivalent if everyone is in the same position relative to everyone else in both arrangements.) `a/h`

2. In how many ways can you embroider n circles of different colors in a row (lengthwise, equally spaced, and centered halfway between the top and bottom edges) on a scarf, as shown?

Figure 1.1

3. Use binomial coefficients to determine the number of ways in which you can line up three identical red apples and two identical golden apples. Use equivalence class counting (in particular, the quotient principle) to determine the same number. `a/h`

4. Use multisets to determine the number of ways to pass out k identical apples to n children. Assume that a child may get more than one apple.

5. In how many ways can n men and n women be seated around a table (as in Problem 1), alternating gender? (Use equivalence class counting!) `a/h`

6. In how many ways can you pass out k identical apples to n children if each child must get at least one apple?

7. In how many ways can you place k distinct books on n shelves of a bookcase (all books pushed to the left as far as possible) if there must be at least one book on each shelf? `a/h`

8. The formula for the number of multisets is $(n + k - 1)!$ divided by a product of two other factorials. We want to use the quotient principle to explain why this formula counts multisets. The formula for the number of multisets is also a binomial coefficient, so it should have an interpretation that involves choosing k items from $n + k - 1$ items. The parts of the problem that follow lead us to these explanations.

    a. In how many ways can you place k red checkers and $n - 1$ black checkers in a row?

    b. How can you relate the number of ways of placing k red checkers and $n - 1$ black checkers in a row to the number of k-element multisets of an n-element set (the set $\{1, 2, \ldots, n\}$ to be specific)?

    c. How can you relate the choice of k items out of $n + k - 1$ items to the placement of red and black checkers, as in parts a and b? Think about how this relates to placing k identical books and $n - k$ identical blocks of wood in a row.

9. How many solutions to the equation $x_1 + x_2 + \cdots + x_n = k$ are there with each x_i a nonnegative integer? a/h

10. How many solutions to the equation $x_1 + x_2 + \cdots + x_n = k$ are there with each x_i a positive integer?

11. In how many ways can n red checkers and $n + 1$ black checkers be arranged in a circle? (This number is a famous number called a Catalan number. While it is not particularly difficult to find the answer to this question, one detail in the proof that the answer is correct is somewhat sophisticated.) a/h

12. A standard notation for the number of partitions of an n-element set into k classes is $S(n, k)$. Because the empty family of subsets of the empty set is a partition of the empty set, $S(0, 0)$ is 1. In addition, $S(n, 0)$ is 0 for $n > 0$, because there are no partitions of a nonempty set into no parts. $S(1, 1)$ is 1.

    a. Explain why $S(n, n)$ is 1 for all $n > 0$. Explain why $S(n, 1)$ is 1 for all $n > 0$.

    b. Explain why $S(n, k) = S(n - 1, k - 1) + kS(n - 1, k)$ for $1 < k < n$.

    c. Make a table like Table 1.1 that shows the values of $S(n, k)$ for values of n and k ranging from 1 to 6.

13. You are given a square that can be rotated 90 degrees at a time (i.e., the square has four orientations). You are also given two red checkers and two black checkers, each to be placed on one corner of the square. How many lists of four letters, two of which are R and two of which are B, are there? Once you choose a starting place on the square, each list represents placing checkers on the square in clockwise order. Consider two lists to be equivalent if they represent the same arrangement of checkers at the corners of the square—that is, if one arrangement can be rotated to create the other. Write the equivalence classes of this equivalence relation. Why can't you apply Theorem 1.5 to compute the number of equivalence classes? a/h

14. The terms *reflexive, symmetric,* and *transitive* were defined in Footnote 9.

    a. Which of these properties is satisfied by the relationship of "greater than"? Which of these properties is satisfied by the relationship of "is a brother of"? Which of these properties is satisfied by "is a sibling of"? (You are not considered to be your own brother or your own sibling.) How about the relationship "is a sibling of or is"? (Note that Charles is a sibling of or is Charles.)

    b. Explain why an equivalence relation (as we have defined it) is a reflexive, symmetric, and transitive relationship.

    c. Suppose you have a reflexive, symmetric, and transitive relationship defined on a set S. For each x in S, let $S_x = \{y | y$ is related to $x\}$. Show that two such sets S_x and S_y are either disjoint or identical. Explain why this means that your relationship is an equivalence relation (as defined in this section of the book, not as defined in the footnote).

    d. Parts b and c prove that a relationship is an equivalence relation if and only if it is symmetric, reflexive, and transitive. Explain why. (A short answer is most appropriate here.)

15. Consider the following C++ function to compute $\binom{n}{k}$.

```
int pascal(int n, int k)
{
   if (n < k)
     {
       cout << "error: n<k" << endl;
       exit(1);
     }
   if ( (k==0) || (n==k))
     return 1;
   return pascal(n-1,k-1) + pascal(n-1,k);
}
```

    Enter this code, compile it, and run it (you will need to create a simple main program that calls it). Run it on larger and larger values of n and k, and observe the running time of the program. It should be surprisingly slow. (For example, try computing $\binom{30}{15}$.) Why is it so slow? Can you write a different function to compute $\binom{n}{k}$ that is *significantly* faster? Why is your new version faster? (Note: An exact analysis of this might be difficult at this point in the course; it will be easier later. However, you should be able to figure out roughly why the original version is so much slower.)

16. Answer the following questions with either n^k, $n^{\underline{k}}$, $\binom{n}{k}$, or $\binom{n+k-1}{k}$.

    a. In how many ways can k different candy bars be distributed to n people (with any person allowed to receive more than one bar)? a/h

    b. In how many ways can k different candy bars be distributed to n people (with nobody receiving more than one bar)?

c. In how many ways can k identical candy bars be distributed to n people (with any person allowed to receive more than one bar)? `a/h`

d. In how many ways can k identical candy bars be distributed to n people (with nobody receiving more than one bar)?

e. How many one-to-one functions f are there from $\{1, 2, \ldots, k\}$ to $\{1, 2, \ldots, n\}$? `a/h`

f. How many functions f are there from $\{1, 2, \ldots, k\}$ to $\{1, 2, \ldots, n\}$?

g. In how many ways can you choose a k-element subset from an n-element set? `a/h`

h. How many k-element multisets can be formed from an n-element set?

i. In how many ways can the top k-ranking officials in the U.S. government be chosen from a group of n people? (We want an ordered list of the people, not a set.) `a/h`

j. In how many ways can k pieces of candy (not necessarily of different types) be chosen from among n different types?

k. In how many ways can k children each choose one piece of candy (all of different types) from among n different types of candy? `a/h`

2 CRYPTOGRAPHY AND NUMBER THEORY

2.1 CRYPTOGRAPHY AND MODULAR ARITHMETIC

Introduction to Cryptography

For thousands of years, people have searched for ways to send messages secretly. There is a story that in ancient times a king needed to send a secret message to his general in battle. The king shaved a servant's head and wrote the message on it. The king waited for the servant's hair to grow back and then sent the servant to the general. The general then shaved the servant's head and read the message. If the enemy had captured the servant, they presumably would not have known to shave his head, and the message would have been safe.

Cryptography is the study of methods for sending and receiving secret messages. In general, there is a **sender** who is trying to send a message to a **receiver**. There is also an **adversary** who wants to steal the message. The method used is deemed successful if the sender is able to communicate a message to the receiver without the adversary learning what that message was.

Cryptography has been used for military and diplomatic purposes over the centuries. Recently, with the advent of the Internet and electronic commerce, cryptography has become vital for the the global economy and is used daily by millions of people. Sensitive information, such as bank records, credit card reports, passwords, or private communication, is (and should be) **encrypted**—modified in such a way that it should be understandable only to people who are allowed to have access to it and undecipherable to others.

Undecipherability by an adversary is, of course, a difficult goal. No code is completely undecipherable. If there is a printed codebook, then the adversary can always steal it. No amount of mathematical sophistication can prevent this possibility. More likely, an adversary may have extremely large amounts of computing power and human resources devoted to trying to crack a code. Thus, our notion of security is tied to computing power—a code is only as safe as the amount of computing power needed to break it. If we design codes that seem to need exceptionally large amounts of computing power to break them, then we can be relatively confident in their security.

Private-Key Cryptography

Traditional cryptography is known as **private-key cryptography**. The sender and receiver agree in advance on a secret code and then send messages using that code. For example, one of the oldest codes is a **Caesar cipher**. In this code, the letters of the alphabet are shifted by some fixed amount. Typically, the original message is called the **plaintext**. and the encoded text is called the **ciphertext**. The code

```
plaintext  A B C D E F G H I J K L M N O P Q R S T U V W X Y Z
ciphertext E F G H I J K L M N O P Q R S T U V W X Y Z A B C D
```

is an example of a Caesar cipher. Thus, if we wanted to send the plaintext message

ONE IF BY LAND AND TWO IF BY SEA ,

we would send the ciphertext

SRI MJ FC PERH ERH XAS MJ FC WIE .

A Caesar cipher is especially easy to implement on a computer using a scheme known as arithmetic mod 26. The symbolism

$$m \bmod n$$

means the remainder we get when we divide m by n. To be more precise, we give the following definition.

Definition 2.1	For an integer m and a positive integer n, $m \bmod n$ is the smallest nonnegative integer r such that

$$m = nq + r \qquad (2.1)$$

for some integer q.

A theorem we call Euclid's division theorem[1] tells us that there is always such an r. We prove this theorem in Section 2.2.

---

[1] In an unfortunate historical evolution of terminology, the statement we call Euclid's division theorem is often called the "division algorithm" and occasionally "Euclid's division algorithm." Because the theorem is not a computational procedure, we have chosen not to call it an algorithm.

| Theorem 2.1 | **(Euclid's Division Theorem)** Let n be a positive integer. Then for every integer m, there exist unique integers q and r such that $m = nq + r$ and $0 \leq r < n$. |

| Exercise 2.1-1 | Use the definition of m mod n to compute 10 mod 7 and -10 mod 7. What are q and r in each case? Does $(-m)$ mod $n = -(m \bmod n)$? |

| Exercise 2.1-2 | Using 0 for A, 1 for B, and so on, let the numbers from 0 to 25 stand for the letters of the alphabet. In this way, convert a message to a sequence of strings of numbers. For example, SEA becomes 18 4 0. What does the numerical representation of this word become if we shift every letter 2 places to the right? What if we shift every letter 13 places to the right? How can we use the idea of m mod n to implement a Caesar cipher? |

| Exercise 2.1-3 | Have someone use a Caesar cipher to encode a message of a few words in your favorite natural language without telling you how far they are shifting the letters of the alphabet. How can you figure out what the message is? Is this something a computer could do quickly? |

In Exercise 2.1-1, $10 = 7(1) + 3$. Thus, 10 mod 7 is 3. Also, $-10 = 7(-2) + 4$. Therefore, -10 mod 7 is 4. These two calculations show that $(-m)$ mod n *does not necessarily equal* $-(m \bmod n)$. (In fact, they are unequal unless m mod $n = 0$.) Note that -3 mod 7 is also 4. Furthermore, $(-10 + 3)$ mod $7 = 0$, suggesting that -10 is essentially the same as -3 when we are considering integers mod 7.

In Exercise 2.1-2, to shift each letter 2 places to the right, we replace each number n in our message by $(n + 2)$ mod 26, so that SEA becomes 20 6 2. To shift 13 places to the right, we replace each number n in our message with $(n + 13)$ mod 26, so that SEA becomes 5 17 13. Similarly, to implement a shift of s places, we replace each number n in our message by $(n + s)$ mod 26. Because most computer languages give us simple ways to handle strings of numbers and a mod function, it is easy to implement a Caesar cipher on a computer.

Exercise 2.1-3 considers the complexity of encoding, decoding, and cracking a Caesar cipher. Even by hand, it is easy for the sender to encode the message and for the receiver to decode the message. The disadvantage of this scheme is that it is also easy for the adversary to try the 26 different possible Caesar ciphers to decode the message. (It is very likely that only one will decode the message into plain English.) Of course, there is no reason to use such a simple code; we can use any arbitrary permutation of the alphabet as the ciphertext. For example,

```
plaintext  A B C D E F G H I J K L M N O P Q R S T U V W X Y Z
ciphertext H D I E T J K L M X N Y O P F Q R U V W G Z A S B C
```

shows one arbitrary permutation we could use. If we encode a short message with a code like this, it would be hard for the adversary to decode it. However, with a message of any reasonable length (greater than about 50 letters), an adversary with a knowledge of the relative frequency of alphabet letters in the English language can easily crack the code. (These codes appear in many newspapers and puzzle books under the name "cryptograms." The fact that many people are able to solve these puzzles is compelling evidence of the lack of security in such a code.)

We do not have to use simple mappings of letters to letters. For example, our coding algorithm could be to do the following:

Step 1: Take three consecutive letters.

Step 2: Reverse their order.

Step 3: Interpret each triple as a base 26 integer (with $A = 0$; $B = 1$, etc.) and then convert that base 26 integer to base 10.

Step 4: Multiply that number by 37 (in base 10).

Step 5: Add 95.

Step 6: Convert that number to base 8.

We continue this processing with each block of three consecutive letters. We append the blocks, using either an 8 or a 9 to separate them. When we are done, we reverse the number and replace each digit 5 with two 5s. Here is an example of this method:

```
plaintext: ONEIFBYLANDTWOIFBYSEA

block and reverse: ENO BFI ALY TDN IOW YBF AES
base 26 integer converted to base 10: 3056 814 310 12935 5794 16255 122
*37 +95 converted to base 8: 335017 73005 26455 1646742 642711 2226672 11001
appended: 3350178730059264559164674296427118222667811001
reverse, 5rep: 10011827662228117246924764619555546295500378710 5533
```

As Problem 19 shows, a receiver who knows the code can decode this message. However, a casual reader of the message, without knowledge of the encryption algorithm, would have no hope of decoding the message. So it seems that with a complicated enough code, we can have secure cryptography. Unfortunately, there are at least two flaws with this method. The first is that if the adversary somehow learns what the code is, she can easily decode it. Second, if this coding scheme is repeated often enough, and if the adversary has enough time, money, and computing power, she could break this code. In the field of cryptography, some entities, such as a government or a large corporation, have all these resources. The infamous German Enigma code is an example of a much more complicated coding scheme; yet successive versions of it were broken, which helped the Allies win World War II. (Breaking the code was aided by the Enigma machine recreated by Polish code breakers and later by machines from captured German ships. However, even with the machines, it was not easy to break the code.) In general, any scheme that uses a **codebook**—a secretly agreed upon (possibly complicated) code—suffers from these drawbacks.

Public-Key Cryptosystems

A **public-key cryptosystem** overcomes the problems associated with using a codebook. In a public-key cryptosystem, the sender and receiver (often called Alice and

Bob, respectively) don't have to agree in advance on a secret code. In fact, they each publish part of their code in a public directory. Yet, an adversary with access to the encoded message and the public directory still cannot decode the message.

More precisely, Alice and Bob each have two keys, a **public key** and a **secret key.** We denote Alice's public and secret keys by KP_A and KS_A and Bob's by KP_B and KS_B. They each keep their secret keys to themselves but can publish their public keys and make them available to anyone, including the adversary. Although the published key is likely to be a symbol string of some sort, the key is used in some standardized way (we shall see examples soon) to create a function from the set $\mathcal{D}$ of possible messages onto itself. (In complicated cases, the key might be the actual function.) We denote the functions associated with KS_A, KP_A, KS_B, and KP_B by S_A, P_A, S_B, and P_B, respectively. We require that the public and secret keys be chosen so that the corresponding functions are inverses of each other, that is, for any message $M \in \mathcal{D}$, we have that

$$M = S_A(P_A(M)) = P_A(S_A(M)) \text{ , and}$$
$$M = S_B(P_B(M)) = P_B(S_B(M)) \text{ .}$$
(2.2)

We also assume that for Alice, S_A and P_A are easily computable. However, it is essential that S_A is hard for everyone except Alice to compute, even if P_A is known. At first glance, this may seem an impossible task: Alice must create a function, P_A, that is public and easy for everyone to compute, yet this function has an inverse, S_A, that is hard for everyone except Alice to compute. It is not at all clear how to design such a function. In fact, when the idea for public-key cryptography was proposed (by Diffie and Hellman[15]), no one knew of any such functions. The first complete public-key cryptosystem is the now-famous RSA cryptosystem, which is widely used in many contexts. This system was developed by Ronald Rivest, Adi Shamir, and Leonard Adleman[28], hence its name. To understand how such a cryptosystem is possible requires some knowledge of number theory and computational complexity. We develop the necessary number theory in the next few sections. Before doing so, however, let us assume that we have a function that is easily computed but that can be inverted only by Alice. This will show us how to make use of it.

If Bob wants to send Alice a message M, he takes the following steps:

1. Bob obtains Alice's public key P_A.

2. Bob applies Alice's public key to M to create ciphertext $C = P_A(M)$.

Bob then sends C to Alice. Alice can decode the message by using her secret key to compute $S_A(C)$, which is identical to $S_A(P_A(M))$, which by Equation 2.2 is identical to the original message M. The beauty of the scheme is that even if the adversary has C and knows P_A, she cannot decode the message without S_A, because S_A is a secret that only Alice has. Even though the adversary knows that S_A is the inverse of P_A, she cannot easily compute this inverse.

Because it is difficult, at this point, to describe an example of a public-key cryptosystem that is hard to decode, we now give an example of one that is easy to decode. Imagine our messages are numbers in the range 1 to 999. We can then imagine that Bob's public key yields the function P_B given by $P_B(M) = rev(1000 - M)$, where $rev()$ is a function that reverses the digits of a number. So, to encrypt the message 167, Alice would compute $1000 - 167 = 833$, reverse the digits, and send Bob $C = 338$. In this case, $S_B(C) = 1000 - rev(C)$, so Bob can easily decode the message. This code is not secure, because if you know P_B, you can figure out S_B. The challenge is to design a function P_B so that even if you know P_B and $C = P_B(M)$, it is exceptionally difficult to figure out what M is.

Arithmetic Modulo n

The RSA encryption scheme is built on the idea of arithmetic mod n, which we now introduce. Our goal is to understand how the basic arithmetic operations of addition, subtraction, multiplication, division, and exponentiation behave when all arithmetic is done mod n. As we shall see, some of the operations, such as addition, subtraction, and multiplication, are straightforward. Others, such as division and exponentiation, behave very differently from how they behave for normal arithmetic.

Exercise 2.1-4   Compute   21 mod 9,   38 mod 9, $(21 \cdot 38)$ mod 9, $(21 \bmod 9) \cdot (38 \bmod 9)$, $(21 + 38)$ mod 9, $(21 \bmod 9) + (38 \bmod 9)$.

Exercise 2.1-5   True or false: $i \bmod n = (i + 2n) \bmod n$; $i \bmod n = (i - 3n) \bmod n$.

In Exercise 2.1-4, the point to notice is that

$$(21 \cdot 38) \bmod 9 = (21 \bmod 9) \cdot (38 \bmod 9)$$

and

$$(21 + 38) \bmod 9 = (21 \bmod 9) + (38 \bmod 9) .$$

These equations are very suggestive, though the general equations they first suggest aren't true! As we shall soon see, some closely related equations are true.

Exercise 2.1-5 is true in both cases, because adding multiples of n to i does not change the value of $i \bmod n$. In general, we have the following:

Lemma 2.2 $i \bmod n = (i + kn) \bmod n$ for any integer k.

Proof: By Theorem 2.1, for unique integers q and r, with $0 \le r < n$, we have

$$i = nq + r . \tag{2.3}$$

Adding kn to both sides of Equation 2.3, we obtain

$$i + kn = n(q + k) + r . \tag{2.4}$$

Applying the definition of $i \bmod n$ to Equation 2.3, we have that $r = i \bmod n$; applying the same definition to Equation 2.4 we have that $r = (i + kn) \bmod n$. The lemma follows. ∎

Now we can go back to the equations of Exercise 2.1-4; the correct versions are stated below. Informally, we are showing that if we have a computation involving addition and multiplication and if we plan to take the end result mod n, then we are free to take any of the intermediate results mod n also.

Lemma 2.3	$(i + j) \bmod n = (i + (j \bmod n)) \bmod n$

$$(i + j) \bmod n = (i + (j \bmod n)) \bmod n$$
$$= ((i \bmod n) + j) \bmod n$$
$$= ((i \bmod n) + (j \bmod n)) \bmod n$$
$$(i \cdot j) \bmod n = (i \cdot (j \bmod n)) \bmod n$$
$$= ((i \bmod n) \cdot j) \bmod n$$
$$= ((i \bmod n) \cdot (j \bmod n)) \bmod n .$$

Proof: We prove that the first and last terms in the sequence of equations for sums are equal; the other equalities for sums follow by similar computations. The proofs of the equalities for products are similar.

By Theorem 2.1, we have that for unique integers q_1 and q_2,

$$i = (i \bmod n) + q_1 n \text{ and } j = (j \bmod n) + q_2 n .$$

Adding these two equations together mod n and using Lemma 2.2, we obtain

$$(i + j) \bmod n = ((i \bmod n) + q_1 n + (j \bmod n) + q_2 n) \bmod n$$
$$= ((i \bmod n) + (j \bmod n) + n(q_1 + q_2)) \bmod n$$
$$= ((i \bmod n) + (j \bmod n)) \bmod n . \ ∎$$

We now introduce a convenient notation for modular arithmetic. We use the notation Z_n to represent the integers $0, 1, \ldots, n - 1$ together with a redefinition of addition, which we denote $+_n$, and a redefinition of multiplication, which we denote $\cdot_n$. The redefinitions are

$$i +_n j = (i + j) \bmod n$$
$$i \cdot_n j = (i \cdot j) \bmod n .$$
(2.5)

The expression $x \in Z_n$ means that x is a variable that can take on any of the integral values between 0 and $n - 1$. In addition, $x \in Z_n$ is a signal that if we do algebraic operations with x, we will use $+_n$ and $\cdot_n$ rather than the usual addition and multiplication.

In ordinary algebra, it is traditional to use letters near the beginning of the alphabet to stand for constants—numbers that are fixed throughout a problem and that would be known in advance in any one instance of that problem. This allows us to describe the solution to many different variations of a problem all at once. Thus, we might say, "For all integers a and b, there is one and only one integer x that is a solution to the equation $a + x = b$, namely, $x = b - a$." We adopt the same system for Z_n. When we say, "Let a be a member of Z_n," we mean the same thing as, "Let a be an integer between 0 and $n - 1$," but we are also signaling that the value of a does not change during the problem and that in equations involving a, we will use $+_n$ and $\cdot_n$.

We call these new operations addition mod n and multiplication mod n. We must now verify that all the "usual" rules of arithmetic that normally apply to addition and multiplication still apply with $+_n$ and $\cdot_n$. In particular, we wish to verify the commutative, associative, and distributive laws.

Theorem 2.4	Addition and multiplication mod n satisfy the commutative and associative laws, and multiplication distributes over addition.

Proof: Commutativity of $+_n$ and $\cdot_n$ follows immediately from the commutativity of ordinary addition and multiplication. We prove the associative law for addition in the following equations; the other laws follow similarly.

$$
\begin{aligned}
a +_n (b +_n c) &= (a + (b +_n c)) \bmod n & \text{(Equation 2.5)} \\
&= (a + ((b + c) \bmod n)) \bmod n & \text{(Equation 2.5)} \\
&= (a + (b + c)) \bmod n & \text{(Lemma 2.3)} \\
&= ((a + b) + c) \bmod n & \text{(Associative law for ordinary sums)} \\
&= \Big(((a + b) \bmod n) + c\Big) \bmod n & \text{(Lemma 2.3)} \\
&= ((a +_n b) + c) \bmod n & \text{(Equation 2.5)} \\
&= (a +_n b) +_n c & \text{(Equation 2.5)} . \blacksquare
\end{aligned}
$$

Notice that $0 +_n i = i$ and $1 \cdot_n i = i$ (these equations are called the **additive identity properties** and the **multiplicative identity properties**). Also, $0 \cdot_n i = 0$. Thus, we can use 0 and 1 in algebraic expressions in Z_n (which we may also refer to as algebraic expressions mod n) the same way we use them in ordinary algebraic expressions. We use $a -_n b$ to stand for $a +_n (-b)$.

We conclude this section by observing that repeated applications of Lemma 2.3 and Theorem 2.4 are useful when computing sums or products mod n in which the numbers are large. For example, suppose you had m integers $x_1, \ldots, x_m$ and you wanted to compute $\left(\sum_{j=1}^{m} x_i\right) \bmod n$. One natural way to do so would be to compute the sum and then take the result modulo n. It is possible, however, that on the computer you are using, even though $\left(\sum_{j=1}^{m} x_i\right) \bmod n$ is a number that can be stored in an integer and each x_i can be stored in an integer, $\sum_{j=1}^{m} x_i$ might be too large to be stored in an integer. (Recall that integers are typically stored as four or eight bytes and thus have a maximum value of roughly $2 \cdot 10^9$ or $9 \cdot 10^{18}$.) Lemma 2.3 tells us that if we are computing a result mod n, we may do all our calculations in Z_n using $+_n$ and $\cdot_n$. Thus, we never have to compute an integer that has significantly more digits than any of the numbers with which we are working.

Cryptography Using Addition mod n

One natural way to use addition of a number a mod n in encryption is first to convert the message to a sequence of digits—say, concatenating all the ASCII codes for all the symbols in the message—and then to add a to the message mod n. Thus,

$$P(M) = M +_n a \quad \text{and} \quad S(C) = C +_n (-a) = C -_n a \,.$$

If n happens to be larger in numerical value than the message, then it is simple for someone who knows a to decode the encrypted message. However, an adversary who sees the encrypted message has no special knowledge. Therefore, unless a was ill-chosen (for example, having all or most of the digits be 0 would be a silly choice), the adversary who knows what system you are using, even including the value of n, but who does not know a, is essentially reduced to trying all possible a values. (In effect, adding a appears to the adversary much like changing digits at random.) Because you use a only once, there is virtually no way for the adversary to collect any data that will aid in guessing a. Thus, if only you and your intended recipient know a, this kind of encryption is quite secure: guessing a is just as hard as guessing the message.

It is possible that once n has been chosen, you will find you have a message that translates to a number larger than n. Normally you would then break the message into segments, each with no more digits than n, and send the segments individually. It might seem that as long as you were not sending a large number of segments, it would still be quite difficult for your adversary to guess a by observing the encrypted information. However, if your adversary knew n but not a and knew you were adding a mod n, he could take two messages and subtract them in Z_n, thus getting the difference of two unencrypted messages. (Problem 13 asks you to explain why, even if your adversary didn't know n but believed you were adding some secret number a mod some other secret number n, he could use three encoded messages to find three differences in the integers, instead of in Z_n, one of which was the difference of two messages.) This difference could contain valuable information for your adversary.[2] Even worse, if your adversary could trick you into sending just one message z that he knows, intercepting the message and subtracting z would give your adversary a. Thus, adding a mod n is not an encoding method you would want to use more than once.

---

[2]If each segment of a message were equally likely to be any number between 0 and n, and if any second (or third, etc.) segment were equally likely to follow any first segment, then knowing the difference between two segments would yield no information about the two segments. However, because language is structured and most information is structured, these two conditions are highly unlikely to hold, in which case your adversary could apply structural knowledge to deduce information about your two messages from their difference.

Cryptography Using Multiplication mod n

We now explore whether multiplication is a good method for encryption. In particular, we could encrypt by multiplying a message mod n by a prechosen value a. We would then expect to decrypt by dividing by a, except that we have not yet defined division in this context. What exactly does division mod a mean? Informally, we think of division as the inverse of multiplication—that is, if we take a number x, multiply by a, and then divide by a, we should get back to x. Clearly, with normal arithmetic, this is the case. However, with modular arithmetic, division is trickier.

Exercise 2.1-6	One possibility for encryption is to take a message x and compute $a \cdot_n x$ for some value a that the sender and receiver both know. You could then decrypt by dividing by a in Z_n, if you knew how to divide in Z_n. How well does this work? In particular, consider the following three cases. First, consider $n = 12$, $a = 4$, and $x = 3$. Second, consider $n = 12$, $a = 3$, and $x = 6$. Third, consider $n = 12$, $a = 5$, and $x = 7$. In each case, if your recipient knows a, could he or she figure out the message x? For this question, you don't need to know what dividing means. There is at least one other way to try to figure out the message.

When we encoded a message by adding a in Z_n, we could decode the message simply by subtracting a in Z_n. By analogy, if we encode by multiplying by a in Z_n, we would expect to decode by dividing by a in Z_n. However, Exercise 2.1-6 shows that division in Z_n doesn't always make much sense. Suppose your value of n is 12 and the value of a is 4. You send the message 3 as $4 \cdot_{12} 3 = 0$. Thus, you send the encoded message 0. Your recipient sees 0 and says the message might have been 0; after all, $4 \cdot_{12} 0 = 0$. On the other hand, $4 \cdot_{12} 3 = 0$, $4 \cdot_{12} 6 = 0$, and $4 \cdot_{12} 9 = 0$. Therefore, your recipient has four different choices for the original message, which is almost as bad as having to guess the original message itself!

It might appear that special problems arose only because the encoded message was 0. The second case in Exercise 2.1-6 gives an encoded message that is not 0. Suppose $a = 3$ and $n = 12$. You encode the message 6 by computing $3 \cdot_{12} 6 = 6$. Straightforward calculation shows that $3 \cdot_{12} 2 = 6$, $3 \cdot_{12} 6 = 6$, and $3 \cdot_{12} 10 = 6$. Thus, the message 6 can be decoded in three possible ways: 2, 6, or 10.

The final case in Exercise 2.1-6 provides some hope. Let $a = 5$ and $n = 12$. The message 7 is encoded as $5 \cdot_{12} 7 = 11$. Simple checking of $5 \cdot_{12} 1$, $5 \cdot_{12} 2$, $5 \cdot_{12} 3$, and so on shows that 7 is the unique solution in Z_{12} to the equation $5 \cdot_{12} x = 11$. Thus, in this case, the recipient can correctly decode the message.

One key point that this exercise shows is that our system of encrypting messages must be one-to-one. That is, each unencrypted message must correspond to a different encrypted message.

As we shall see in Section 2.2, the kinds of problems in Exercise 2.1-6 happen only when a and n have a common divisor that is greater than 1. Thus, when a and n have no common factors greater than 1, all our receiver needs to know to decrypt the message is how to divide by a in Z_n. If you don't know how to divide by a in Z_n, then you can begin to understand the idea of public-key cryptography. The message is there to find for anyone who knows how to divide by a. However, if nobody but our receiver can divide by a, we can tell everyone what a and n are and our messages will still be secret. This is the second point our system illustrates: If we have some knowledge that nobody else has, such as how to divide by a mod n, then we have a possible public-key cryptosystem. As we shall soon see, however, dividing by a is not particularly difficult, so a better trick is needed for public-key cryptography to work.

Important Concepts, Formulas, and Theorems

1. *Cryptography. Cryptography* is the study of methods for sending and receiving secret messages.

    a. The *sender* wants to send a message to a *receiver*.

    b. The *adversary* wants to steal the message.

    c. In *private-key cryptography*, the sender and receiver agree in advance on a *secret code* and then send messages using that code.

    d. In *public-key cryptography*, the encoding method can be published. Each person has a *public key* used to encrypt messages and a *secret key* used to decode an encrypted message.

    e. The original message is called the *plaintext*.

    f. The encoded text is called the *ciphertext*.

2. A *Caesar cipher* is a code in which each letter of the alphabet is shifted by a fixed amount.

3. *Euclid's division theorem.* Let n be a positive integer. Then for every integer m, there exist unique integers q and r such that $m = nq + r$ and $0 \leq r < n$. By definition, r is equal to m mod n.

4. *Adding multiples of n does not change values* mod n. This means that $i \bmod n = (i + kn) \bmod n$ for any integer k.

5. *Mods (by* n) *can be taken anywhere in calculation involving only addition and multiplication, as long as we take the final result* mod n.

$$(i + j) \bmod n = (i + (j \bmod n)) \bmod n$$
$$= ((i \bmod n) + j) \bmod n$$
$$= ((i \bmod n) + (j \bmod n)) \bmod n$$

$$(i \cdot j) \bmod n = (i \cdot (j \bmod n)) \bmod n$$
$$= ((i \bmod n) \cdot j) \bmod n$$
$$= ((i \bmod n) \cdot (j \bmod n)) \bmod n$$

6. *Commutative, associative, and distributive laws.* Addition and multiplication mod n satisfy the commutative and associative laws, and multiplication distributes over addition.

7. Z_n. Use the notation Z_n to represent the integers $0, 1, \ldots, n - 1$ together with a redefinition of addition, which we denote $+_n$, and a redefinition of multiplication, which we denote $\cdot_n$. The redefinitions are

$$i +_n j = (i + j) \bmod n \text{ and}$$

$$i \cdot_n j = (i \cdot j) \bmod n .$$

The expression $x \in Z_n$ means that x is a variable that can take on any of the integral values between 0 and $n - 1$ and that, in algebraic expressions involving x, we will use $+_n$ and $\cdot_n$. We use the expression $a \in Z_n$ to mean that a is a constant between 0 and $n - 1$ and that in algebraic expressions involving a, we will use $+_n$ and $\cdot_n$.

Problems

1. What is 14 mod 9? What is -1 mod 9? What is -11 mod 9? a/h

2. Encrypt the message HERE IS A MESSAGE using a Caesar cipher in which each letter is shifted three places to the right.

3. Encrypt the message HERE IS A MESSAGE using a Caesar cipher in which each letter is shifted three places to the left. a/h

4. How many places has each letter been shifted in the Caesar cipher used to encode the message XNQQD RJXXFLJ?

5. What is $16 +_{23} 18$? What is $16 \cdot_{23} 18$? a/h

6. A short message was encoded by converting it to an integer by replacing each "a" with 1, each "b" with 2, and so on, and then concatenating the integers. The result had six or fewer digits. An unknown number a was added to the message mod 913,647, giving 618,232. Without knowledge of a, what can you say about the message? With knowledge of a, what could you say about the message?

7. What would it mean to say there is an integer x equal to $(1/4)$ mod 9? If it is meaningful to say there is such an integer, what is it? Is there an integer equal to $(1/3)$ mod 9? If so, what is it? a/h

8. Multiplying a number x by 487 in Z_{30031} gives 13,008. If you know how to find the number x, do so. If not, explain why the problem seems difficult to do by hand.

9. Write the addition table for $+_7$ addition. Why is the table symmetric? Why does every number appear in every row? a/h

10. It is straightforward to solve for x any equation of the form

$$x +_n a = b$$

in Z_n and to see that the result will be a unique value of x. However, in the discussion of Exercise 2.1-6, we saw that 0, 3, 6, and 9 are all solutions to the equation

$$4 \cdot_{12} x = 0.$$

    a. Are there any integral values of a and b, with 1 less than or equal to a and b, which are both less than 12, for which the equation $a \cdot_{12} x = b$ does not have any solutions in Z_{12}? If there are, give one set of values for a and b. If there are not, explain how you know this.

    b. Are there any integers a, with $1 < a < 12$, such that for every integral value of b, with $1 \le b < 12$, the equation $a \cdot_{12} x = b$ has a solution? If so, give one and explain why it works. If not, explain how you know this.

11. Does every equation of the form $a \cdot_n x = b$, with $a, b \in Z_n$ and $a \neq 0$, have a solution in Z_5? In Z_7? In Z_9? In Z_{11}? a/h

12. Recall that if a prime number divides a product of two integers, then it divides one of the factors.

    a. Use this to show that as b runs though the integers from 0 to $p-1$, with p prime, the products $a \cdot_p b$ are all different (for each fixed choice of a between 1 and $p-1$).

    b. Explain why every integer greater than 0 and less than p has a unique multiplicative inverse in Z_p if p is prime.

13. Explain why, if you were encoding messages x_1, x_2, and x_3 to obtain y_1, y_2, and y_3 by adding an arbitrary number a and taking the sum mod n, your adversary would know that at least one of the differences $y_1 - y_2$, $y_1 - y_3$, or $y_2 - y_3$ taken in the integers, not in Z_n, would be the difference of two unencoded messages. (*Note:* We are not saying that your adversary would know which of the three was such a difference.) a/h

14. Write the $\cdot_7$ multiplication table for Z_7.

15. Prove the equalities for multiplication in Lemma 2.3.

16. State and prove the associative law for $\cdot_n$ multiplication. a/h

17. State and prove the distributive laws for $\cdot_n$ multiplication over $+_n$ addition.

18. Write pseudocode to take m integers $x_1, x_2, \ldots, x_m$ and an integer n and return $(\Pi_i^m x_i)$ mod n. Be careful about overflow; in this context, being careful about overflow means that at no point should you ever compute a value that is greater than n^2.

19. Write pseudocode to decode a message that has been encoded using the following algorithm.

    Step 1: Take three consecutive letters.

    Step 2: Reverse their order.

    Step 3: Interpret each as a base 26 integer (with A = 0, B = 1, etc.) and convert back to base 10.

Step 4: Multiply that number by 37 (in base 10).

Step 5: Add 95.

Step 6: Convert that number to base 8.

Continue this process with each block of three consecutive letters. Append the blocks, using either an 8 or a 9 to separate them. Finally, reverse the number, and replace each digit 5 with two 5s.

2.2 INVERSES AND GREATEST COMMON DIVISORS

Solutions to Equations and Inverses mod n

In Section 2.1, we explored multiplication in Z_n. In the special case of $n = 12$ and $a = 4$, if we used multiplication by a in Z_n to encrypt a message, then our receiver would need to be able to solve the equation $4 \cdot_n x = b$ for x to decode a received message b. We saw that if the encrypted message was 0, then there were four possible values for x. More generally, Exercise 2.1-6 and some of the problems in Section 2.1 showed that for certain values of n, a, and b, equations of the form $a \cdot_n x = b$ have a unique solution, while for other values of n, a, and b, the equation could have no solutions or more than one solution.

To decide whether an equation of the form $a \cdot_n x = b$ has a unique solution in Z_n, it helps to know whether a has a **multiplicative inverse** in Z_n, that is, whether there is another number a' such that $a' \cdot_n a = 1$. For example, in Z_9, the inverse of 2 is 5 because $2 \cdot_9 5 = 1$. On the other hand, 3 does not have an inverse in Z_9, because the equation $3 \cdot_9 x = 1$ does not have a solution. (This can be verified by checking the nine possible values for x.) If a does have an inverse a', then we can find a solution to the equation

$$a \cdot_n x = b .$$

To do so, we multiply both sides of the equation by a', obtaining

$$a' \cdot_n (a \cdot_n x) = a' \cdot_n b .$$

By the associative law, this gives us

$$(a' \cdot_n a) \cdot_n x = a' \cdot_n b .$$

But $a' \cdot_n a = 1$ by definition, so we have that

$$x = a' \cdot_n b .$$

Because this computation is valid for any x that satisfies the equation, we conclude that the only x that satisfies the equation is $a' \cdot_n b$. We summarize this discussion in the following lemma.

Lemma 2.5	Suppose a has a multiplicative inverse a' in Z_n. Then for any $b \in Z_n$, the equation	

$$a \cdot_n x = b$$

has the unique solution

$$x = a' \cdot_n b.$$

Note that this lemma holds for *any* value of $b \in Z_n$.

Lemma 2.5 tells us that whether or not a number has an inverse mod n is important for the solution of modular equations. We therefore wish to understand exactly when a member of Z_n has an inverse.

Inverses mod n

We now consider some of the examples related to Problem 11 from Section 2.1.

Exercise 2.2-1 Determine whether every nonzero element a of Z_n has an inverse for $n = 5, 6, 7, 8,$ and 9.

Exercise 2.2-2 If an element of Z_n has a multiplicative inverse, can it have two different multiplicative inverses?

The following table gives multiplicative inverses for each nonzero element a of Z_5. We created the table by multiplying each number in the top row by all nonzero members of Z_5. For example, the products $2 \cdot_5 1 = 2$, $2 \cdot_5 2 = 4$, $2 \cdot_5 3 = 1$, and $2 \cdot_5 4 = 3$ tell us that 3 is the unique multiplicative inverse for 2 in Z_5, which is why we put 3 below 2 in the table. We could make the same kinds of computations with 3 or 4 instead of 2 on the left side of the products to get the rest of the table.

a	1	2	3	4
a'	1	3	2	4

Similarly, for Z_7, we have the following table.

a	1	2	3	4	5	6
a'	1	4	5	2	3	6

For Z_9, we have already said that $3 \cdot_9 x = 1$ does not have a solution, so by Lemma 2.5, the number 3 does not have an inverse. (Notice how we are using the lemma: Lemma 2.5 says that if 3 had an inverse, then the equation $3 \cdot_9 x = 1$ *would* have a solution, and this would contradict the fact that $3 \cdot_9 x = 1$ does not have a solution. Thus, assuming that 3 had an inverse would lead us to a contradiction. Therefore, 3 has no multiplicative inverse.)

This computation is a special case of the following corollary[3] to Lemma 2.5.

Corollary 2.6 Suppose there is a b in Z_n such that the equation

$$a \cdot_n x = b$$

does not have a solution. Then a does not have a multiplicative inverse in Z_n.

Proof: Suppose that $a \cdot_n x = b$ has no solution. Suppose further that a does have a multiplicative inverse a' in Z_n. By Lemma 2.5, $x = a'b$ is a solution to the equation $a \cdot_n x = b$. This contradicts the hypothesis given in the corollary that the equation does not have a solution. Thus, some supposition we made must be incorrect. One of the assumptions—namely, that $a \cdot_n x = b$ has no solution—was the hypothesis given in the corollary's statement. The only other supposition we made was that a has an inverse a' in Z_n. This supposition must be incorrect, because it led to the contradiction. Therefore, it must be the case that a does not have a multiplicative inverse in Z_n. ■

Our proof of the corollary is a classical example of the use of the principle of **proof by contradiction**.

Principle 2.1 **(Proof by Contradiction)** If by assuming a statement we want to prove is false we are led to a contradiction, then the statement we are trying to prove must be true.

We now complete our discussion of Exercise 2.2-1. The following table shows an X for the nonzero elements of Z_9 that do not have inverses and gives an inverse for each element that has one.

a	1	2	3	4	5	6	7	8
a'	1	5	X	7	2	X	4	8

In Z_6, the number 1 has an inverse—namely, 1—but the equations

$$2 \cdot_6 1 = 2, \quad 2 \cdot_6 2 = 4, \quad 2 \cdot_6 3 = 0, \quad 2 \cdot_6 4 = 2, \quad \text{and} \quad 2 \cdot_6 5 = 4$$

tell us that 2 does not have an inverse. Less directly, but with less work, we see that the equation $2 \cdot_6 x = 3$ has no solution because $2x$ will always be even, so $2x \bmod 6$

---

[3]In Section 2.3, we show that this corollary is actually equivalent to part of Lemma 2.5.

will always be even. Corollary 2.6 tells us that 2 has no inverse. Once again, we give a table that shows exactly which nonzero elements of Z_6 have inverses.

a	1	2	3	4	5
a'	1	X	X	X	5

A similar set of equations shows that 2 does not have an inverse in Z_8. The following table shows which nonzero elements of Z_8 have inverses.

a	1	2	3	4	5	6	7
a'	1	X	3	X	5	X	7

We see that every nonzero element in Z_5 and Z_7 does have a multiplicative inverse, but in Z_6, Z_8, and Z_9, some elements do not have a multiplicative inverse. Notice that 5 and 7 are prime, while 6, 8, and 9 are not. Further, notice that, in all our examples, the elements in Z_n that do not have a multiplicative inverse are exactly those that share a common factor with n.

We showed that 2 has exactly one inverse in Z_5 by checking each multiple of 2 in Z_5 and showing that exactly one multiple of 2 equals 1. In fact, for any element that has an inverse in Z_5, Z_6, Z_7, Z_8, and Z_9, you can check in the same way that it has exactly one inverse. The following theorem explains why.

Theorem 2.7 If an element of Z_n has a multiplicative inverse, then it has exactly one inverse.

Proof: Suppose an element a of Z_n has an inverse a'. Suppose that a^* is also an inverse of a. Then a' is a solution to $a \cdot_n x = 1$, and a^* is a solution to $a \cdot_n x = 1$. By Lemma 2.5, however, the equation $a \cdot_n x = 1$ has a unique solution. Therefore, $a' = a^*$. ∎

Just as we use a^{-1} to denote the inverse of a in the real numbers, we use a^{-1} to denote the unique inverse of a in Z_n when a has an inverse. We define what we mean by *dividing* a member of Z_n by a in the case that a has an inverse a^{-1} mod n. If a has a multiplicative inverse, we define dividing b by a in Z_n to be the same as multiplying b by a^{-1} mod n. We were led to our discussion of inverses because of their role in solving equations. We observed that in our examples, an element of Z_n that has an inverse mod n has no factors greater than 1 in common with n. This is a statement about a and n as integers with ordinary multiplication rather than multiplication mod n. Thus, to prove that a has an inverse mod n if and only if a and n have no common factors other than 1 and -1, we have to convert the equation $a \cdot_n x = 1$ into an equation involving ordinary multiplication.

Converting Modular Equations to Normal Equations

We can reexpress the equation

$$a \cdot_n x = 1$$

as

$$ax \bmod n = 1 .$$

But $ax \bmod n$ is defined as the remainder r that we get when we write $ax = qn + r$, with $0 \leq r < n$. This means that $ax \bmod n = 1$ if and only if there is an integer q with $ax = qn + 1$, or

$$ax - qn = 1 .$$

Thus, we have shown the following:

Lemma 2.8	The equation $$a \cdot_n x = 1$$ has a solution in Z_n if and only if there exist integers x and y such that $$ax + ny = 1 . \qquad (2.6)$$

Proof: We simply take $y = -q$ in the equation $ax - qn = 1$. ∎

We make the change from $-q$ to y for two reasons. First, if you read a book on number theory, you are more likely to see the equation with y in this context. Second, to solve this equation, we must find both x and y. Using a letter near the end of the alphabet in place of $-q$ emphasizes that this is a variable for which we need to solve.

It appears that we have made our work harder, not easier. We have converted the problem of solving the equation $a \cdot_n x = 1$ in Z_n, an equation with just one variable x that could only have $n - 1$ different values, to a problem of solving Equation 2.6, which has two variables, x and y. Further, in this second equation, x and y can take on any integer values, even negative values.

However, we will see that this equation is exactly what we need in order to prove that a has an inverse mod n if and only if a and n have no common factors larger than 1.

Greatest Common Divisors

Exercise 2.2-3	Suppose a and n are integers such that n is positive and $ax + ny = 1$ for some integers x and y. What does that tell us about being able to find a multiplicative inverse for a (mod n)? In this situation, if a has an inverse in Z_n, what is the inverse?

| Exercise 2.2-4 | If $ax + ny = 1$ for integers x and y, can a and n have any common divisors other than 1 and -1? |

In Exercise 2.2-3, because Lemma 2.8 tells us that the equation $a \cdot_n x = 1$ has a solution in Z_n if and only if there exist integers x and y such that $ax + ny = 1$, we conclude the following theorem.

| Theorem 2.9 | A number a has a multiplicative inverse in Z_n if and only if there are integers x and y such that $ax + ny = 1$. |

We answer the rest of Exercise 2.2-3 with a corollary.

| Corollary 2.10 | If $a \in Z_n$ and x and y are integers such that $ax + ny = 1$, then the multiplicative inverse of a in Z_n is x mod n. |

Proof: Because $n \cdot_n y = 0$ in Z_n, we have $a \cdot_n x = 1$ in Z_n. Therefore, x is the inverse of a in Z_n. ∎

Now let's consider Exercise 2.2-4. If a and n have a common divisor k, then there must exist integers s and q such that

$$a = sk$$

and

$$n = qk .$$

Substituting these into $ax + ny = 1$, we obtain

$$1 = ax + ny$$
$$= skx + qky$$
$$= k(sx + qy) .$$

But then k is a divisor of 1. Because the only integer divisors of 1 are ± 1, we must have $k = \pm 1$. Therefore, a and n can have no common divisors other than 1 and -1.

In general, the **greatest common divisor** (GCD) of two numbers j and k is the largest number d that is a factor of both j and k.[4] We denote the greatest common divisor of j and k by $\gcd(j, k)$. When two integers j and k have $\gcd(j, k) = 1$, we say that j and k are **relatively prime**.

---

[4]There is one common factor of j and k for sure—namely, 1. No common factor can be larger than $|j|$ or $|k|$, so there are finitely many factors. Therefore, there must be a largest common factor.

We can now restate Exercise 2.2-4 as the following lemma.

Lemma 2.11

Given a and n, if there exist integers x and y such that $ax + ny = 1$, then $\gcd(a, n) = 1$—that is, a and n are relatively prime.

If we combine Theorem 2.9 and Lemma 2.11, we see that if a has a multiplicative inverse mod n, then $\gcd(a, n) = 1$. It is natural to ask whether the statement "If $\gcd(a, n) = 1$, then a has a multiplicative inverse" is true as well.[5] If it were, this would give a way to test whether a has a multiplicative inverse mod n by computing the greatest common divisor of a and n. For this purpose, we would need an algorithm to find $\gcd(a, n)$. It turns out that there is such an algorithm, and a byproduct of the algorithm is a proof of our conjectured converse statement.

Euclid's Division Theorem

One of the important tools in understanding greatest common divisors is Euclid's division theorem, a result that has already been important to us in defining what we mean by m mod n. Although it appears obvious, Euclid's division theorem follows from one simple principle, and the proof of it helps us understand how the greatest common divisor algorithm works. Thus, we restate the theorem and present a proof of it here. Our proof uses the method of proof by contradiction, which you first saw in Corollary 2.6. Notice that we are assuming m is nonnegative, which we didn't assume in our earlier statement of Euclid's division theorem (Theorem 2.1). Problem 16 explores how we can remove this additional assumption. By definition, the r in the following theorem is equal to m mod n.

Theorem 2.12

(Euclid's Division Theorem, Restricted Version) Let n be a positive integer. Then for every nonnegative integer m, there exist unique integers q and r such that $m = nq + r$ and $0 \le r < n$.

Proof: First we show that for each m, there is at least one pair of integers q and r with $0 \le r < n$ so that $m = qn + r$. Assume, for the sake of proof by contradiction, that there is a nonnegative integer m for which no such q and r exist. Choose the smallest[6] such nonnegative integer m. If $m < n$, then $m = n \cdot 0 + m$ with $0 \le m < n$, so $m \ge n$. Thus, $m - n$ is a nonnegative integer smaller than m. Therefore, there exist integers q' and r' such that $m - n = nq' + r'$ with $0 \le r' < n$. But then $m = n(q' + 1) + r'$. So, by taking $q = q' + 1$ and $r = r'$, we obtain $m = qn + r$ with $0 \le r < n$. This is a contradiction to the assumption

---

[5]Notice that this statement is *not* equivalent to the statement in the lemma. This statement is called the "converse" of the lemma; we explain the idea of converse statements more in Chapter 3.

[6]There is a principle called the well-ordering principle for nonnegative integers that says any set of nonnegative integers has a smallest element. Thus, for any given positive integer n, the set of numbers m that make the statement of the theorem false has a smallest element.

that there are not integers q and r with $0 \leq r < n$ such that $m = qn + r$. Thus, by the principle of proof by contradiction, such integers q and r exist.

Next, we show that the integers q and r are unique by showing that any two pairs (q, r) satisfying the theorem are identical. For this purpose, we suppose that $m = nq + r$ and $m = nq^* + r^*$ with $0 \leq r < n$ and $0 \leq r^* < n$. By subtraction, $0 = n(q - q^*) + r - r^*$, so that $n(q - q^*) = r^* - r$. Because r^* and r are both between 0 and $n - 1$ (inclusive), the absolute value of their difference is less than n, giving us

$$|n(q - q^*)| = |r^* - r| < n .$$

Because n is a factor of the left side, the only way the inequality can hold is if $|n(q - q^*)| = |r^* - r| = 0$. Therefore, $q = q^*$ and $r = r^*$, proving that q and r are unique. ∎

Here we have used a special case of proof by contradiction that we call **proof by smallest counterexample**. In this method, we assume, as in all proofs by contradiction, that the theorem is false, which implies that there must be a **counterexample** that does not satisfy the theorem's conditions. For a given n, the counterexample would consist of a nonnegative integer m such that there are *not* unique integers q and r, with $0 \leq r < n$, that satisfy $m = qn + r$. Further, if there are counterexamples, then there must be one having the smallest m. We assume we have chosen a counterexample with such a smallest m. Then we reason that if such an m exists, then every example with a smaller m satisfies the theorem's conclusion. If we can then use a smaller true example to show that our supposedly false example is true as well, then we have created a contradiction. The only thing this can contradict is our assumption that the theorem was false. Therefore, this assumption has to be invalid, and the theorem has to be true. As we will see in Section 4.1, this method is closely related both to a proof method called **proof by induction** and to recursive algorithms. In essence, the proof of Theorem 2.12 describes a recursive program to find q and r in Theorem 2.12 so that $0 \leq r < n$.

Exercise 2.2-5

Suppose that $k = jq + r$, as in Euclid's division theorem. Is there a relationship between $\gcd(j, k)$ and $\gcd(r, j)$?

In this exercise, if $r = 0$, then $\gcd(r, j)$ is j, because any number is a divisor of 0. But this is also the GCD of k and j, because, in this case, $k = jq$. The answer to the remainder of Exercise 2.2-5 appears in the following lemma.

Lemma 2.13

If j, k, q, and r are positive integers such that $k = jq + r$, then

$$\gcd(j, k) = \gcd(r, j) . \tag{2.7}$$

Proof: To prove that both sides of Equation 2.7 are equal, we will show that j and k have exactly the same common factors as r and j. That is, we will first show that if d is a factor j and k, then it is a factor of r and j. Second, we will show that if d is a factor of both r and j, then it is a factor of both j and k.

If d is a factor of j and k, then there must be integers i_1 and i_2 so that $k = i_1 d$ and $j = i_2 d$. Thus, d is also a factor of

$$r = k - jq$$
$$= i_1 d - i_2 dq$$
$$= (i_1 - i_2 q)d \ . \qquad (2.8)$$

Because d is a factor of j (by supposition) and r (by Equation 2.8), it is a common factor of r and j.

Similarly, if d is a factor of r and j, then we can write $j = i_3 d$ and $r = i_4 d$. Therefore,

$$k = jq + r$$
$$= i_3 dq + i_4 d$$
$$= (i_3 q + i_4)d \ ,$$

and d is a factor of k and, therefore, a common factor of j and k.

Because j and k have the same common factors as r and j, their greatest common factors (their GCDs) must be equal. ∎

Although we did not need to assume $r < j$ to prove the lemma, Theorem 2.1 tells us we may assume $r < j$. The assumption in the lemma that j, q, and r are positive implies that $j < k$. Thus, this lemma reduces our problem of finding $\gcd(j, k)$ to the simpler problem (in a recursive sense) of finding $\gcd(r, j)$.

Euclid's GCD Algorithm

Exercise 2.2-6

Using Lemma 2.13, write a recursive algorithm to find $\gcd(j, k)$, given that $0 < j \le k$. Use it to find, by hand, the GCD of 24 and 14 and the GCD of 252 and 189.

Our algorithm for Exercise 2.2-6 is based on Lemma 2.13 and the observation that if $k = jq$ for any q, then $j = \gcd(j, k)$. Notice from the statement of the exercise that we are assuming that j and k are both positive and $j \le k$. We first write $k = jq + r$ in the usual way. If $r = 0$, then we return j as the greatest common divisor. Otherwise, j and r are both positive with $r \le j$, and we apply our algorithm to find the GCD of j and r. Finally, we return the result as the GCD of j and k. This is called **Euclid's GCD algorithm**.

To find $\gcd(14, 24)$, we write

$$24 = 14(1) + 10 .$$

In this case, $k = 24$, $j = 14$, $q = 1$, and $r = 10$. Thus, we can apply Lemma 2.13 and conclude that

$$\gcd(14, 24) = \gcd(10, 14) .$$

We therefore continue our computation of $\gcd(10, 14)$ by writing $14 = 10 \cdot 1 + 4$ and have that

$$\gcd(10, 14) = \gcd(4, 10) .$$

Because

$$10 = 4 \cdot 2 + 2 ,$$

we have that

$$\gcd(4, 10) = \gcd(2, 4) .$$

Because

$$4 = 2 \cdot 2 + 0 ,$$

we know that $k = 4$, $j = 2$, $q = 2$, and $r = 0$. In this case, our algorithm tells us that our current value of j is the GCD of the original j and k. This step is the base case of our recursive algorithm.

Thus,

$$\gcd(14, 24) = \gcd(2, 4) = 2 .$$

It turns out to be even easier to find the GCD of 252 and 189, even though the numbers are larger.

We write

$$252 = 189 \cdot 1 + 63 ,$$

so that $\gcd(189, 252) = \gcd(63, 189)$, and

$$189 = 63 \cdot 3 + 0 .$$

This tells us that $\gcd(189, 252) = \gcd(189, 63) = 63$.

Extended GCD Algorithm

By analyzing our process in a bit more detail, we can return not only the greatest common divisor but also numbers x and y such that $\gcd(j, k) = jx + ky$. This solves the problem we have been working on, because it proves that if $\gcd(a, n) = 1$, then there are integers x and y such that $ax + ny = 1$. Further, it tells us how to find x and, therefore, the multiplicative inverse of a.

In the case that $k = jq$ and we want to return j as our greatest common divisor, we also want to return 1 for the value of x and 0 for the value of y. Suppose we are now in the case that $k = jq + r$, with $0 < r < j$ (that is, the case that $k \neq jq$). We recursively compute $\gcd(r, j)$ and, in the process, get an x' and a y' such that $\gcd(r, j) = rx' + jy'$. Because $r = k - jq$, we get by substitution that

$$\gcd(r, j) = (k - jq)x' + jy' = kx' + j(y' - qx') .$$

Thus, when we return $\gcd(r, j)$ as $\gcd(j, k)$, we want to return the value of x' as y and the value of $y' - qx'$ as x.

We refer to this process as **Euclid's extended GCD algorithm**.

Exercise 2.2-7	Apply Euclid's extended GCD algorithm to find numbers x and y such that the GCD of 14 and 24 is $14x + 24y$.

For Exercise 2.2-7, we give pseudocode for the extended GCD algorithm. We expressed the algorithm more concisely earlier by using recursion; we now give an iterative version that is longer but that can make the computational process clearer. Instead of using the variables q, j, k, r, x, and y, we use six arrays, where $q[i]$ is the value of q computed on the ith iteration and so forth. We use the index 0 for the input values—that is, $j[0]$ and $k[0]$ will be the numbers whose GCD we wish to compute. Eventually, $x[0]$ and $y[0]$ will become the x and y we want. (In Line 8, the notation $\lfloor x \rfloor$ stands for the floor of x, the largest integer less than or equal to x.)

```
gcd(j,k)

// Assume that j ≤ k and that j and k are positive integers.
(1)   if (j==k)
(2)        return j as gcd
(3)        return 1 as x
(4)        return 0 as y
(5)   else
(6)        i = 0; k[i] = k; j[i] = j
(7)   Repeat
(8)        q[i] = ⌊k[i]/j[i]⌋
(9)        r[i] = k[i]-q[i]j[i]
           // Now r[i] = k[i] mod j[i].
(10)       k[i + 1] = j[i]; j[i + 1] = r[i]
(11)       i = i + 1
(12) Until  (r[i-1] = 0)
(13) i = i-1
(14) gcd = j[i]
// we have found the value of the gcd, now we compute the x and y
(15) y[i] = 0; x[i] = 1
(16) i = i-1
```

```
(17) While (i ≥ 0)
(18)     y[i] = x[i + 1]
(19)     x[i] = y[i + 1] − q[i] * x[i + 1]
(20)     i = i − 1
(21) Return gcd
(22) Return x[0] as x
(23) Return y[0] as y
```

Table 2.1 shows the details of how this algorithm applies to gcd(24, 14). In a row, the $q[i]$ and $r[i]$ values are computed from the $j[i]$ and $k[i]$ values. Then the $j[i]$ and $r[i]$ are passed down to the next row as $k[i + 1]$ and $j[i + 1]$, respectively. This process continues until we finally reach a case where $k[i] = q[i]j[i]$, and we can answer $j[i]$ for the GCD. We can then begin computing $x[i]$ and $y[i]$. In the row of Table 2.1 with $i = 3$, we have that $x[i] = 1$ and $y[i] = 0$. As i decreases, we compute $x[i]$ and $y[i]$ for a row by setting $y[i]$ to $x[i + 1]$ and $x[i]$ to $y[i + 1] − q[i]x[i + 1]$. We note that in every row, we have the property that $j[i]x[i] + k[i]y[i] = \gcd(j, k)$.

Table 2.1: *The computation of* $\gcd(14, 24)$ *by algorithm* $\gcd(j, k)$

i	$j[i]$	$k[i]$	$q[i]$	$r[i]$	$x[i]$	$y[i]$
0	14	24	1	10		
1	10	14	1	4		
2	4	10	2	2		
3	2	4	2	0	1	0
2	4	10	2	2	−2	1
1	10	14	1	4	3	−2
0	14	24	1	10	−5	3

Results: gcd = 2, $x = -5$, $y = 3$.

We summarize Euclid's extended GCD algorithm in the following theorem.

Theorem 2.14

Given two integers j and k, Euclid's extended GCD algorithm computes $\gcd(j, k)$ and two integers x and y such that $\gcd(j, k) = jx + ky$.

We now use Eculid's extended GCD algorithm to extend Lemma 2.11.

Theorem 2.15

Two positive integers j and k have greatest common divisor 1 (and thus are relatively prime) if and only if there are integers x and y such that $jx + ky = 1$.

Proof: The statement that if there are integers x and y such that $jx + ky = 1$, then $\gcd(j, k) = 1$ is proved in Lemma 2.11. In other words, $\gcd(j, k) = 1$ if there are integers x and y such that $jx + ky = 1$.

On the other hand, we just showed, by Euclid's extended GCD algorithm, that given positive integers j and k, there are integers x and y such that $\gcd(j, k) = jx + ky$. Therefore, $\gcd(j, k) = 1$ only if there are integers x and y such that $jx + ky = 1$. ∎

Combining Lemma 2.8 and Theorem 2.15, we obtain the following corollary.

Corollary 2.16   For any positive integer n, an element a of Z_n has a multiplicative inverse if and only if $\gcd(a, n) = 1$.

Using the fact that if n is prime, then $\gcd(a, n) = 1$ for all nonzero $a \in Z_n$, we obtain the following corollary.

Corollary 2.17   For any prime p, every nonzero element a of Z_p has an inverse.

Computing Inverses

Not only does Euclid's extended GCD algorithm tell us whether an inverse exists, but it also computes it for us, as we saw in Exercise 2.2-3. Combining Exercise 2.2-3 with Theorem 2.15, we get the following:

Corollary 2.18   If an element a of Z_n has an inverse, we can compute it by running Euclid's extended GCD algorithm to determine integers x and y so that $ax + ny = 1$. The inverse of a in Z_n is $x \bmod n$.

For completeness, we now give pseudocode that determines whether an element a in Z_n has an inverse and that computes the inverse if it exists.

```
inverse(a,n)

(1)    Run procedure gcd(a,n) to obtain gcd(a,n), x, and y
(2)    if (gcd(a,n) == 1)
(3)          return x mod n
(4)    else
(5)          print "no inverse exists"
```

The correctness of the algorithm follows immediately from the fact that $\gcd(a, n) = ax + ny$; so, if $\gcd(a, n) = 1$, then $ax \bmod n$ must be equal to 1.

Important Concepts, Formulas, and Theorems

1. *Multiplicative inverse.* The element a' of Z_n is a multiplicative inverse of a in Z_n if $a \cdot_n a' = 1$. If a has a multiplicative inverse, then it has a unique multiplicative inverse, which we denote a^{-1}.

2. *An important way to solve modular equations.* Suppose a has a multiplicative inverse mod n, and this inverse is a^{-1}. Then for any $b \in Z_n$, the unique solution to the equation

$$a \cdot_n x = b$$

is

$$x = a^{-1} \cdot_n b .$$

3. *Converting modular to regular equations.* The equation

$$a \cdot_n x = 1$$

has a solution in Z_n if and only if there exist integers x and y such that

$$ax + ny = 1 .$$

4. *When do inverses exist in Z_n?* A number a has a multiplicative inverse in Z_n if and only if there are integers x and y such that $ax + ny = 1$.

5. *Greatest common divisor (GCD).* The *greatest common divisor* of two numbers j and k is the largest number d that is a factor of both j and k.

6. *Relatively prime.* When two numbers j and k have $\gcd(j, k) = 1$, we say that j and k are *relatively prime*.

7. *Connecting inverses to GCD.* Given a and n, if there exist integers x and y such that $ax + ny = 1$, then $\gcd(a, n) = 1$.

8. *GCD recursion lemma.* If j, k, q, and r are positive integers such that $k = jq + r$, then $\gcd(j, k) = \gcd(r, j)$.

9. *Euclid's GCD algorithm.* Given two numbers j and k, Euclid's GCD algorithm returns $\gcd(j, k)$.

10. *Euclid's extended GCD algorithm.* Given two numbers j and k, Euclid's extended GCD algorithm returns $\gcd(j, k)$ and two integers x and y such that $\gcd(j, k) = jx + ky$.

11. *Relating GCD of 1 to Euclid's extended GCD algorithm.* Two positive integers j and k have greatest common divisor 1 if and only if there are integers x and y such that $jx + ky = 1$. One of the integers x and y could be negative.

12. *Condition for multiplicative inverse in Z_n.* For any positive integer n, an element a of Z_n has an inverse if and only if $\gcd(a, n) = 1$.

13. *Multiplicative inverses in Z_p, with p prime.* For any prime p, every nonzero element a of Z_p has a multiplicative inverse.

14. *A way to solve some modular equations $a \cdot_n x = b$.* Use Euclid's extended GCD algorithm to compute a^{-1} (if it exists), and multiply both sides of the equation by a^{-1}. (If a has no inverse, the equation might or might not have a solution.)

Problems

1. If $a \cdot 133 - m \cdot 277 = 1$, does this guarantee that a has an inverse mod m? If so, what is it? If not, why not? `a/h`

2. If $a \cdot 133 - 2m \cdot 277 = 1$, does this guarantee that a has an inverse mod m? If so, what is it? If not, why not?

3. Determine whether every nonzero element of Z_n has a multiplicative inverse for $n = 10$ and $n = 11$. `a/h`

4. How many elements a are there such that $a \cdot_{31} 22 = 1$? How many elements a are there such that $a \cdot_{10} 2 = 1$?

5. Given an element b in Z_n, what can you say in general about the possible number of elements a such that $a \cdot_n b = 1$ in Z_n? `a/h`

6. If $a \cdot 133 - m \cdot 277 = 1$, what can you say about all possible common divisors of a and m?

7. Compute the GCD of 210 and 126 by using Euclid's GCD algorithm. `a/h`

8. If $k = jq + r$, as in Euclid's division theorem, is there a relationship between $\gcd(q, k)$ and $\gcd(r, q)$? If so, what is it?

9. Bob and Alice want to choose a key that they can use for cryptography, but all they have to communicate is a bugged phone line. Bob proposes that they each choose a secret number, a for Alice and b for Bob. They also choose, over the phone, a prime number p, with more digits than any key they want to use, and one more number q. Bob will send Alice bq mod p, and Alice will send Bob aq mod p. Their key (which they will keep secret) will then be abq mod p. (In this case, don't worry about the details of how they use their key; only worry about how they choose it.) As Bob explains, their wiretapper will know p, q, aq mod p, and bq mod p but will not know a or b, so their key should be safe.

    Is this scheme safe—that is, can the wiretapper compute abq mod p? If so, how?

    Alice says, "You know, the scheme sounds good, but wouldn't it be more complicated for the wiretapper if I send you q^a mod p, you send me q^b mod p, and we use q^{ab} mod p as our key?" In this case, can you think of a way for the wiretapper to compute q^{ab} mod p? If so, how? If not, what is the stumbling block? (It is fine for the stumbling block to be that you don't know how to compute something; you don't need to prove that you can't compute it.) `a/h`

10. Write pseudocode for a recursive version of the extended GCD algorithm.

11. Run Euclid's extended GCD algorithm to compute $\gcd(576, 486)$. Show all the steps. `a/h`

12. Use Euclid's extended GCD algorithm to compute the multiplicative inverse of 16 mod 103.

13. Solve the equation $16 \cdot_{103} x = 21$ in Z_{103}. `a/h`

14. Which elements of Z_{35} do not have multiplicative inverses in Z_{35}?

15. If $k = jq + r$, as in Euclid's division theorem, is there a relationship between $\gcd(j, k)$ and $\gcd(r, k)$? If so, what is it? `a/h`

16. Notice that if m is negative, then $-m$ is positive. Thus, by Theorem 2.12, $-m = qn + r$ for $0 \leq r < n$. This gives $m = -qn - r$. If $r = 0$, then $m = q'n + r'$ for $0 \leq r' \leq n$ and $q' = -q$. However, if $r > 0$, then you cannot take $r' = -r$ and have $0 \leq r' < n$. Notice, though, that because you have already finished the case in which $r = 0$, you may assume that $0 \leq n - r < n$. This suggests that if you were to take r' to be $n - r$, you might be able to find a q' so that $m = q'n + r'$, with $0 \leq r' \leq n$, which would let you conclude that Euclid's division theorem is valid for negative values m as well as for nonnegative values m. Find a q' that works, and explain how you have extended Euclid's division theorem from the version in Theorem 2.12 to the version in Theorem 2.1.

17. The Fibonacci numbers F_i are defined as follows:

$$F_i = \begin{cases} 1 & \text{if } i \text{ is 1 or 2}, \\ F_{i-1} + F_{i-2} & \text{otherwise}. \end{cases}$$

What happens when you run Euclid's extended GCD algorithm on F_i and F_{i+1}? (This problem is asking not only for the answer but also about the execution of the algorithm.) `a/h`

18. Write (and run on several different inputs) a program to implement Euclid's extended GCD algorithm. Be sure to return x and y in addition to the GCD. About how many times does your program have to make a recursive call to itself? What does that say about how long you should expect the program to run as you increase the size of the j and k whose GCD you are computing?

19. The least common multiple (LCM) of two positive integers x and y is the smallest positive integer z such that z is an integer multiple of both x and y. Give a formula for the least common multiple that involves the GCD. `a/h`

20. Write pseudocode that, given integers a, b, and n in Z_n, either computes an x such that $a \cdot_n x = b$ or concludes that no such x exists.

21. Give an example of an equation of the form $a \cdot_n x = b$ that has a solution even though a and n are not relatively prime, or show that no such equation exists. `a/h`

22. Either find an equation of the form $a \cdot_n x = b$ in Z_n that has a unique solution even though a and n are not relatively prime, or prove that no such equation exists. In other words, either prove the statement that if $a \cdot_n x = b$ has a unique solution in Z_n, then a and n are relatively prime, or find a counterexample.

23. Prove Theorem 2.14. `a/h`

2.3 THE RSA CRYPTOSYSTEM

Exponentiation mod n

In the previous sections, we considered encryption using modular addition and multiplication and saw the shortcomings of both. In this section, we consider using exponentiation for encryption and see that it can provide a much greater level of security.

The idea behind RSA encryption is **exponentiation** in Z_n. By Lemma 2.3, if $a \in Z_n$, then

$$a^j \bmod n = \underbrace{a \cdot_n a \cdot_n \cdots \cdot_n a}_{j \text{ factors}} . \qquad (2.9)$$

In other words, $a^j \bmod n$ is the product in Z_n of j factors, each equal to a.

The Rules of Exponents

From Lemma 2.3 and the rules of exponents for the integers, we have Lemma 2.19.

Lemma 2.19	For any $a \in Z_n$ and any nonnegative integers i and j,

$$(a^i \bmod n) \cdot_n (a^j \bmod n) = a^{i+j} \bmod n \qquad (2.10)$$

and

$$(a^i \bmod n)^j \bmod n = a^{ij} \bmod n . \qquad (2.11)$$

Exercise 2.3-1	Compute the powers of 2 mod 7. What do you observe? Now compute the powers of 3 mod 7. What do you observe?

Exercise 2.3-2	Compute the sixth powers of the nonzero elements of Z_7. What do you observe?

Exercise 2.3-3	Compute the numbers $1 \cdot_7 2$, $2 \cdot_7 2$, $3 \cdot_7 2$, $4 \cdot_7 2$, $5 \cdot_7 2$, and $6 \cdot_7 2$. What do you observe? Now compute the numbers $1 \cdot_7 3$, $2 \cdot_7 3$, $3 \cdot_7 3$, $4 \cdot_7 3$, $5 \cdot_7 3$, and $6 \cdot_7 3$. What do you observe?

Exercise 2.3-4	Suppose we choose an arbitrary nonzero number a between 1 and 6. Are the numbers $1 \cdot_7 a$, $2 \cdot_7 a$, $3 \cdot_7 a$, $4 \cdot_7 a$, $5 \cdot_7 a$, and $6 \cdot_7 a$ all different? Why or why not?

In Exercise 2.3-1, we have that

$$2^0 \bmod 7 = 1$$
$$2^1 \bmod 7 = 2$$
$$2^2 \bmod 7 = 4$$
$$2^3 \bmod 7 = 1$$
$$2^4 \bmod 7 = 2$$
$$2^5 \bmod 7 = 4$$
$$2^6 \bmod 7 = 1$$
$$2^7 \bmod 7 = 2$$
$$2^8 \bmod 7 = 4 \, .$$

Continuing, we see that the powers of 2 will cycle through the list of three values—1, 2, and 4—again and again. Performing the same computation for 3, we have

$$3^0 \bmod 7 = 1$$
$$3^1 \bmod 7 = 3$$
$$3^2 \bmod 7 = 2$$
$$3^3 \bmod 7 = 6$$
$$3^4 \bmod 7 = 4$$
$$3^5 \bmod 7 = 5$$
$$3^6 \bmod 7 = 1$$
$$3^7 \bmod 7 = 3$$
$$3^8 \bmod 7 = 2 \, .$$

In this case, we cycle through the list of six values—1, 3, 2, 6, 4, and 5—again and again.

Now observe that in Z_7, we have $2^6 = 1$ and $3^6 = 1$. This suggests an answer to Exercise 2.3-2. Is it the case that $a^6 \bmod 7 = 1$ for all $a \in Z_7$? We can compute that $1^6 \bmod 7 = 1$, and

$$4^6 \bmod 7 = (2 \cdot_7 2)^6 \bmod 7$$
$$= (2^6 \cdot_7 2^6) \bmod 7$$
$$= (1 \cdot_7 1) \bmod 7$$
$$= 1 \, .$$

What about 5^6? Notice that $3^5 = 5$ in Z_7 by the computations made above. Using Equation 2.11 twice gives us

$$5^6 \bmod 7 = (3^5)^6 \bmod 7$$
$$= 3^{5 \cdot 6} \bmod 7$$
$$= 3^{6 \cdot 5} \bmod 7$$
$$= (3^6)^5 = 1^5 = 1$$

in Z_7. Finally, because $-1 \bmod 7 = 6$, Lemma 2.3 tells us that $6^6 \bmod 7 = (-1)^6 \bmod 7 = 1$. Thus, the sixth power of each element of Z_7 is 1.

In Exercise 2.3-3, we see that

$$1 \cdot_7 2 = 1 \cdot 2 \bmod 7 = 2$$
$$2 \cdot_7 2 = 2 \cdot 2 \bmod 7 = 4$$
$$3 \cdot_7 2 = 3 \cdot 2 \bmod 7 = 6$$
$$4 \cdot_7 2 = 4 \cdot 2 \bmod 7 = 1$$
$$5 \cdot_7 2 = 5 \cdot 2 \bmod 7 = 3$$
$$6 \cdot_7 2 = 6 \cdot 2 \bmod 7 = 5 \ .$$

These numbers are a permutation of the set $\{1, 2, 3, 4, 5, 6\}$. Similarly,

$$1 \cdot_7 3 = 1 \cdot 3 \bmod 7 = 3$$
$$2 \cdot_7 3 = 2 \cdot 3 \bmod 7 = 6$$
$$3 \cdot_7 3 = 3 \cdot 3 \bmod 7 = 2$$
$$4 \cdot_7 3 = 4 \cdot 3 \bmod 7 = 5$$
$$5 \cdot_7 3 = 5 \cdot 3 \bmod 7 = 1$$
$$6 \cdot_7 3 = 6 \cdot 3 \bmod 7 = 4 \ .$$

Again, we get a permutation of $\{1, 2, 3, 4, 5, 6\}$.

In Exercise 2.3-4, we are asked whether this is always the case. Notice that because 7 is a prime, Corollary 2.17 tells us that each nonzero number between 1 and 6 has a mod 7 multiplicative inverse a^{-1}. Thus, if i and j are integers in Z_7, with $i \cdot_7 a = j \cdot_7 a$, we multiply mod 7 on the right by a^{-1} to get

$$(i \cdot_7 a) \cdot_7 a^{-1} = (j \cdot_7 a) \cdot_7 a^{-1} \ .$$

After using the associative law, we get

$$i \cdot_7 (a \cdot_7 a^{-1}) = j \cdot_7 (a \cdot_7 a^{-1}) \ . \tag{2.12}$$

Because $a \cdot_7 a^{-1} = 1$, Equation 2.12 simply becomes $i = j$. Thus, we have shown that the only way for $i \cdot_7 a$ to equal $j \cdot_7 a$ is for i to equal j. Therefore, all the values $i \cdot_7 a$ for $i = 1, 2, 3, 4, 5, 6$ must be different. Because we have six different values that must be integers between 1 and 6, we have that the values ia for $i = 1, 2, 3, 4, 5, 6$ are a permutation of $\{1, 2, 3, 4, 5, 6\}$.

As you can see, the only fact we used in our analysis of Exercise 2.3-4 is that if p is a prime, then any number between 1 and $p - 1$ has a multiplicative inverse in Z_p. In other words, we have proved the following lemma.

| Lemma 2.20 | Let p be a prime number. For any fixed nonzero number a in Z_p, the numbers $(1 \cdot a) \bmod p$, $(2 \cdot a) \bmod p$, $\ldots$, $((p - 1) \cdot a) \bmod p$ are a permutation of the set $\{1, 2, \ldots, p - 1\}$. |

With this lemma in hand, we can prove a famous theorem that explains the phenomenon we saw in Exercise 2.3-2.

Fermat's Little Theorem

Theorem 2.21 | **(Fermat's Little Theorem)** Let p be a prime number. Then $a^{p-1} \bmod p = 1$ in Z_p for each nonzero a in Z_p.

Proof: Because p is a prime, Lemma 2.20 tells us that the numbers $1 \cdot_p a$, $2 \cdot_p a, \ldots, (p-1) \cdot_p a$ are a permutation of the set $\{1, 2, \ldots, p-1\}$. But then

$$1 \cdot_p 2 \cdot_p \cdots \cdot_p (p-1) = (1 \cdot_p a) \cdot_p (2 \cdot_p a) \cdot_p \cdots \cdot_p ((p-1) \cdot_p a) .$$

Using Equation 2.9 and the commutative and associative laws for multiplication in Z_p, we get

$$1 \cdot_p 2 \cdot_p \cdots \cdot_p (p-1) = 1 \cdot_p 2 \cdot_p \cdots \cdot_p (p-1) \cdot_p (a^{p-1} \bmod p) .$$

Now we multiply both sides of the equation by the multiplicative inverses in Z_p of $2, 3, \ldots, p-1$; the left side of our equation becomes 1, and the right side becomes $a^{p-1} \bmod p$, which is exactly the conclusion of our theorem. ∎

Corollary 2.22 | **(Fermat's Little Theorem, Version 2)** For every positive integer a and prime p, if a is not a multiple of p, then

$$a^{p-1} \bmod p = 1 .$$

Proof: This is a direct application of Lemma 2.3, because if we replace a with $a \bmod p$, then Theorem 2.21 applies. ∎

The RSA Cryptosystem

Fermat's Little Theorem is at the heart of the RSA cryptosystem, a system that allows Bob to tell the world how to encode a message to send to him so that he, and only he, can read it. In other words, even though he tells everyone how to encode the message, nobody except Bob has a significant chance of figuring out what the message is from looking at the encoded message. What Bob is giving out is called a **one-way function**—a function f that has an inverse f^{-1}; but even though $y = f(x)$ is reasonably easy to compute, nobody but Bob (who has some extra information that he keeps secret) can compute $f^{-1}(y)$. Thus, when Alice wants to send a message x to Bob, she computes $f(x)$ and sends it to Bob, who uses his secret information to compute $f^{-1}(f(x)) = x$.

In the RSA cryptosystem, Bob chooses two prime numbers p and q (which, in practice, each have at least 150 digits) and computes the number $n = pq$. He also chooses a number $e \neq 1$, which need not have a large number of digits but which is relatively prime to $(p-1)(q-1)$. Thus, e has an inverse d in $Z_{(p-1)(q-1)}$, and Bob computes $d = e^{-1}$ mod $(p-1)(q-1)$. Bob publishes e and n. The number e is called his public key. The number d is called his private key. To summarize what we just said, here is a pseudocode outline of what Bob does:

Bob's RSA Key Choice Algorithm

(1) Choose 2 large prime numbers p and q
(2) $n = p \star q$
(3) Choose $e \neq 1$ so that e is relatively prime to $(p-1)(q-1)$
(4) Compute $d = e^{-1}$ mod $(p-1)(q-1)$
(5) Publish e and n
(6) Keep d secret

People who want to send a message x to Bob compute $y = x^e$ mod n and send that to him instead. (We assume x has fewer digits than n so that it is in Z_n. If not, the sender has to break the message into blocks of size less than the number of digits of n and send each block individually.)

To decode the message, Bob will compute $z = y^d$ mod n. The following pseudocode summarizes this process.

Alice-send-message-to-Bob(x)

(1) Alice does the following:
(2) Read the public directory for Bob's keys e and n
(3)   Compute $y = x^e$ mod n
(4)   Send y to Bob

(5) Bob does the following:
(6) Receive y from Alice
(7) Compute $z = y^d$ mod n, using secret key d
(8) Read z

Each step in these algorithms can be computed using methods from this chapter. Section 2.4 deals with computational issues in more detail.

To show that the RSA cryptosystem works—that is, it allows us to encode and then correctly decode messages—we must show that $z = x$. In other words, we must show that when Bob decodes, he gets back the original message. To show that the RSA cryptosystem is secure, we must argue that an eavesdropper, who knows n, e, and y but does not know p, q, or d, cannot easily compute x.

Exercise 2.3-5 To show that the RSA cryptosystem works, we first show a simpler fact. Why is

$$y^d \bmod p = x \bmod p \ ?$$

Does this tell us what x is?

Plugging in the value of y, we have

$$y^d \bmod p = x^{ed} \bmod p \ . \tag{2.13}$$

But, in Lines 3 and 4 of Bob's key choice algorithm, we chose e and d so that $e \cdot_m d = 1$, where $m = (p-1)(q-1)$. In other words,

$$ed \bmod (p-1)(q-1) = 1 .$$

Therefore, for some integer k,

$$ed = k(p-1)(q-1) + 1 .$$

Plugging this into Equation 2.13, we obtain

$$
\begin{aligned}
x^{ed} \bmod p &= x^{k(p-1)(q-1)+1} \bmod p \\
&= x^{(k(q-1))(p-1)} x \bmod p .
\end{aligned}
\tag{2.14}
$$

For any number a that is not a multiple of p, however, $a^{p-1} \bmod p = 1$, by Fermat's Little Theorem (Corollary 2.22). We could simplify Equation 2.14 by applying Fermat's Little Theorem to $x^{k(q-1)}$, as you will see below. However, we can do this only when $x^{k(q-1)}$ is not a multiple of p. This gives us two cases: the case in which $x^{k(q-1)}$ is not a multiple of p (we'll call this Case 1) and the case in which $x^{k(q-1)}$ is a multiple of p (we'll call this Case 2). In Case 1, we apply Equation 2.11 and Fermat's Little Theorem, with $a = x^{k(q-1)}$. We have that

$$
\begin{aligned}
x^{(k(q-1))(p-1)} \bmod p &= \left(x^{k(q-1)} \right)^{(p-1)} \bmod p \\
&= 1 .
\end{aligned}
\tag{2.15}
$$

Combining Equations 2.13, 2.14, and 2.15, we have that

$$
\begin{aligned}
y^d \bmod p &= x^{k(q-1)(p-1)} x \bmod p \\
&= 1 \cdot x \bmod p \\
&= x \bmod p ;
\end{aligned}
$$

hence, $y^d \bmod p = x \bmod p$.

We still have to deal with Case 2. In this case, x is a multiple of p as well because x is an integer and p is prime. Thus, $x \bmod p = 0$. Combining Equations 2.13 and 2.14 with Lemma 2.3, we get

$$y^d \bmod p = \left(x^{k(q-1)(p-1)} \bmod p \right)(x \bmod p) = 0 = x \bmod p .$$

Hence, in this case as well, we have $y^d \bmod p = x \bmod p$.

Although this will turn out to be useful information, it does not tell us what x is, because x may or may not equal $x \bmod p$. The same reasoning shows us that $y^d \bmod q = x \bmod q$. What remains to show is what these two facts tell us about $y^d \bmod pq = y \bmod n$, which is what Bob computes.

Notice that by Lemma 2.3, we have proved that

$$(y^d - x) \bmod p = 0 \tag{2.16}$$

and

$$(y^d - x) \bmod q = 0 \, . \tag{2.17}$$

Exercise 2.3-6 Equation 2.16 says that $(y^d - x) \bmod p = 0$. Write an equivalent equation, using only integers and addition, subtraction, and multiplication in the integers but perhaps additional variables. (Do not use mods.)

Exercise 2.3-7 Equation 2.17 says that $(y^d - x) \bmod q = 0$. Write an equivalent equation, using only integers and addition, subtraction, and multiplication in the integers but perhaps additional variables. (Do not use mods.)

Exercise 2.3-8 If a number is a multiple of a prime p and a different prime q, then what else is it a multiple of? What does this tell us about y^d and x?

The statement "$y^d - x \bmod p = 0$" is equivalent to the statement "$y^d - x = ip$ for some integer i." The statement "$y^d - x \bmod q = 0$" is equivalent to the statement "$y^d - x = jq$ for some integer j." If something is a multiple of the prime p and of the prime q, then it is a multiple of pq. Thus, $(y^d - x) \bmod pq = 0$. Lemma 2.3 tells us that $(y^d - x) \bmod pq = ((y^d \bmod pq) - x) \bmod pq = 0$. But x and $y^d \bmod pq$ are both integers between 0 and $pq - 1$, so their difference is between $-(pq - 1)$ and $pq - 1$. The only integer between these two values that is 0 mod pq is 0 itself. Thus, $(y^d \bmod pq) - x = 0$. In other words,

$$x = y^d \bmod pq$$
$$= y^d \bmod n \, ,$$

which means that Bob will, in fact, get the correct answer.

Theorem 2.23 **(Rivest, Shamir, and Adleman)** The RSA procedure for encoding and decoding messages works correctly.

Proof: Proved above. ∎

One might ask, given that Bob published e and n and messages are encrypted by computing $x^e \bmod n$, why can't any adversary who learns $x^e \bmod n$ simply compute eth roots mod n and break the code? At present, nobody knows a quick scheme for computing eth roots mod n for an arbitrary n. Someone who does not know p and q cannot duplicate Bob's work and discover x. Thus, as far as we know, modular exponentiation is an example of a one-way function.

The Chinese Remainder Theorem

The method we used to do the last step of the proof of Theorem 2.23 also proves a theorem known as the Chinese remainder theorem.

Exercise 2.3-9 For each number in $x \in Z_{15}$, write x mod 3 and x mod 5. Is x uniquely determined by these values? If so, explain why.

As we see from Table 2.2, each of the $3 \cdot 5 = 15$ pairs (i, j) of integers i and j, with $0 \leq i \leq 2$ and $0 \leq j \leq 4$, occurs exactly once as x ranges through the 15 integers from 0 to 14. Thus, the function f given by $f(x) = (x \bmod 3, x \bmod 5)$ is a one-to-one function from a 15-element set to a 15-element set; thus, each x is uniquely determined by its pair of remainders.

Table 2.2: *The values of x mod 3 and x mod 5 for each x between 0 and 14*

x	x mod 3	x mod 5
0	0	0
1	1	1
2	2	2
3	0	3
4	1	4
5	2	0
6	0	1
7	1	2
8	2	3
9	0	4
10	1	0
11	2	1
12	0	2
13	1	3
14	2	4

The Chinese remainder theorem tells us that this observation always holds.

Theorem 2.24 **(Chinese Remainder Theorem)** If m and n are relatively prime integers and $a \in Z_m$ and $b \in Z_n$, then the equations

$$x \bmod m = a \tag{2.18}$$

$$x \bmod n = b \tag{2.19}$$

have one and only one solution for an integer x between 0 and $mn - 1$.

Proof: If we show that as x ranges over the integers from 0 to $mn - 1$, the ordered pairs $(x \bmod m, x \bmod n)$ are all different, then we will have shown that the function given by $f(x) = (x \bmod m, x \bmod n)$ is a one-to-one function from an mn-element set to an mn-element set; thus, it is onto as well.[7] In other words, we will have shown that each pair of Equations 2.18 and 2.19 has one and only one solution.

To show that f is one-to-one, we must show that if x and y are different numbers between 0 and $mn - 1$, then $f(x)$ and $f(y)$ are different. To do so, assume instead that we have an x and a y with $f(x) = f(y)$. Then $x \bmod m = y \bmod m$ and $x \bmod n = y \bmod n$, so that $(x - y) \bmod m = 0$ and $(x - y) \bmod n = 0$. That is, $x - y$ is a multiple of both m and n. Then, as Problem 11 shows, $x - y$ is a multiple of mn; that is, $x - y = dmn$ for some integer d. We assumed x and y were different, which means x and y cannot both be between 0 and $mn - 1$, because their difference is mn or more. This contradicts our hypothesis that x and y were different numbers between 0 and $mn - 1$, so our assumption must be incorrect; that is, f must be one-to-one. This completes the proof of the theorem. ∎

Important Concepts, Formulas, and Theorems

1. *Exponentiation in Z_n.* For each $a \in Z_n$ and each positive integer j,

$$a^j \bmod n = \underbrace{a \cdot_n a \cdot_n \cdots \cdot_n a}_{j \text{ factors}} .$$

2. *Rules of exponents.* For each $a \in Z_n$ and any nonnegative integers i and j,

$$(a^i \bmod n) \cdot_n (a^j \bmod n) = a^{i+j} \bmod n$$

and

$$(a^i \bmod n)^j \bmod n = a^{ij} \bmod n .$$

3. *Multiplication by a fixed nonzero a in Z_p is a permutation.* Let p be a prime number. For any fixed nonzero number a in Z_p, the numbers $(1 \cdot a) \bmod p$, $(2 \cdot a) \bmod p, \ldots, ((p-1) \cdot a) \bmod p$ are a permutation of the set $\{1, 2, \ldots, p - 1\}$.

4. *Fermat's Little Theorem.* If we let p be a prime number, then $a^{p-1} \bmod p = 1$ for each nonzero a in Z_p.

5. *Fermat's Little Theorem, version 2.* For every positive integer a and prime p, if a is not a multiple of p, then

$$a^{p-1} \bmod p = 1 .$$

---

[7]If the function weren't onto, then two values of x would have to map to the same pair, because the number of pairs is the same as the number of possible values of x. So, the function wouldn't be one-to-one after all.

6. *RSA cryptosystem* (*the first implementation of a public-key cryptosystem*). In the RSA cryptosystem, Bob chooses two prime numbers p and q (which, in practice, each have at least 150 digits) and computes the number $n = pq$. He also chooses a number $e \neq 1$, which need not have a large number of digits but which is relatively prime to $(p-1)(q-1)$. Thus, e has an inverse d, and Bob computes $d = e^{-1} \bmod (p-1)(q-1)$. Bob publishes e and n. To send a message x to Bob, Alice sends $y = x^e \bmod n$. Bob decodes by computing $y^d \bmod n$.

7. *Chinese remainder theorem.* If m and n are relatively prime integers and $a \in Z_m$ and $b \in Z_n$, then the equations

$$x \bmod m = a$$
$$x \bmod n = b$$

have one and only one solution for an integer x between 0 and $mn - 1$.

Problems

1. Compute the positive powers of 4 in Z_7. Compute the positive powers of 4 in Z_{10}. What is the most striking similarity? What is the most striking difference? a/h

2. Compute the numbers $1 \cdot_{11} 5$, $2 \cdot_{11} 5$, $3 \cdot_{11} 5, \ldots, 10 \cdot_{11} 5$. Do you get a permutation of the set $\{1, 2, 3, 4, 5, 6, 7, 8, 9, 10\}$? Would you get a permutation of the set $\{1, 2, 3, 4, 5, 6, 7, 8, 9, 10\}$ if you used another nonzero member of Z_{11} in place of 5?

3. Compute the fourth power mod 5 of each element of Z_5. What do you observe? What general principle explains this observation? a/h

4. The numbers 29 and 43 are primes. What is $(29-1)(43-1)$? What is $199 \cdot 1111$ in Z_{1176}? What is $(23^{1111})^{199}$ in Z_{29}? In Z_{43}? In Z_{1247}?

5. The numbers 29 and 43 are primes. What is $(29-1)(43-1)$? What is $199 \cdot 1111$ in Z_{1176}? What is $(105^{1111})^{199}$ in Z_{29}? In Z_{43}? In Z_{1247}? How does this answer the second question in Exercise 2.3-5? a/h

6. How many solutions with x between 0 and 34 are there to the system of equations

$$x \bmod 5 = 4$$
$$x \bmod 7 = 5 ?$$

What are these solutions?

7. Compute each of the following. Show or explain your work. Do not use a calculator or computer. a/h

   a. 15^{96} in Z_{97}.

   b. 67^{72} in Z_{73}.

   c. 67^{73} in Z_{73}.

8. Show that in Z_p, if $a^i \bmod p = 1$, then $a^n \bmod p = a^{n \bmod i} \bmod p$ when p is prime.

9. Show that there are $p^2 - p$ elements with multiplicative inverses in Z_{p^2} when p is prime. If x has a multiplicative inverse in Z_{p^2}, what is $x^{p^2-p} \bmod p^2$? Is the same statement true for an element without an inverse? (Working out an example might help here.) Can you find something interesting that is true about x^{p^2-p} when x does not have an inverse? a/h

10. How many elements have multiplicative inverses in Z_{pq} when p and q are primes?

11. The paragraph preceding the proof of Theorem 2.23 says that if a number is a multiple of the prime p and the prime q, then it is a multiple of pq. This is proved here.

    a. What equation in the integers does Euclid's extended GCD algorithm solve for when m and n are relatively prime? a/h

    b. Suppose that m and n are relatively prime and that k is a multiple of each—that is, $k = bm$ and $k = cn$ for integers b and c. If you multiply both sides of the equation in part a by k, you get an equation expressing k as a sum of two products. By making appropriate substitutions in these terms, you can show that k is a multiple of mn. Do so. Does this justify the assertion made in the paragraph preceding the proof of Theorem 2.23? a/h

12. (Requires Section 1.4.) The relation of "congruence modulo n" is denoted by $\equiv$ and defined by $x \equiv y \pmod{n}$ if and only if $x \bmod n = y \bmod n$.

    a. Show that congruence modulo n is an equivalence relation by showing that it defines a partition of the integers into equivalence classes.

    b. Show that congruence modulo n is an equivalence relation by showing that it is reflexive, symmetric, and transitive.

    c. Express the Chinese remainder theorem in the notation of congruence modulo n.

13. Write and implement code to do RSA encryption and decryption. Use it to send a message to someone else in class. (For the sake of efficiency, you may use smaller numbers than are usually used in implementing the RSA algorithm. In other words, you may choose your numbers so that your computer can multiply them without overflow.)

14. Show that if $x^{n-1} \bmod n = 1$ for all integers x that are not multiples of n, then n is prime. (The slightly weaker statement "$x^{n-1} \bmod n = 1$ for all x relatively prime to n" does not imply that n is prime. There is a famous infinite family of numbers called Carmichael numbers that are counterexamples.[2],[13]) a/h

2.4 DETAILS OF THE RSA CRYPTOSYSTEM

This section deals with some issues related to implementing the RSA cryptosystem: exponentiating large numbers, finding primes, and factoring.

Practical Aspects of Exponentiation mod n

Suppose you are going to raise a 150-digit number a to the 10^{120}th power modulo a 300-digit integer n. Note that the exponent is a 121-digit number.

Exercise 2.4-1 Propose an algorithm to compute $a^{10^{120}} \bmod n$, where a is a 150-digit number and n is a 300-digit number.

Exercise 2.4-2 What can we say about how long the algorithm in Exercise 2.4-1 would take on a computer that can do one infinite precision arithmetic operation in constant time?

Exercise 2.4-3 What can we say about how long the algorithm in Exercise 2.4-1 would take on a computer that can multiply integers in time proportional to the product of the number of digits in the two numbers, that is, multiplying an x-digit number by a y-digit number takes roughly xy time?

Notice that if we form the sequence a, a^2, a^3, a^4, a^5, a^6, a^7, a^8, a^9, a^{10}, and a^{11}, we are modeling the process of forming a^{11} by successively multiplying by a. If, on the other hand, we form the sequence a, a^2, a^4, a^8, a^{16}, a^{32}, a^{64}, a^{128}, a^{256}, a^{512}, a^{1024}, we are modeling the process of successive squaring, and in the same number of multiplications, we are able to get a raised to a four-digit number. Each time we square, we double the exponent; so, every ten steps or so we will increase the number of digits in the exponent by three. Thus, in a bit under 400 multiplications, we will get $a^{10^{120}}$. This suggests that our algorithm should be to square a some number of times until the result is almost $a^{10^{120}}$ and then multiply by some smaller powers of a until we get exactly what we want. More precisely, we square a and continue squaring the result until we get the largest $a^{2^{k_1}}$ such that 2^{k_1} is less than 10^{120}. Then we multiply $a^{2^{k_1}}$ by the largest $a^{2^{k_2}}$ such that $2^{k_1} + 2^{k_2}$ is less than 10^{120}. We continue until we have

$$10^{120} = 2^{k_1} + 2^{k_2} + \cdots + 2^{k_r}$$

for some integer r. (Can you connect this with the binary representation of 10^{120}?) Then we get

$$a^{10^{120}} = a^{2^{k_1}} a^{2^{k_2}} \cdots a^{2^{k_r}}.$$

Notice that all these powers of a have been computed in the process of discovering k_1. Thus, it makes sense to save them as you compute them.

To be more concrete, let's see how to compute a^{43}. We may write $43 = 32 + 8 + 2 + 1$, and thus

$$a^{43} = a^{2^5} a^{2^3} a^{2^1} a^{2^0}. \tag{2.20}$$

So, we first compute $a^{2^0}, a^{2^1}, a^{2^2}, a^{2^3}, a^{2^4}, a^{2^5}$, using five multiplications. Then we can compute a^{43} via Equation 2.20, using three additional multiplications. This saves a large number of multiplications.

On a machine that could do infinite precision arithmetic in constant time, we would need about $\log_2(10^{120})$ steps to compute all the powers a^{2^i}, and perhaps equally many steps to do the multiplications of the appropriate powers. At the end, we could take the result mod n. Thus, the length of time it would take to do these computations would be more or less $2\log_2(10^{120}) = 240\log_2 10$ times the time needed to do one operation. Because $\log_2 10$ is about 3.32, it will take at most 800 times the amount of time for one operation to compute $a^{10^{120}}$.

You may not be used to thinking about how large the numbers get when you are doing computation. Computers hold fairly large numbers (4-byte integers in the range roughly -2^{31} to 2^{31} are typical). This suffices for most purposes. Because of the way computer hardware works, as long as numbers fit into one 4-byte integer, the time to do simple arithmetic operations doesn't depend on the value of the numbers involved. (A standard way to say this is that the time to do a simple arithmetic operation is constant.) However, when we talk about numbers that are much larger than 2^{31}, we have to take special care to implement our arithmetic operations correctly; we also have to be aware that operations are slower.

Because $2^{10} = 1024$, we have that 2^{31} is twice as big as $2^{30} = (2^{10})^3 = (1024)^3$; thus, it is somewhat more than two billion, or $2 \cdot 10^9$. In particular, it is less than 10^{10}. Because 10^{120} is a 1 followed by 120 zeros, raising a positive integer other than 1 to the 10^{120}th power takes us completely out of the realm of the numbers with which we are used to making exact computations. For example, the decimal representation of $10^{(10^{120})}$ has 119 more zeros following the 1 in the exponent than does 10^{10}.

It is accurate to assume that when multiplying large numbers, the time it takes is roughly proportional to the product of the number of digits in each. If we computed our 150-digit number to the 10^{120}th power, we would be computing a number with more than 10^{120} digits. We clearly do *not* want to compute with such numbers, as no computer can store such a number, even using all its memory (including disks)!

Fortunately, because the number we are computing will ultimately be taken modulo some 300-digit number, we can make all our multiplications modulo that number (see Lemma 2.3). By doing so, we ensure that the two numbers we are multiplying have at most 300 digits, and so the time needed for the problem proposed in Exercise 2.4-1 would be a proportionality constant multiplied by 90,000 multiplied by $\log_2(10^{120})$ multiplied by the time needed for a basic operation plus the time needed to figure out which powers of a are multiplied together, which would be quite small in comparison.

This algorithm on 300-digit numbers could be on the order of a million times slower than an algorithm on simple integers.[8] This is a noticeable effect, and if you use or write an encryption program, you can see this effect when you run it. However, we can still typically do this calculation in less than a second—a small price to pay for secure communication.

How Long Does It Take to Use the RSA Algorithm?

Encoding and decoding messages according to the RSA algorithm requires many calculations. How long will all this arithmetic take? Let's assume for now that Bob has already chosen p, q, e, and d; so he knows n as well. When Alice wants to send Bob the message x, she sends $x^e \bmod n$. By our analyses in Exercises 2.4-2 and 2.4-3, we see that the amount of time needed to compute this number is more or less proportional to $\log_2 e$, which is itself proportional to the number of digits of e, though the first constant of proportionality depends on the method our computer uses to multiply numbers. Because e has no more than 300 digits, this should not be too time consuming for Alice if she has a reasonable computer. (On the other hand, if she wants to send a message consisting of many segments of 300 digits each, she might want to use the RSA system to send a key for another simpler secret key system and then use that simpler system for the message.)

It takes Bob a similar amount of time to decode, as he has to take the message to the dth power mod n.

We commented already that nobody knows a fast way to find x from $x^e \bmod n$. In fact, nobody knows that there isn't a fast way either, which means that it is possible that the RSA cryptosystem could be broken some time in the future. You may have heard about the family of **NP**-complete problems (see Chapter 6). These are a family of problems that people believe to be reasonably difficult; that is, no one has yet come up with an efficient algorithm for any problem in the class, and the set of problems in the class are all roughly equivalent. We would be happy if cryptography were based on an **NP**-complete problem; unfortunately, it is not. The problem of extracting eth roots mod n is not an **NP**-complete problem although it is known to be no more difficult than the **NP**-complete problems. It is also true that no one has yet designed an efficient algorithm for eth roots; the security of our cryptosystems rests on the hope that no one will develop such an algorithm.

To get around the RSA system, however, someone is not restricted to extracting roots to discover x. Someone who knows n and knows that Bob is using the RSA system could presumably factor n, discover p and q, use the extended GCD algorithm to compute d, and then decode all of Bob's messages. But nobody knows how to factor integers quickly either. In fact, we don't know if factoring is as hard as **NP**-complete problems, but we do know that it is no harder than the **NP**-complete problems. However,

---

[8]Let us assume that our computer can multiply four-digit integers, but not five-digit numbers, exactly. Then efficiently multiplying two 300-digit numbers is like multiplying 75 integers times 75 integers, or 5625 products. Also, $\log_2(10^{120}) \approx \log_2(2^{10})^{40} = \log_2(2^{400}) = 400$. Because each of the approximately 400 computations we need to do to compute $10^{10^{120}}$ is like 5625 integer multiplications, we would have something like two million steps, each equivalent to multiplying two integers, in executing our algorithm.

enough people have worked on the factoring problem that most computer scientists are confident that it is in fact difficult. In this case, the RSA system is safe, as long as we use keys that are long enough.

How Hard Is Factoring?

Exercise 2.4-4 Factor 225,413. (The idea is to try to do this without resorting to computers; but if you give up by hand and calculator, using a computer is fine.)

Unless you know some special factoring techniques, it probably took you a while to discover that 225,413 is 431 times 523. In other words, factoring is real work. With current technology, keys with roughly 100 digits are not that hard to crack. In other words, people can factor numbers that are roughly 100-digits long using methods that are a little more sophisticated than the obvious approach of trying all possible divisors. However, when the numbers get longer, say more than 300 digits, they become very hard to factor. As of the year 2004, the record size of numbers that can be factored is roughly 174 digits. Factoring a "test" number of this size takes months, even when using many computers. Given the current technology, RSA with a 300-digit key seems to be very secure.

Finding Large Primes

There is one more issue to consider in implementing the RSA system for Bob. We said that Bob chose two primes of about 150 digits each. But how did he choose them? A theorem called the prime number theorem tells us that if we choose a number m at random and check about $\log_e m$ numbers around m for primality, we would expect one of these numbers to be prime. Thus, we shouldn't have to guess many numbers, even with hundreds of digits, before we find a prime. So, if we have a quick way to check if a number is prime, finding one shouldn't take too long.

However, we have just mentioned that nobody knows a quick way to find any or all factors of a number. The standard way to prove a number is prime is to show that its only factors are the number and 1. For the same reasons that factoring is hard, the simple approach to primality testing—testing all possible divisors—is much too slow. If we did not have a faster way to check whether a number is prime, the RSA system would be useless.

In 2002, Agrawal, Kayal, and Saxena[1] announced an algorithm for testing whether an integer n is prime. They showed that the algorithm takes no more than the 12th power of the number of digits of n to determine whether n is prime. Lenstra and Pomerance[14] have improved the algorithm in a way that reduces the exponent to six. In practice, the algorithm seems to take significantly less time. Although the algorithm requires more than the background we are able to provide in this book, its description and the proof that it works in the specified time uses only results that one might find in an undergraduate course on abstract algebra or number theory. The central theme of the algorithm is the use of a variation of Fermat's Little Theorem.

In 1976, Miller[26] was able to use Fermat's Little Theorem to show that if a conjecture called the "extended Riemann hypothesis" was true, then an algorithm he developed would determine whether a number n was prime in a time bounded above by a polynomial in the number of digits of n. In 1980, Rabin[27] modified Miller's method to determine, in polynomial time, whether a number was prime without the extra hypothesis but with a probability of error that could be made as small a positive number as one might desire, though not 0. We describe the general idea behind all of these advances in the context of what people now call the Miller-Rabin primality test. As of the writing of this book, variations on this kind of algorithm are being used to provide primes for cryptography.

By Fermat's Little Theorem, we know that in Z_p with p prime, $x^{p-1} \bmod p = 1$ for every x between 1 and $p - 1$. What about x^{m-1} in Z_m when m is not prime?

Exercise 2.4-5 Suppose x is a member of Z_m that has no multiplicative inverse. Is it possible that $x^{m-1} \bmod m = 1$?

Our next lemma answers the question of this exercise.

Lemma 2.25 Let m be a nonprime and let x be a number in Z_m that has no multiplicative inverse. Then $x^{m-1} \bmod m \neq 1$.

Proof: Assume for the purpose of contradiction that

$$x^{m-1} \bmod m = 1 .$$

Then

$$x \cdot x^{m-2} \bmod m = 1 .$$

But then $x^{m-2} \bmod m$ is the inverse of x in Z_m, which contradicts the fact that x has no multiplicative inverse. Thus, it must be the case that $x^{m-1} \bmod m \neq 1$. ∎

This distinction between primes and nonprimes suggests an idea we could use to create an algorithm to test for primality. Suppose we have some number m and are not sure whether it is prime. We can run the following algorithm:

```
PrimeTest(m)

(1)    choose a random number x, 2 ≤ x ≤ m−1
(2)    compute y = xᵐ⁻¹ mod m
(3)    if (y==1)
(4)        output "m might be prime"
(5)    else
(6)        output "m is definitely not prime"
```

Note the asymmetry here. If $y \neq 1$, then m is definitely not prime, and we are done. If $y = 1$, however, then m might be prime, and we will probably want to do some other calculations. In fact, we can repeat the algorithm PrimeTest(m) t times, with a different random number x each time. If, on any of the t runs, the algorithm outputs "`m is definitely not prime,`" then the number m is definitely not prime, because we have an x for which $x^{m-1} \neq 1$. If, on all t runs, the algorithm PrimeTest(m) outputs "`m might be prime`", however, then we can say with reasonable certainty that the number m is prime. This is actually an example of a **randomized algorithm**. We will be studying these in greater detail in Chapter 5. For now, let's informally estimate how likely it is that we will make a mistake.

We can see that for a particular nonprime m, the chance of making a mistake depends on exactly how many numbers a have the property that $a^{m-1} = 1$. If very few do, then our algorithm is very likely to give the correct answer. If most of them do, however, then we are more likely to give an incorrect answer.

Problem 12 shows that the number of elements in Z_m without inverses is at least $\sqrt{m}$. In fact, even many numbers that do have inverses will fail the test $x^{m-1} = 1$. For example, in Z_{12}, only 1 passes the test; in Z_{15}, only 1 and 14 pass the test. (Z_{12} is not really typical. Can you explain why? See Problem 15 for a hint.)

In fact, the Miller-Rabin algorithm modifies the test slightly (in a way that we won't describe here[13]) so that for any nonprime m, at least three-quarters of the possible values we could choose for x will fail the modified test and, hence, will show that m is composite. This suggests intuitively that if we repeat the test t times and assert that an x that passes these t tests is prime, then the probability of being wrong is actually 4^{-t}. So, if we repeat the test 5 times, we seem to have only about a one in a thousand chance of making a mistake, and if we repeat it 50 times, we seem to have only about a one in 2^{100} (a little less than one in a nonillion) chance of making a mistake! As you may have guessed from our careful phrasing, the matter is not quite this simple, but in Chapter 5 we will see that we still have a highly effective test for primality.

Numbers we have chosen by this algorithm are sometimes called **pseudoprimes**. (They are called this because they are very likely to be prime.) In practice, pseudo-primes are used instead of primes in implementations of the RSA cryptosystem. The worst that can happen when a pseudoprime is not prime is that a message may be garbled. In this case, we know that our pseudoprime is not really prime. As a result, we choose new pseudoprimes and ask our sender to send the message again. (Note that we do not change p and q with each use of the system; unless we were to receive a garbled message, we would have no reason to change them.)

Recall that we said the prime number theorem tells us that if we check about $\log_e n$ numbers near n, we can expect one of them to be prime. A d-digit number is at least 10^{d-1} and is less than 10^d, so its natural logarithm is between $(d-1)\log_e 10$ and $d \log_e 10$. If we want to find a d-digit prime, we can take any d-digit number and test about $d \log_e 10$ numbers near it for primality. In Chapter 5, we will see that it is reasonable for us to expect that one of them will turn out to be prime. The number $\log_e 10$ is 2.3 to two decimal places. Thus, it does not take a really large amount of time to find two prime numbers with 150 (or so) digits each.

Important Concepts, Formulas, and Theorems

1. *Exponentiation.* To perform exponentiation mod n efficiently, we use repeated squaring and take mods after each arithmetic operation.

2. *Security of RSA.* The security of RSA rests on the fact that no one has developed an efficient algorithm factoring or finding x given x^e mod n.

3. *Fermat's Little Theorem does not hold for composites.* Let m be a nonprime and x be a number in Z_n that has no multiplicative inverse. Then x^{m-1} mod $m \neq 1$.

4. *Testing numbers for primality.* The randomized Miller-Rabin algorithm will tell you almost surely if a given number is prime.

5. *Finding prime numbers.* If we apply the randomized Miller-Rabin algorithm to numbers with d digits until we find a pseudoprime, then we expect to test about $d \ln 10$ (which is about $2.3d$) numbers.

Problems

1. What is 3^{1024} in Z_7? (This is a straightforward problem to do by hand.) `a/h`

2. Suppose you have computed a^2, a^4, a^8, a^{16}, and a^{32}. What is the most efficient way to compute a^{53}?

3. A gigabyte is one billion bytes; a terabyte is one trillion bytes. A byte is eight bits, each a 0 or a 1. Because $2^{10} = 1024$, which is about 1000, you can store about three digits (any number between 0 and 999) in ten bits. About how many decimal digits could you store in five gigabytes of memory (a gigabyte is 2^{30}, or approximately one billion bytes)? About how many decimal digits could you store in five terabytes of memory (a terabyte is 2^{40}, or approximately one trillion bytes)? How does this compare with the number 10^{120}? (To do this problem, it is reasonable to continue to assume that 1024 is about 1000.) `a/h`

4. Find all numbers, if any, $a \in Z_9$ different from 1 and 8 (notice that -1 mod $9 = 8$) such that a^8 mod $9 = 1$.

5. Use a spreadsheet, programmable calculator, or computer to find all numbers a different from 1 and 32 (which equals -1 mod 33) with a^{32} mod $33 = 1$. (This problem is relatively straightforward to do with a spreadsheet that can compute mods and that will let you "fill in" rows and columns with formulas. However, you do have to know how to use the spreadsheet in this way to make it straightforward!) `a/h`

6. How many digits does the 10^{120}th power of 10^{100} have?

7. If a is a 100-digit number, is the number of digits of $a^{10^{120}}$ closer to 10^{120} or 10^{240}? Is it a lot closer? Does the answer depend on what a actually is rather than the number of digits it has? `a/h`

8. Explain what our outline of the solution to Exercise 2.4-1 has to do with the binary representation of 10^{120}.

9. Suppose you want to compute $a^{e_1 e_2 \cdots e_m}$ mod n. Discuss whether it makes sense to reduce the exponents mod n as you compute their product. In particular, what rule of exponents would allow you to do this, and do you think this rule of exponents makes sense? a/h

10. Give careful pseudocode to compute a^x mod n. Make your algorithm as efficient as possible.

11. Number theorists use $\varphi(n)$ to stand for the number of elements of Z_n that have inverses. Suppose you want to compute $a^{e_1 e_2 \cdots e_m}$ mod n. Would it make sense to reduce the exponents mod $\varphi(n)$ as you compute their product? Why? (*Hint:* The answer might be different in different cases.) a/h

12. Show that if m is not prime, then at least $\sqrt{m}$ elements of Z_m do not have multiplicative inverses.

13. Suppose for applying RSA, $p = 11$, $q = 19$, and $e = 7$. What is the value of d? Show how to encrypt the message 100, and then show how to decrypt the resulting message. a/h

14. Suppose for applying RSA, $p = 11$, $q = 23$, and $e = 13$. What is the value of d? Show how to encrypt the message 100 and then how to decrypt the resulting message.

15. Show that in Z_{p+1}, where p is an odd prime, only one element passes the primality test x^{m-1} mod $m = 1$. (In this case, $m = p + 1$.)

16. A digital signature is a way to sign a document securely. In other words, it is a way to put your "signature" on a document so that anyone reading it knows that it is you who have signed it: no one else can "forge" your signature. The document itself may be public; it is your signature that we are trying to protect. Digital signatures are, in a way, the opposite of encryption: If Bob wants to sign a message, he first applies his signature to it (think of this as encryption) and then the rest of the world can easily read it (think of this as decryption). Explain, in detail, how to achieve digital signatures by using ideas similar to those used for RSA. In particular, anyone who has the document and has your signature of the document (and knows your public key) should be able to determine that you signed it. a/h

3 REFLECTIONS ON LOGIC AND PROOF

In this chapter, we cover some basic principles of logic and describe some methods for constructing proofs. This chapter is not meant to be a complete enumeration of all possible proof techniques. The philosophy of this book is that most people learn more about proofs by reading, watching, and attempting proofs than by an extended study of the logical rules behind proofs. However, now that we have some examples of proofs, reflecting on their structure and discussing what constitutes a proof will help you read and do proofs. We first develop a language that will allow us to talk about proofs, and then we use this language to describe the logical structure of a proof.

3.1 EQUIVALENCE AND IMPLICATION

Equivalence of Statements

Exercise 3.1-1

A group of students is working on a project that involves writing a merge sort program. Joe and Mary have each written an algorithm for a function that takes two lists, List1 and List2, of lengths p and q, respectively, and merges them into a third list, List3. Part of Mary's algorithm is as follows:

```
(1)  if ((i+j ≤ p+q) && (i ≤ p) && ((j > q)||(List1[i] ≤ List2[j])))
(2)       List3[k] = List1[i]
(3)       i = i+1
(4)  else
(5)       List3[k] = List2[j]
(6)       j = j+1
(7)  k = k+1
```

The corresponding part of Joe's algorithm is

```
(1)   if ((( i+j ≤ p+q) && ( i ≤ p) && ( j > q))
         || (( i+j ≤ p+q) && ( i ≤ p) && (List1[i] ≤ List2[j])))
(2)        List3[k] = List1[i]
(3)        i = i+1
(4)   else
(5)        List3[k] = List2[j]
(6)        j = j+1
(7)   k = k+1
```

Do Joe's and Mary's algorithms do the same thing?

Notice that Joe's and Mary's algorithms are exactly the same except for the if statement in Line 1 (how convenient; they even used the same local variables!). In Mary's algorithm, we put entry i of List1 into position k of List3 if

$$i + j \leq p + q \text{ and } i \leq p \text{ and } (j > q \text{ or } List1[i] \leq List2[j]) ,$$

whereas in Joe's algorithm, we put entry i of List1 into position k of List3 if

$$(i + j \leq p + q \text{ and } i \leq p \text{ and } j > q) \text{ or } (i + j \leq p + q \text{ and } i \leq p \text{ and } List1[i] \leq List2[j]) .$$

Joe's and Mary's statements are both built from the same constituent parts (namely, comparison statements), so we can name these constituent parts and rewrite the statements. We use

- s to stand for $i + j \leq p + q$,

- t to stand for $i \leq p$,

- u to stand for $j > q$, and

- v to stand for $List1[i] \leq List2[j]$.

The condition in Mary's if statement on Line 1 of her code becomes

$$s \text{ and } t \text{ and } (u \text{ or } v) ,$$

while Joe's if statement on Line 1 of his code becomes

$$(s \text{ and } t \text{ and } u) \text{ or } (s \text{ and } t \text{ and } v) .$$

By recasting the statements in this symbolic form, we see that s and t always appear together as "s and t." We can thus simplify the expressions by substituting w for "s and t." Mary's condition now has the form

$$w \text{ and } (u \text{ or } v) ,$$

and Joe's has the form

$$(w \text{ and } u) \text{ or } (w \text{ and } v) .$$

Although we can argue, based on our knowledge of the structure of the English language, that Joe's statement and Mary's statement are saying the same thing, it will help us understand logic if we formalize the idea of "saying the same thing." If you look closely at Joe's and Mary's statements, you can see that we are saying that the word "and" distributes over the word "or," just as set intersection distributes over set union and multiplication distributes over addition. To analyze when statements mean the same thing, and to explain more precisely what it means to say something like "'and' distributes over 'or,'" logicians have adopted a standard notation for writing symbolic versions of compound statements. We use the symbol $\land$ to stand for "and" and $\lor$ to stand for "or." In this notation, Mary's condition becomes

$$w \land (u \lor v) \,,$$

and Joe's becomes

$$(w \land u) \lor (w \land v) \,.$$

We now have a nice notation (which makes our compound statements look a lot like the two sides of the distributive law for intersection of sets over union), but we have not yet explained why two statements with these symbolic forms mean the same thing. We must therefore give a precise definition of "meaning the same thing" and develop a tool for analyzing when two statements satisfy this definition. We are going to consider symbolic compound statements that may be built up from the following notation:

- Symbols (s, t, etc.), which we call *variables,* standing for statements
- The symbol $\land$, standing for "and"
- The symbol $\lor$, standing for "or"
- The symbol $\oplus$, standing for "exclusive or"
- The symbol $\neg$, standing for "not"
- Left and right parentheses

Truth Tables

We now develop a theory for deciding when a compound statement is true based on the truth or falsity of its component statements. Using this theory, we can determine for a particular setting of variables, such as s, t, and u, whether a particular compound statement, such as

$$(s \oplus t) \land (\neg u \lor (s \land t)) \land \neg(s \oplus (t \lor u)) \,,$$

is true or false. Our technique uses truth tables, which you have probably seen before. We will soon see why truth tables are the proper tool for determining whether two statements are equivalent.

As with arithmetic, the order of operations in a logical statement is important. Our sample compound statement used parentheses to make it clear which operation to do first, with one exception: the use of the symbol $\neg$. The symbol $\neg$ always has the highest priority, which means that $\neg u \lor (s \land t)$ means $(\neg u) \lor (s \land t)$ rather than

$\neg(u \lor (s \land t))$. The principle is simple—the symbol $\neg$ applies to either the symbol or the parenthesized expression immediately following it. This is the same principle used with negative numbers in algebraic expressions. With this one exception, we will always use parentheses to make the order in which we are to perform operations clear; you should do the same.

The operators $\land$, $\lor$, $\oplus$, and $\neg$ are called **logical connectives**. The truth table for a logical connective tells us, in terms of the possible truth or falsity of the component parts, when the compound statement made by connecting those parts is true and when it is false. The truth tables for the connectives we have mentioned so far are in Figure 3.1.

Figure 3.1: *Truth tables for the basic logical connectives*

AND			OR			XOR			NOT	
s	t	$s \land t$	s	t	$s \lor t$	s	t	$s \oplus t$	s	$\neg s$
T	T	T	T	T	T	T	T	F	T	F
T	F	F	T	F	T	T	F	T	F	T
F	T	F	F	T	T	F	T	T		
F	F	F	F	F	F	F	F	F		

These truth tables define the words "and," "or," "exclusive or" ("xor" for short), and "not" in the context of symbolic compound statements. For example, the truth table for $\lor$—"or"—tells us that when s and t are both true, then so is "s or t." It tells us that when s is true and t is false, or s is false and t is true, then "s or t" is true. Finally, it tells us that when s and t are both false, then so is "s or t." Is this how we use the word "or" in English? The answer is "sometimes." The word "or" is used ambiguously in English. When a teacher says, "Each question on the test will be short answer or multiple choice," the teacher is presumably not intending that a question could be both. Thus, the word "or" is being used here in the sense of "exclusive or"—the $\oplus$ in Figure 3.1. When someone says, "Let's see, this afternoon I could take a walk or I could shop for some new gloves," he probably does not mean to preclude the possibility of doing both—perhaps even taking a walk downtown and then shopping for new gloves before walking back. Thus, in English, we determine the way in which someone uses the word "or" from context. In mathematics and computer science, because we don't always have context, we agree to say "exclusive or," or "xor" for short, when that is what we mean; otherwise, we mean the "or" whose truth table is given by $\lor$. In the case of "and" and "not," the truth tables are exactly what we would expect.

We have been thinking of s and t as variables that stand for statements. The purpose of a truth table is to define when a compound statement is true or false in terms of when its component statements are true and false. Because we focus on just the truth and falsity of our statements when we are giving truth tables, we can also think of s and t as variables that can take on the values "true" (T) and "false" (F). We refer to these values as the **truth values** of s and t. A truth table, then, gives us the truth values of a compound statement in terms of the truth values of the component parts of the compound statement. The statements $s \land t$, $s \lor t$, and $s \oplus t$ each have two component parts, s and t. Notice that there are two values we can assign to s, and for each value we assign to s, there are two values we can assign to t. By the product principle, there are $2 \cdot 2 = 4$ ways to assign truth values to s and t. Thus, we have four rows in our truth table, one for each way of assigning truth values to s and t.

For a more complex compound statement, such as the one in Line 1 in Joe's and Mary's programs, we still want to describe situations in which the statement is true and situations in which the statement is false. We do this by working out a truth table for the compound statement from the truth tables of its symbolic statements and its connectives. We use a variable to represent the truth value of each symbolic statement. The truth table has one column for each of the original variables and one column for each of the pieces we use to build up the compound statement. The truth table has one row for each possible way of assigning truth values to the original variables. Thus, if we have two variables, we have four rows, as in the "AND," "OR," and "XOR" tables in Figure 3.1. If we have just one variable, then we have just two rows, as in the "NOT" table in Figure 3.1. If we have three variables, then we have $2^3 = 8$ rows, and so on.

Table 3.1 gives the truth table for the symbolic statement derived from Line 1 of Joe's algorithm. The columns to the left of the dark blue line contain the possible truth values of the variables. The columns to the right correspond to various subexpressions whose truth values we need to compute. The truth table has as many columns as we need in order to compute the final result correctly. As a general rule, each column should be easily computed from one or two previous columns.

Table 3.1: *The truth table for Joe's statement*

w	u	v	$u \vee v$	$w \wedge (u \vee v)$
T	T	T	T	T
T	T	F	T	T
T	F	T	T	T
T	F	F	F	F
F	T	T	T	F
F	T	F	T	F
F	F	T	T	F
F	F	F	F	F

Table 3.2 gives the truth table for the statement derived from Line 1 of Mary's algorithm.

Notice that the pattern of T's and F's used to the left of the dark blue line in both Joe's and Mary's truth tables are the same—namely, they are in reverse alphabetical order.[1] Thus, row i of Table 3.1 represents exactly the same assignment of truth values to u, v, and w as row i of Table 3.2. The final columns of Joe's and Mary's truth tables are identical, which means that Joe's symbolic statement and Mary's symbolic statement

---

[1] Alphabetical order is sometimes called *lexicographic order*. Lexicography is the study of the principles and practices used in making dictionaries. Thus, the order we used for the T's and F's is called reverse lexicographic order, or reverse lex order for short.

Table 3.2: *The truth table for Mary's statement*

w	u	v	$w \wedge u$	$w \wedge v$	$(w \wedge u) \vee (w \wedge v)$
T	T	T	T	T	T
T	T	F	T	F	T
T	F	T	F	T	T
T	F	F	F	F	F
F	T	T	F	F	F
F	T	F	F	F	F
F	F	T	F	F	F
F	F	F	F	F	F

are true in exactly the same cases. Therefore, the two statements must say the same thing, and Mary's and Joe's program segments return exactly the same values. We say that two symbolic compound statements are **equivalent** if they are true in exactly the same cases. Alternatively, two statements are equivalent if their truth tables have the same final column (assuming both tables assign truth values to the original symbolic statements in the same pattern).

Tables 3.1 and 3.2 actually prove a **distributive law**:

Lemma 3.1   The statements

$$w \wedge (u \vee v)$$

and

$$(w \wedge u) \vee (w \wedge v)$$

are equivalent.

DeMorgan's Laws

Exercise 3.1-2   **DeMorgan's laws** say that $\neg(p \vee q)$ is equivalent to $\neg p \wedge \neg q$ and that $\neg(p \wedge q)$ is equivalent to $\neg p \vee \neg q$. Use truth tables to demonstrate that DeMorgan's laws are correct.

Exercise 3.1-3   Show that $p \oplus q$ (the exclusive or of p and q) is equivalent to $(p \vee q) \wedge \neg(p \wedge q)$. Apply one of DeMorgan's laws to $\neg(\neg(p \vee q)) \wedge \neg(p \wedge q)$ to find another symbolic statement equivalent to the exclusive or.

To verify the first DeMorgan's law, we create a "double truth table" by (mentally) condensing two truth tables into one (see Table 3.3). The left sides of the two truth

Table 3.3: *Proving the first DeMorgan's law*

p	q	$p \vee q$	$\neg(p \vee q)$	$\neg p$	$\neg q$	$\neg p \wedge \neg q$
T	T	T	F	F	F	F
T	F	T	F	F	T	F
F	T	T	F	T	F	F
F	F	F	T	T	T	T

tables we are condensing are identical, so we give just one left side to the left of the first dark blue line. The second dark blue line separates the right sides of the two truth tables we are condensing. In this way, we can still see the computation of the truth values of $\neg(p \vee q)$ and $\neg p \wedge \neg q$. We see that the fourth and the last columns are identical; therefore, the first DeMorgan's law is correct. We can verify the second DeMorgan's law by a similar process.

To show that $p \oplus q$ is equivalent to $(p \vee q) \wedge \neg(p \wedge q)$, we use the double truth table in Table 3.4. Now we deal with the second question in Exercise 3.1-3. Notice first that $\neg(\neg(p \vee q))$ is equivalent to $p \vee q$; thus, the statement $\neg(\neg(p \vee q)) \wedge \neg(p \wedge q)$ is equivalent to $p \oplus q$. By applying DeMorgan's first law[2] to $\neg(\neg(p \vee q)) \wedge \neg(p \wedge q)$, we see that $p \oplus q$ is also equivalent to $\neg(\neg(p \vee q) \vee (p \wedge q))$. It was easier to use DeMorgan's law to show this equivalence than to use another double truth table.

Table 3.4: *An equivalent statement to $p \oplus q$*

p	q	$p \oplus q$	$p \vee q$	$p \wedge q$	$\neg(p \wedge q)$	$(p \vee q) \wedge \neg(p \wedge q)$
T	T	F	T	T	F	F
T	F	T	T	F	T	T
F	T	T	T	F	T	T
F	F	F	F	F	T	F

Implication

Another kind of compound statement occurs frequently in mathematics and computer science. Recall Fermat's Little Theorem (Theorem 2.21):

If p is a prime, then $a^{p-1} \bmod p = 1$ for each nonzero $a \in Z_p$.

Fermat's Little Theorem combines two constituent statements:

- p is a prime, and
- $a^{p-1} \bmod p = 1$ for each nonzero $a \in Z_p$.

---

[2]Notice that we are applying the law to a statement of the form $\neg s \wedge \neg t$ and getting one of the form $\neg(s \vee t)$.

We can also restate Fermat's Little Theorem (a bit clumsily) as

- p is a prime only if $a^{p-1} \bmod p = 1$ for each nonzero $a \in Z_p$, or
- p is a prime implies $a^{p-1} \bmod p = 1$ for each nonzero $a \in Z_p$, or
- $a^{p-1} \bmod p = 1$ for each nonzero $a \in Z_p$ if p is prime.

Using s to stand for "p is a prime" and t to stand for "$a^{p-1} \bmod p = 1$ for every nonzero $a \in Z_p$," we can express any of the four statements of Fermat's Little Theorem in symbols as

$$s \Rightarrow t ,$$

which most people read as "s implies t." When we translate from symbolic language to English, it is often clearer to say, "If s, then t."

We summarize this discussion in the following definition.

Definition 3.1	The following four English phrases are intended to mean the same thing. In other words, they are defined by the same truth table.

- s implies t.
- If s, then t.
- t if s.
- s only if t.

Observe that the use of "only if" may seem a little different from the normal usage in English. Also observe that there are still other ways of making an "if ... then" statement in English. A number of our lemmas, theorems, and corollaries (for example, Lemma 2.5 and Corollary 2.6) have had two sentences. The first says, "Suppose ...." The second says, "Then ...." The two sentences "Suppose s." and "Then t." are equivalent to the single sentence "$s \Rightarrow t$." When we have a statement equivalent to $s \Rightarrow t$, we call the statement s the **hypothesis** of the implication, and we call the statement t the **conclusion** of the implication.

If and Only If

The word "if" and the phrase "only if" frequently appear together in mathematical statements. For example, Theorem 2.9 stated:

A number a has a multiplicative inverse in Z_n if and only if there are integers x and y such that $ax + ny = 1$.

Using s to stand for the statement "a number a has a multiplicative inverse in Z_n" and t to stand for the statement "there are integers x and y such that $ax + ny = 1$," we can write this statement symbolically as

$$s \text{ if and only if } t .$$

Referring to Definition 3.1, we parse this as

$$s \text{ if } t, \text{ and } s \text{ only if } t,$$

which by the definition above is the same as

$$t \Rightarrow s \text{ and } s \Rightarrow t.$$

We denote the statement "s if and only if t" by $s \Leftrightarrow t$. Statements of the form $s \Rightarrow t$ and $s \Leftrightarrow t$ are called **conditional statements**, and the connectives $\Rightarrow$ and $\Leftrightarrow$ are called **conditional connectives**.

Exercise 3.1-4 Use truth tables to explain the difference between $s \Rightarrow t$ and $s \Leftrightarrow t$.

To analyze the truth and falsity of statements involving "implies" and "if and only if," we need to understand exactly how they are different. By constructing truth tables for these statements, we see that there is only one case in which they could have different truth values. In particular, if s is true and t is true, then we would say that both $s \Rightarrow t$ and $s \Leftrightarrow t$ are true. If s is true and t is false, we would say that both $s \Rightarrow t$ and $s \Leftrightarrow t$ are false. In the case that both s and t are false, we would say that $s \Leftrightarrow t$ is true. What about $s \Rightarrow t$? Let's try an example. Suppose s is the statement "It is supposed to rain" and t is the statement "I carry an umbrella." If, on a given day, it is not supposed to rain and I do not carry an umbrella, we would say that the statement "If it is supposed to rain, then I carry an umbrella" is true on that day. This suggests that we also want to say $s \Rightarrow t$ is true if s is false and t is false.[3] Thus, the truth tables are identical in Rows 1, 2, and 4. For "implies" and "if and only if" to mean different things, the truth tables must therefore be different in Row 3. (Row 3 is the case where s is false and t is true.) Clearly, in this case, we would want $s \Leftrightarrow t$ to be false, so our only choice is to say that $s \Rightarrow t$ is true. This gives us the truth tables in Figure 3.2.

Figure 3.2: *The truth tables for "implies" and for "if and only if"*

	IMPLIES			IF AND ONLY IF	
s	t	$s \Rightarrow t$	s	t	$s \Leftrightarrow t$
T	T	T	T	T	T
T	F	F	T	F	F
F	T	T	F	T	F
F	F	T	F	F	T

[3]Note that we are making this conclusion on the basis of one example. Why can we do so? We are not trying to prove something; rather, we are trying to figure out what the appropriate definition is for the $\Rightarrow$ connective. Because we have said that the truth or falsity of $s \Rightarrow t$ depends only on the truth or falsity of s and t, one example serves to lead us to an appropriate definition. If a different example led us to a different definition, then we would want to define two different kinds of implications, just as we have two different kinds of "ors," $\vee$ and $\oplus$. Fortunately, the only kinds of conditional statements we need for doing mathematics and computer science are "implies" and "if and only if."

Here is another place where English usage is sometimes inconsistent. Suppose a parent says, "I will take the family to McDougall's for dinner if you get an A on this test," and even though the student gets a C, the parent still takes the family to McDougall's for dinner. Although this is something we didn't expect, was the parent's statement still true? Some people would say "yes;" others would say "no." Those who would say "no" mean, in effect, that in this context, the parent's statement meant the same as, "I will take the family to dinner at McDougall's if and only if you get an A on this test." In other words, to some people and in certain contexts, "if" and "if and only if" mean the same thing. Fortunately, questions of child rearing aren't part of mathematics or computer science (at least not this kind of question!). In mathematics and computer science, we adopt the two truth tables in Figure 3.2 as the meaning of the compound statement $s \Rightarrow t$ (or "if s, then t" or "t if s") and the compound statement $s \Leftrightarrow t$ (or "s if and only if t"). In particular, the truth table for "implies" in Figure 3.2 is the one referred to in Definition 3.1, and thus it defines the mathematical meaning of s implies t or any of the other three statements referred to in that definition.

Some people have difficulty using the truth table for $s \Rightarrow t$ because of this ambiguity in English. The following example can be helpful in resolving this ambiguity: Suppose a classmate holds an ordinary playing card (with its back to you) and says, "If this card is a heart, then it is a queen." In which of the following four circumstances would you say your classmate lied?

1. The card is a heart and a queen.

2. The card is a heart and a king.

3. The card is a diamond and a queen.

4. The card is a diamond and a king.

You would certainly say she lied in the case that the card is the king of hearts, and you would certainly say she didn't lie if the card is the queen of hearts. In this example, the inconsistency of the English language should seem out of place to you, and you would not say your classmate is a liar in either of the other cases. Now we apply the **principle of the excluded middle**.

Principle 3.1	**(The Principle of the Excluded Middle)** A statement is true exactly when it is not false.

This principle tells us that the statement is true in the three cases where you wouldn't say your classmate lied. We used this principle implicitly when we introduced proof by contradiction (Principle 2.1). We were explaining Corollary 2.6, which states:

Suppose there is a b in Z_n such that the equation

$$a \cdot_n x = b$$

does not have a solution. Then a does not have a multiplicative inverse in Z_n.

We had assumed that the hypothesis of the corollary was true so that $a \cdot_n x = b$ does not have a solution. Then we assumed that the conclusion that a does not have a multiplicative inverse was false. We saw that these two assumptions led to a contradiction; thus, it was impossible for both of them to be true. We concluded that whenever the first assumption was true, the second had to be false. Why could we conclude this? Because the principle of the excluded middle says that the second assumption has to be either true or false. We didn't introduce the principle of the excluded middle at that point for two reasons. First, we expected that you would agree with our proof even if we didn't mention the principle, and second, we didn't want to confuse your understanding of proof by contradiction by talking about two principles at once.

Important Concepts, Formulas, and Theorems

1. *Logical statements.* Logical statements may be built up from the following notation:

   - Symbols (s, t, etc.), which we call variables, standing for statements
   - The symbol $\land$, standing for "and"
   - The symbol $\lor$, standing for "or"
   - The symbol $\oplus$, standing for "exclusive or"
   - The symbol $\neg$, standing for "not"
   - The symbol $\Rightarrow$, standing for "implies"
   - The symbol $\Leftrightarrow$, standing for "if and only if"
   - Left and right parentheses

   The operators $\land$, $\lor$, $\oplus$, $\Rightarrow$, $\Leftrightarrow$, and $\neg$ are called *logical connectives*. The operators $\Rightarrow$ and $\Leftrightarrow$ are called *conditional connectives*.

2. *Truth tables.* The following are truth tables for the basic logical connectives.

	AND			OR			XOR			NOT
s	t	$s \land t$	s	t	$s \lor t$	s	t	$s \oplus t$	s	$\neg s$
T	T	T	T	T	T	T	T	F	T	F
T	F	F	T	F	T	T	F	T	F	T
F	T	F	F	T	T	F	T	T		
F	F	F	F	F	F	F	F	F		

3. *Equivalence of logical statements.* We say that two symbolic compound statements are *equivalent* if they are true in exactly the same cases.

4. *Distributive law.* The statements $w \land (u \lor v)$ and $(w \land u) \lor (w \land v)$ are equivalent.

5. *DeMorgan's laws.* DeMorgan's laws say that $\neg(p \lor q)$ is equivalent to $\neg p \land \neg q$ and that $\neg(p \land q)$ is equivalent to $\neg p \lor \neg q$.

6. *Implication.* The following four English phrases are equivalent:

- *s* implies *t*.
- If *s*, then *t*.
- *t* if *s*.
- *s* only if *t*.

7. *Truth tables for "implies" and "if and only if."*

	IMPLIES			IF AND ONLY IF	
s	t	$s \Rightarrow t$	s	t	$s \Leftrightarrow t$
T	T	T	T	T	T
T	F	F	T	F	F
F	T	T	F	T	F
F	F	T	F	F	T

8. *Principle of the excluded middle.* A statement is true exactly when it is not false.

Problems

1. Give truth tables for the following expressions.

   a. $(s \vee t) \wedge (\neg s \vee t) \wedge (s \vee \neg t)$ a/h

   b. $(s \Rightarrow t) \wedge (t \Rightarrow u)$ a/h

   c. $(s \vee t \vee u) \wedge (s \vee \neg t \vee u)$ a/h

2. Find at least two more examples of the use of some word or phrase equivalent to "implies" in lemmas, theorems, or corollaries in Chapters 1 or 2.

3. Find at least two more examples of the use of the phrase "if and only if" in lemmas, theorems, and corollaries in Chapters 1 or 2.

4. Show that the statements $s \Rightarrow t$ and $\neg s \vee t$ are equivalent. a/h

5. Prove the DeMorgan law that states $\neg(p \wedge q) = \neg p \vee \neg q$. a/h

6. Show that $p \oplus q$ is equivalent to $(p \wedge \neg q) \vee (\neg p \wedge q)$.

7. Give a simplified form of each of the following expressions (using T to stand for a statement that is always true and F to stand for a statement that is always false).[4] a/h

   a. $s \vee s$

   b. $s \wedge s$

---

[4]A statement that is always true is called a *tautology;* a statement that is always false is called a *contradiction.*

c. $s \lor \neg s$

d. $s \land \neg s$

8. Using T to stand for a statement that is always true and F to stand for a statement that is always false, give a simplified form of each of the following statements.

    a. $T \land s$

    b. $F \land s$

    c. $T \lor s$

    d. $F \lor s$

9. Use DeMorgan's law, the distributive law, and Problems 7 and/or 8 to show that

$$\neg(s \lor t) \lor \neg(s \lor \neg t)$$

is equivalent to $\neg s$. a/h

10. Give an example in English where "or" seems to mean "exclusive or" (or where you think it would for many people) and an example in English where "or" seems to mean "inclusive or" (or where you think it would for many people).

11. Give an example in English where "if ... then" seems to mean "if and only if" (or where you think it would to many people) and an example in English where it seems not to mean "if and only if" (or where you think it would not to many people).

12. Find a statement involving only $\land$, $\lor$, and $\neg$ (and s and t) equivalent to $s \Leftrightarrow t$. Does your statement have as few symbols as possible? If you think it doesn't, try to find one with fewer symbols. a/h

13. Suppose that for each line of a two-variable truth table, you are told whether the final column in that line should evaluate to true or to false. (For example, you might be told that the final column should contain T, F, F, and T, in that order. Notice that Problem 12 can be interpreted as asking for this pattern.) Explain how to create a logical statement using the symbols s, t, $\land$, $\lor$, and $\neg$ that has that pattern as its final column. Can you extend this procedure to an arbitrary number of variables?

14. In Problem 13, your solution may have used $\land$, $\lor$, and $\neg$. Is it possible to give a solution using only one of these symbols? Is it possible to give a solution using only two of these symbols? a/h

15. We proved that $\land$ distributes over $\lor$ in the sense of giving two equivalent statements that represent the two "sides" of the distributive law. Answer each question that follows, and explain why your answer is correct.

    a. Does $\lor$ distribute over $\land$?

    b. Does $\lor$ distribute over $\oplus$?

    c. Does $\land$ distribute over $\oplus$?

3.2 VARIABLES AND QUANTIFIERS

Variables and Universes

Statements we use in computer languages to control loops or conditionals are statements about variables. When we declare these variables, we give the computer information about their possible values. For example, in some programming languages, we may declare a variable to be a Boolean or an integer or a real number.[5] In English and in mathematics, we also make statements about variables, but it is not always clear which words are being used as variables and what values these variables may take on. We use the phrase **varies over** to describe the set of values a variable may take on. For example, in English we might say, "If someone's umbrella is up, then it must be raining." In this case, the word "someone" is a variable, and presumably it *varies over* the people who happen to be in a given place at a given time. In mathematics, we might say, "For every pair of positive integers m and n, there are nonnegative integers q and r, with $0 \leq r < n$, such that $m = nq + r$." In this case, m, n, q, and r are clearly variables; our statement itself suggests that two variables range over the positive integers and two range over the nonnegative integers. We call the set of possible values for a variable the **universe** of that variable.

In the statement "m is an even integer," it is clear that m is a variable, but the universe is not given. The universe might be the integers, only the even integers, the rational numbers, or one of many other sets. The choice of the universe is crucial for determining the truth or falsity of a statement. If we choose the set of integers as the universe for m, then the statement is true for some integers and false for others. On the other hand, if we choose integer multiples of 10 as our universe, then the statement is always true. In the same way, when we control a `while` loop with a statement such as "$i < j$," there are some values of i and j for which the statement is true and some for which it is false. In statements like "m is an even integer" and "$i < j$," our variables are not constrained, and so they are called **free variables**. For each possible value of a free variable, we have a new statement, which might be either true or false, determined by substituting the possible value for the variable. The truth value of the statement is determined only after such a substitution.

Exercise 3.2-1   For what values of m is the statement $m^2 > m$ a true statement and for what values is it a false statement? Because a universe is not specified, our answer will depend on what universe we choose to use.

For the universe of positive integers, the statement is true for every value of m but 1. For the universe of the real numbers, the statement is true for every value of m except for those in the closed interval $[0, 1]$. There are really two points to make here. First, a statement about a variable can often be interpreted as a statement about more than one

---

[5]Note that to declare a variable x as an integer in, say, a C program does not mean the same thing as saying that x is an integer. In a C program, an integer may really be a 32-bit integer, so it is limited to values between $2^{31} - 1$ and -2^{31}. Similarly, a real has some fixed precision; hence, a real variable y may not be able to take on a value of, say, 10^{-985}.

universe; so, to make the statement unambiguous, we must clearly state the universe we have in mind. Second, a statement about a variable can be true for some values of a variable and false for others.

Quantifiers

In contrast, the statement

$$\text{For every integer } m, m^2 > m. \tag{3.1}$$

is false; we do not need to qualify our answer by saying that it is true some of the time and false at other times. To determine whether Statement 3.1 is true or false, we could substitute various values for m into the simpler statement $m^2 > m$ and decide, for each of these values, whether the statement $m^2 > m$ is true or false. Doing so, we see that the statement $m^2 > m$ is true for values such as $m = -3$ or $m = 9$ but false for $m = 0$ or $m = 1$. Thus, it is not the case that $m^2 > m$ for every integer m. Therefore, Statement 3.1 is false, because it is an assertion that the simpler statement $m^2 > m$ holds for each integer value of m we substitute. A phrase like "for every integer m," which converts a symbolic statement about potentially any member of our universe into a statement about the universe instead, is called a **quantifier**. A quantifier that asserts that a statement about a variable is true for every value of the variable in its universe, for example, "for every integer," is called a **universal quantifier**. This example illustrates a very important point.

> If a statement asserts something for every value of a variable, then to show the statement is false, we need only give one value of the variable for which the assertion is untrue.

Another example of a quantifier is the phrase "There is an integer m" in the sentence "There is an integer m such that $m^2 > m$." This statement is also about the universe of integers, and as such, it is true—there are plenty of integers m we can substitute into the symbolic statement $m^2 > m$ to make it true. This is an example of an **existential quantifier**, which asserts that a certain element of our universe exists. A second important point similar to the one we made above is as follows:

> To show that a statement with an existential quantifier is true, we need only exhibit one value of the variable being quantified that makes the statement true.

As the more complex statement

> For every pair of positive integers m and n, there are nonnegative integers q and r with $0 \leq r < n$ such that $m = qn + r$

shows, statements of mathematical interest abound with quantifiers. Mathematical statements of theorems, lemmas, and corollaries often have quantifiers. For example, in Lemma 2.5, the phrase "for any" is a quantifier, and in Corollary 2.6, the phrase "there is" is a quantifier. Quantifiers often occur in definitions as well. Recall the following definition of the big O notation, which you have probably used in earlier computer science courses.

| Definition 3.2 | For a function $g : R \rightarrow R$ with nonnegative values, we say that $f(x) = O(g(x))$ if there are positive numbers c and n_0 such that $f(x) \leq cg(x)$ for every $x > n_0$. |

| Exercise 3.2-2 | Quantification is present in our everyday language. The sentences "Every child wants a pony" and "No child wants a toothache" are two different examples of quantified sentences. Give ten examples of everyday sentences that use quantifiers, but use different words to indicate the quantification. |

| Exercise 3.2-3 | Convert the sentence "No child wants a toothache" into a sentence of the form "It is not the case that. . . ." Find an existential quantifier in your sentence. |

| Exercise 3.2-4 | What would you have to do to show that a statement about one variable with an existential quantifier is false? Correspondingly, what would you have to do to show that a statement about one variable with a universal quantifier is true? |

As Exercise 3.2-2 points out, English has many different ways to express quantifiers. For example, the sentences, "All hammers are tools," "Each sandwich is delicious," "No one in their right mind would do that," "Somebody loves me," and "Yes, Virginia, there is a Santa Claus" all contain quantifiers. For Exercise 3.2-3, we can say, "It is not the case that there is a child who wants a toothache." Our quantifier is the phrase "there is."

To show that a statement about one variable with an existential quantifier is false, we have to show that every element of the universe makes the statement (such as $m^2 > m$) false. Thus, to show that the statement "There is an x in [0, 1] with $x^2 > x$" is false, we have to show that every x in the interval makes the statement "$x^2 > x$" false. Similarly, to show that a statement with a universal quantifier is true, we have to show that the statement being quantified is true for every member of our universe. Later in this section, we give more details about how to show that a statement about a variable is true or false for every member of our universe.

Standard Notation for Quantification

Each of the many variants of a language that describe quantification describe one of two situations. A quantified statement about a variable x asserts either that

- the statement is true for all x in the universe, or
- there exists an x in the universe that makes the statement true.

All quantified statements have one of these two forms. We use the standard shorthand of $\forall$ for the phrase "for all" and the standard shorthand of $\exists$ for the phrase "there exists." We also adopt the convention of putting parentheses around the expression that is subject to the quantification. For example, using Z to stand for the universe of all integers, we write

$$\forall n \in Z(n^2 \geq n)$$

as a shorthand for the statement "For all integers n, $n^2 \geq n$." It is perhaps more natural to read the notation as "For all n in Z, $n^2 \geq n$," which is how we recommend reading the symbolism. We similarly use

$$\exists n \in Z(n^2 \not\geq n)$$

to stand for "There exists an n in Z such that $n^2 \not\geq n$." Notice that to cast our symbolic form of an existence statement into grammatical English, we have included the supplementary word "an" and the supplementary phrase "such that." People often leave out the "an" as they read an existence statement, but they rarely leave out the "such that." Such supplementary language is not needed with $\forall$.

As another example, we use these symbols to rewrite the definition of the big O notation. We use the letter R to stand for the universe of real numbers and the symbol R^+ to stand for the universe of positive real numbers. We assume implicitly that the function $g : R \to R$ takes nonnegative values.

$f = O(g)$ means that
$\exists c \in R^+(\exists n_0 \in R^+(\forall x \in R(x > n_0 \Rightarrow f(x) \leq cg(x))))$.

We would read this literally as

"f is big O of g" means that there exists a c in R^+ such that there exists an n_0 in R^+ such that for all x in R, if $x > n_0$, then $f(x) \leq cg(x)$.

Clearly, this has the same meaning (when we translate it into more idiomatic English) as

"f is big O of g" means that there exist positive real numbers c and n_0 such that for all real numbers $x > n_0$, $f(x) \leq cg(x)$.

This statement is identical to the definition of big O that we gave in Definition 3.2, except it is more precise in describing what c and n_0 actually are.

Exercise 3.2-5

Using the shorthand notation for quantifiers, how would you rewrite the part of Euclid's division theorem (Theorem 2.12), "for every positive integer n and every nonnegative integer m, there are integers q and r, with $0 \leq r < n$, such that $m = qn + r$"? Use Z^+ to stand for the positive integers and N to stand for the nonnegative integers.

We can rewrite Euclid's division theorem as

$$\forall n \in Z^+ \left(\forall m \in N \left(\exists q \in N \left(\exists r \in N ((r < n) \wedge (m = qn + r)) \right) \right) \right).$$

Statements about Variables

To discuss a statement about a variable, it is helpful to have a notation for referring to the statement. For example, we can use $p(n)$ to stand for the statement $n^2 > n$. Now we can say that $p(4)$ and $p(-3)$ are true, while $p(1)$ and $p(0.5)$ are false. In effect, we are introducing variables that stand for statements about other variables. We use symbols like $p(n)$, $q(x)$, and so forth to stand for statements about a variable *n* or *x*. The statement "For all *x* in *U* $p(x)$" can thus be written as $\forall x \in U(p(x))$, and the statement "There exists an *n* in *U* such that $q(n)$" can be written as $\exists n \in U(q(n))$. Sometimes we have statements about more than one variable. For example, our definition of big *O* notation had the form $\exists c(\exists n_0(\forall x(p(c, n_0, x))))$, where $p(c, n_0, x)$ stands for $(x > n_0 \Rightarrow f(x) \leq cg(x))$. (We have left out mention of the universes for our variables to emphasize the form of the statement.)

Exercise 3.2-6

Use the notation for statements about variables to rewrite the part of Euclid's division theorem we gave in Exercise 3.2-5. Leave out the references to universes so that you can see clearly the order in which the quantifiers occur. Use $p(m, n, q, r)$ to stand for "$m = nq + r$ with $0 \leq r < n$."

The form of Euclid's division theorem is $\forall n(\forall m(\exists q(\exists r(p(m, n, q, r)))))$.

Rewriting Statements to Encompass Larger Universes

It is sometimes useful to rewrite a quantified statement so that the universe is larger while the statement itself focuses on a subset of the new universe.

Exercise 3.2-7

Let *R* stand for the real numbers and R^+ stand for the positive real numbers. Consider the following two statements.

   a. $\forall x \in R^+(x > 1)$

   b. $\exists x \in R^+(x > 1)$

Rewrite these statements so that the universe is all the real numbers but the statements say the same thing in everyday English that they did before.

For Exercise 3.2-7, there are potentially many ways to rewrite the statements. Two particularly simple ways are $\forall x \in R(x > 0 \Rightarrow x > 1)$ and $\exists x \in R(x > 0 \wedge x > 1)$. Notice that we translated one of these statements with "implies" and one with "and." We can state this rule as a general theorem.

Theorem 3.2

Let U_1 be a universe and let U_2 be another universe, with $U_1 \subseteq U_2$. Suppose that $q(x)$ is a statement such that

$$U_1 = \{x \mid q(x) \text{ is true}\} . \tag{3.2}$$

Then, if $p(x)$ is a statement about U_2, it may also be interpreted as a statement about U_1, and

   a. $\forall x \in U_1(p(x))$ is equivalent to $\forall x \in U_2(q(x) \Rightarrow p(x))$, and

   b. $\exists x \in U_1(p(x))$ is equivalent to $\exists x \in U_2(q(x) \wedge p(x))$.

Proof: By Equation 3.2, the statement $q(x)$ must be true for all $x \in U_1$ and false for all x in U_2 but not U_1. To prove part a, we must show that $\forall x \in U_1(p(x))$ is true in exactly the same cases as the statement $\forall x \in U_2(q(x) \Rightarrow p(x))$. For this purpose, suppose first that $\forall x \in U_1(p(x))$ is true. Then $p(x)$ is true for all x in U_1. Therefore, by the truth table for "implies" and our remark about Equation 3.2, the statement $\forall x \in U_2(q(x) \Rightarrow p(x))$ is true. Now suppose $\forall x \in U_1(p(x))$ is false. Then there exists an x in U_1 such that $p(x)$ is false. By the truth table for "implies," the statement $\forall x \in U_2(q(x) \Rightarrow p(x))$ is false. Thus, the statement $\forall x \in U_1(p(x))$ is true if and only if the statement $\forall x \in U_2(q(x) \Rightarrow p(x))$ is true. Therefore, the two statements are true in exactly the same cases. Part a of the theorem follows.

Similarly, for part b, we observe that if $\exists x \in U_1(p(x))$ is true, then $p(x')$ is true for some $x' \in U_1$. For that x', $q(x')$ is also true. Hence, $p(x') \wedge q(x')$ is true so that $\exists x \in U_2(q(x) \wedge p(x))$ is true as well. On the other hand, if $\exists x \in U_1(p(x))$ is false, then no $x \in U_1$ has $p(x)$ true. Therefore, by the truth table for "and," $q(x) \wedge p(x)$ won't be true either. Thus, the two statements in part b are true in exactly the same cases and, so, are equivalent. ∎

Proving Quantified Statements True or False

Exercise 3.2-8

Let R stand for the real numbers and R^+ stand for the positive real numbers. For each of the following statements, state whether it is true or false and explain why.

    a. $\forall x \in R^+(x > 1)$

    b. $\exists x \in R^+(x > 1)$

    c. $\forall x \in R(\exists y \in R(y > x))$

    d. $\forall x \in R(\forall y \in R(y > x))$

    e. $\exists x \in R(x \geq 0 \wedge \forall y \in R^+(y > x))$

In Exercise 3.2-8, because $1/2$ is not greater than 1, statement a is false. However, because $2 > 1$, statement b is true. Statement c says that for each real number x, there is a real number y bigger than x, which we know is true. Statement d says that every y in R is larger than every x in R, and so it is false. Statement e says that there is a nonnegative number x such that every positive y is larger than x, which is true because $x = 0$ fills the bill.

We can summarize what we know about the meaning of quantified statements as follows.

- The statement $\exists x \in U(p(x))$ is true if there is at least one value of x in U for which the statement $p(x)$ is true.

- The statement $\exists x \in U(p(x))$ is false if there is no $x \in U$ for which $p(x)$ is true.

- The statement $\forall x \in U(p(x))$ is true if $p(x)$ is true for each value of x in U.

- The statement $\forall x \in U(p(x))$ is false if $p(x)$ is false for at least one value of x in U.

Negation of Quantified Statements

An interesting connection between $\forall$ and $\exists$ arises from the negation of statements.

| Exercise 3.2-9 | What does the statement "It is not the case that $n^2 > 0$ for all integers n" mean? |

From our knowledge of English, we see that the statement[6] $\neg \forall n \in Z(n^2 > 0)$ asserts that it is not the case that we have $n^2 > 0$ for all integers n. Therefore, the statement asserts that there must be some integer n such that $n^2 \not> 0$. In other words, it says there is some integer n such that $n^2 \leq 0$. Thus, the negation of our "for all" statement is a "there exists" statement. We can make this idea more precise by recalling the notion of equivalence of statements. We have said that two symbolic statements are equivalent if they are true in exactly the same cases. By considering the case where $p(x)$ is true for all $x \in U$ (we call this case "always true") and the case where $p(x)$ is false for at least one $x \in U$ (we call this case "not always true"), we can analyze the equivalence. The theorem that follows formalizes the example above in which $p(x)$ was the statement $x^2 > 0$. The theorem is proved by dividing all values of the variables into two possibilities: the case where $p(x)$ is always true and the case where it is not always true.

| Theorem 3.3 | The statements $\neg \forall x \in U(p(x))$ and $\exists x \in U(\neg p(x))$ are equivalent. |

Proof: Consider the following table (which we have set up much like a truth table, except that the relevant cases are not determined by whether $p(x)$ is true or false, but by whether or not $p(x)$ is true for all x in the universe U).

---

[6]The convention is that when $\neg$ appears before a quantifier, the entire quantified statement is negated.

$p(x)$	$\neg p(x)$	$\forall x \in U(p(x))$	$\neg \forall x \in U(p(x))$	$\exists x \in U(\neg p(x))$
always true	always false	true	false	false
not always true	not always false	false	true	true

Because the last two columns are identical, the theorem holds. ∎

Corollary 3.4 The statements $\neg \exists x \in U(q(x))$ and $\forall x \in U(\neg q(x))$ are equivalent.

Proof: Because the two statements in Theorem 3.3 are equivalent, their negations are also equivalent. We then substitute $\neg q(x)$ for $p(x)$ to prove the corollary. ∎

Put another way, when you negate a quantified statement, you switch the quantifier and "push" the negation inside.

To deal with the negation of more complicated statements, we simply take them one quantifier at a time. Recall Definition 3.2, the definition of big O notation:

$$f(x) = O(g(x)) \text{ if } \exists c \in R^+ \left(\exists n_0 \in R^+ \left(\forall x \in R(x > n_0 \Rightarrow f(x) \le cg(x)) \right) \right).$$

What does it mean to say that $f(x)$ is *not* $O(g(x))$? First, we can write

$$f(x) \ne O(g(x)) \text{ if } \neg \exists c \in R^+ \left(\exists n_0 \in R^+ \left(\forall x \in R(x > n_0 \Rightarrow f(x) \le cg(x)) \right) \right).$$

After one application of Corollary 3.4, we get

$$f(x) \ne O(g(x)) \text{ if } \forall c \in R^+ \left(\neg \exists n_0 \in R^+ \left(\forall x \in R(x > n_0 \Rightarrow f(x) \le cg(x)) \right) \right).$$

After another application of Corollary 3.4, we obtain

$$f(x) \ne O(g(x)) \text{ if } \forall c \in R^+ \left(\forall n_0 \in R^+ \left(\neg \forall x \in R(x > n_0 \Rightarrow f(x) \le cg(x)) \right) \right).$$

Now we apply Theorem 3.3 to obtain

$$f(x) \ne O(g(x)) \text{ if } \forall c \in R^+ \left(\forall n_0 \in R^+ \left(\exists x \in R(\neg(x > n_0 \Rightarrow f(x) \le cg(x))) \right) \right).$$

Because $\neg(p \Rightarrow q)$ is equivalent to $p \wedge \neg q$, we can write

$$f(x) \ne O(g(x)) \text{ if } \forall c \in R^+ \left(\forall n_0 \in R^+ \left(\exists x \in R\left((x > n_0) \wedge (f(x) \not\le cg(x))\right) \right) \right).$$

Thus, $f(x)$ is *not* $O(g(x))$ if for every c in R^+ and every n_0 in R, there is an x such that $x > n_0$ and $f(x) \not\leq cg(x)$.

In our next exercise, we will use the big Θ notation, defined as follows:

Definition 3.3 $f(x) = \Theta(g(x))$ means that $f(x) = O(g(x))$ *and* $g(x) = O(f(x))$.

Exercise 3.2-10   Express $\neg(f(x) = \Theta(g(x)))$ in terms similar to those used to describe $f(x) \neq O(g(x))$.

Exercise 3.2-11   Suppose the universe for a statement $p(x)$ is the integers from 1 to 10. Express the statement $\forall x(p(x))$ without any quantifiers. Express the negation in terms of $\neg p$ without any quantifiers. Discuss how negation of "for all" and "there exists" statements corresponds to DeMorgan's law.

By DeMorgan's law, $\neg(f = \Theta(g))$ means $\neg(f = O(g)) \lor \neg(g = O(f))$. Thus, $\neg(f = \Theta(g))$ means either that

- for every c in R^+ and n_0 in R, there is an x in R with $x > n_0$ and $f(x) \not< cg(x)$, or

- for every c in R^+ and n_0 in R, there is an x in R with $x > n_0$ and $g(x) < cf(x)$,

or both. For Exercise 3.2-11, we see that $\forall x(p(x))$ is simply

$$p(1) \land p(2) \land p(3) \land p(4) \land p(5) \land p(6) \land p(7) \land p(8) \land p(9) \land p(10) \,.$$

by DeMorgan's law, the negation of this statement is

$$\neg p(1) \lor \neg p(2) \lor \neg p(3) \lor \neg p(4) \lor \neg p(5) \lor \neg p(6) \lor \neg p(7) \lor \neg p(8) \lor \neg p(9) \lor \neg p(10) \,.$$

Thus, the relationship that negation gives between "for all" and "there exists" statements is the extension of DeMorgan's law from a finite number of statements to potentially infinitely many statements about a potentially infinite universe.

Implicit Quantification

Exercise 3.2-12   Are there any quantifiers in the statement "The sum of even integers is even"?

An elementary fact about numbers is that the sum of even integers is even. Another way to say this is that if m and n are even, then $m + n$ is even. If $p(n)$ stands for the statement "n is even," then this last sentence translates to $p(m) \land p(n) \Rightarrow p(m + n)$. From the logical form of the statement, we see that our variables are free, so we could substitute various integers for m and n to see whether the statement is true. In Exercise 3.2-12, however, we said that we were stating a more general fact about the integers. What we meant to say is that for *every pair of integers m and n*, if m and n

are even, then $m + n$ is even. In symbols, using $p(k)$ for "k is even," we have

$$\forall m \in Z\Big(\forall n \in Z\big(p(m) \wedge p(n) \Rightarrow p(m + n)\big)\Big).$$

This way of representing the statement captures the meaning we originally intended. This is one of the reasons that mathematical statements and their proofs sometimes seem confusing—just as in English, sentences in mathematics have to be interpreted in context. Because mathematics has to be written in some natural language, and because context is used to remove ambiguity in natural language, context must be used to remove ambiguity from mathematical statements made in natural language. In fact, we frequently rely on context when writing mathematical statements with implicit quantifiers, because it makes the statements easier to read. For example, Lemma 2.8 said

> The equation $a \cdot_n x = 1$ has a solution in Z_n if and only if there exist integers x and y such that $ax + ny = 1$.

In context, it was clear that the a we were talking about was an arbitrary member of Z_n. It would simply have made the statement read more clumsily if we had said

> For every $a \in Z_n$, the equation $a \cdot_n x = 1$ has a solution in Z_n if and only if there exist integers x and y such that $ax + ny = 1$.

On the other hand, we were making a transition from talking about Z_n to talking about the integers, so it was important for us to include the quantified statement "there exist integers x and y such that $ax + ny = 1$." More recently, in Theorem 3.3, we also did not feel it was necessary to say "for all universes U and for all statements p about U" at the beginning of the theorem. We felt the theorem would be easier to read if we kept those quantifiers implicit and let you infer them from context (not necessarily consciously).

Proof of Quantified Statements

We said that "the sum of even integers is even" is an elementary fact about numbers. How do we know it is a fact? One answer is that we know it because our teachers told us so (and presumably they knew it because their teachers told them so). But someone had to figure it out in the first place. So we ask, "How would we prove this statement?" A mathematician asked to give a proof that the sum of even numbers is even might write, "If m and n are even, then $m = 2i$ and $n = 2j$ so that $m + n = 2i + 2j = 2(i + j)$, and thus $m + n$ is even."[7] Because mathematicians think and write in natural language, they often rely on context to remove ambiguities. For example, there are no quantifiers in the mathematician's proof. However, the sentence, though technically incomplete as a proof, captures the essence of why the sum of two even numbers is even. A typical complete (but more formal and wordy than usual) proof might go like the following.

---

[7] In the context of this book, a mathematician might simply say that this statement follows from Lemma 2.3 since being even is the same as being 0 mod 2. However, our point in proving elementary statements about even and odd numbers is not that we are learning new facts. Instead, we have chosen facts about numbers because they offer a familiar context for illustrating a variety of different aspects of proof. We do not expect any of the facts to be new to you. In fact, we hope that because they are not new, they will help you focus on the actual proof techniques.

Let m and n be integers. Suppose m and n are even. If m and n are even, then by definition there are integers i and j such that $m = 2i$ and $n = 2j$. Thus, there are integers i and j such that $m = 2i$ and $n = 2j$. Then

$$m + n = 2i + 2j = 2(i + j) \,;$$

so by definition, $m + n$ is an even integer. We have shown that if m and n are even, then $m + n$ is even. Therefore, for every m and n, if m and n are even integers, then so is $m + n$.

We began our proof by assuming that m and n are integers. This gives us symbolic notation for talking about two integers. We then appealed to the definition of an even integer, namely, that an integer h is even if there is another integer k so that $h = 2k$. (Note the use of a quantifier in the definition.) Then we used algebra to show that $m + n$ is also two times another number. Because being two times another integer is the definition of $m + n$ being even, we concluded that $m + n$ is even. This allowed us to say that if m and n are even, then $m + n$ is even. Finally, we asserted that for every pair of integers m and n, if m and n are even, then $m + n$ is even.

There are a number of principles of proof illustrated here. Section 3.3 is devoted to a discussion of principles used in constructing proofs. For now, let us conclude with a remark about the limitations of logic. How did we know that we wanted to write the symbolic equation

$$m + n = 2i + 2j = 2(i + j) \;?$$

It was not logic that told us to do this, but intuition and experience.

Important Concepts, Formulas, and Theorems

1. *Varies over.* We use the phrase "varies over" to describe the set of values a variable may take on.

2. *Universe.* We call the set of possible values for a variable the *universe* of that variable.

3. *Free variables.* Variables that are not constrained in any way are called *free variables.*

4. *Quantifier.* A phrase that converts a symbolic statement about potentially any member of our universe into a statement about the universe instead is called a *quantifier.* There are two types of quantifiers:

   • *Universal quantifiers* assert that a statement about a variable is true for every value of the variable in its universe.

   • *Existential quantifiers* assert that a statement about a variable is true for at least one value of the variable in its universe.

5. *Larger universes.* Let U_1 be a universe, and let U_2 be another universe, with $U_1 \subseteq U_2$. Suppose that $q(x)$ is a statement such that $U_1 = \{x \mid q(x) \text{ is true}\}$. If

$p(x)$ is a statement about U_2, it may also be interpreted as a statement about U_1, and

    a. $\forall x \in U_1(p(x))$ is equivalent to $\forall x \in U_2(q(x) \Rightarrow p(x))$, and

    b. $\exists x \in U_1(p(x))$ is equivalent to $\exists x \in U_2(q(x) \wedge p(x))$.

6. *Proving quantified statements true or false.*

- The statement $\exists x \in U(p(x))$ is true if there is at least one value of x in U for which the statement $p(x)$ is true.

- The statement $\exists x \in U(p(x))$ is false if there is no $x \in U$ for which $p(x)$ is true.

- The statement $\forall x \in U(p(x))$ is true if $p(x)$ is true for each value of x in U.

- The statement $\forall x \in U(p(x))$ is false if $p(x)$ is false for at least one value of x in U.

7. *Negation of quantified statements.* To negate a quantified statement, you switch the quantifier and push the negation inside.

- The statements $\neg \forall x \in U(p(x))$ and $\exists x \in U(\neg p(x))$ are equivalent.

- The statements $\neg \exists x \in U(p(x))$ and $\forall x \in U(\neg p(x))$ are equivalent.

8. *Big O.* We say that $f(x) = O(g(x))$ if there are positive numbers c and n_0 such that $f(x) \le cg(x)$ for every $x > n_0$.

9. *Big Θ.* $f(x) = \Theta(g(x))$ means that $f = O(g(x))$ *and* $g = O(f(x))$.

10. *Some notation for sets of numbers.* We use R to stand for the real numbers, R^+ to stand for the positive real numbers, Z to stand for the integers (positive, negative, and zero), Z^+ to stand for the positive integers, and N to stand for the nonnegative integers.

Problems

1. For what positive integers x is the statement $(x - 2)^2 + 1 \le 2$ true? For what integers is it true? For what real numbers is it true? If you expand the universe for which you are considering a statement about a variable, does this always increase the size of the statement's truth set? a/h

2. Is the statement "There is an integer greater than 2 such that $(x - 2)^2 + 1 \le 2$" true or false? How do you know?

3. Write the statement "The square of every real number is greater than or equal to 0" as a quantified statement about the universe of real numbers. You may use R to stand for the universe of real numbers. a/h

4. A prime number is defined as an integer greater than 1 whose only positive integer factors are itself and 1. Find two ways to write this definition so that all

quantifiers are explicit. (It may be convenient to introduce a variable to stand for the number and perhaps a variable or some variables for its factors.)

5. Write the definition of a greatest common divisor of m and n in such a way that all quantifiers are explicit and expressed explicitly as "for all" or "there exists." Write the part of Euclid's extended greatest common divisor theorem (Theorem 2.14) that relates the greatest common divisor of m and n algebraically to m and n. Again, make sure all quantifiers are explicit and expressed explicitly as "for all" or "there exists."

6. Using $s(x, y, z)$ to be the statement $x = yz$ and $t(x, y)$ to be the statement $x \leq y$, what is the form of the definition of a greatest common divisor d of m and n? (You need not include references to the universes for the variables.)

7. Which of the following statements (in which Z^+ stands for the positive integers and Z stands for all integers) is true and which is false? Explain why.

   a. $\forall z \in Z^+(z^2 + 6z + 10 > 20)$ a/h

   b. $\forall z \in Z(z^2 - z \geq 0)$ a/h

   c. $\exists z \in Z^+(z - z^2 > 0)$ a/h

   d. $\exists z \in Z(z^2 - z = 6)$ a/h

8. Are there any (implicit) quantifiers in the statement "The product of odd integers is odd"? If so, what are they? a/h

9. Rewrite the statement "The product of odd integers is odd" with all quantifiers (including any in the definition of odd integers) explicitly stated as "for all" or "there exist."

10. Rewrite the following statement without any negations: "There is no positive integer n such that for all integers $m > n$, all polynomial equations $p(x) = 0$ of degree m have no real numbers for solutions." a/h

11. Consider the following slight modifications of Theorem 3.2. For each part, either prove that it is true or give a counterexample.

    Let U_1 be a universe, and let U_2 be another universe, with $U_1 \subseteq U_2$. Suppose that $q(x)$ is a statement about U_2 such that $U_1 = \{x \mid q(x) \text{ is true}\}$ and $p(x)$ is a statement about U_2.

    a. $\forall x \in U_1(p(x))$ is equivalent to $\forall x \in U_2(q(x) \wedge p(x))$. a/h

    b. $\exists x \in U_1(p(x))$ is equivalent to $\exists x \in U_2(q(x) \Rightarrow p(x))$. a/h

12. Let $p(x)$ stand for "x is a prime," $q(x)$ for "x is even," and $r(x, y)$ stand for "$x = y$." Use these three symbolic statements and appropriate logical notation to write the statement "There is one and only one even prime." (Use the set Z^+ of positive integers for your universe.)

13. Each of the following expressions represents a statement about the integers. Using $p(x)$ for "x is prime," $q(x, y)$ for "$x = y^2$," $r(x, y)$ for "$x \leq y$,"

$s(x, y, z)$ for "$z = xy$," and $t(x, y)$ for "$x = y$," determine which expressions represent true statements and which represent false statements.

    a. $\forall x \in Z(\exists y \in Z(q(x, y) \lor p(x)))$ a/h

    b. $\forall x \in Z(\forall y \in Z(s(x, x, y) \Leftrightarrow q(x, y)))$ a/h

    c. $\forall y \in Z(\exists x \in Z(q(y, x)))$ a/h

    d. $\exists z \in Z(\exists x \in Z(\exists y \in Z(p(x) \land p(y) \land \neg t(x, y))))$ a/h

14. Why is $(\exists x \in U(p(x))) \land (\exists y \in U(q(y)))$ not equivalent to $\exists z \in U(p(z) \land q(z))$? Are the statements $(\exists x \in U(p(x))) \lor (\exists y \in U(q(y)))$ and $\exists z \in U(p(z) \lor q(z))$ equivalent?

15. Give an example (in English) of a statement that has the form $\forall x \in U(\exists y \in V(p(x, y)))$. (The statement can be a mathematical statement, a statement about everyday life, or whatever you prefer.) Now write (in English) the statement using the same $p(x, y)$ but of the form $\exists y \in V(\forall x \in U(p(x, y)))$. Comment on whether "for all" and "there exist" commute. a/h

3.3 INFERENCE

Direct Inference (Modus Ponens) and Proofs

In this section, we talk about the logical structure of proofs. The examples of proofs we give are chosen to illustrate a concept in a context that we hope will be familiar to you. These examples are not necessarily the only or the best way to prove the results. If you see other ways to do the proofs, that is good, because it means you are putting your prior knowledge to work. It would be useful to try to see how the ideas of this section apply to your alternate proofs.

Section 3.2 concluded with a proof that the sum of two even numbers is even. That proof contained several crucial ingredients. First, it introduced symbols for members of the universe of integers. In other words, rather than saying, "Suppose we have two integers," we used symbols for the two members of our universe by saying, "Let m and n be integers." How did we know to use algebraic symbols? There are many possible answers to this question. In this case, our intuition was probably based on thinking about what an even number is and realizing that the definition itself is essentially symbolic. (You may argue that an even number is just twice another number, and you would be right. Apparently there are no symbols (variables) in that definition. But they really are there in the phrases "even number" and "another number.") Because we all know algebra is easier with symbolic variables than with words, we should recognize that it makes sense to use algebraic notation. Thus, this decision was based on experience, not logic.

Next, we assumed the two integers were even. We then used the definition of even numbers; as our previous parenthetic comment suggests, it was natural to use the definition symbolically. The definition tells us that if m is an even number, then there exists another integer i such that $m = 2i$. We combined this with the assumption that m is even and concluded that, in fact, there does exist an integer i such that $m = 2i$. This is an example of using the principle of **direct inference** (called *modus ponens* in Latin).

Principle 3.3   **(Direct Inference)** From p and $p \Rightarrow q$, we may conclude q.

This common-sense principle is a cornerstone of logical arguments. But why is it valid? In Table 3.5, we take another look at the truth table for implication.

Table 3.5: *Another look at implication*

p	q	$p \Rightarrow q$
T	T	T
T	F	F
F	T	T
F	F	T

In Table 3.5, the only line that has a T in both the p column and the $p \Rightarrow q$ column is the first line. In this line, q is also true; thus, we conclude that if p and $p \Rightarrow q$ hold, then q must hold also. Although this may seem like a somewhat inside-out application of the truth table, it is simply a different way of using a truth table.

There are quite a few rules (called "rules of inference"), such as the principle of direct inference, that people commonly use in proofs without explicitly stating them. Before beginning a formal study of rules of inference, however, let's complete our analysis of which rules we used in the proof that the sum of two even integers is even. After concluding that $m = 2i$ and $n = 2j$, we used algebra to show that because $m = 2i$ and $n = 2j$, there exists a k such that $m + n = 2k$ (our k was $i + j$). Next, we used the definition of even numbers again to say that $m + n$ was even. We then used the following rule of inference.

Principle 3.4   **(Conditional Proof)** If by assuming p we may prove q, then the statement $p \Rightarrow q$ is true.

Using this principle, we reached the conclusion that if m and n are even integers, then $m + n$ is an even integer. To conclude that this statement is true for all integers m and n, we used another rule of inference, one that is more difficult to describe. We originally introduced the variables m and n. We used only well-known consequences of the fact that they were in the universe of integers in our proof. Thus, we felt justified in asserting that what we concluded about m and n is true for any pair of integers. We might say that we were treating m and n as generic members of our universe. Thus, our rule of inference says:

Principle 3.5   **(Universal Generalization)** If we can prove a statement about x by assuming only that x is a member of our universe, then we can conclude the statement is true for every member of our universe.

Perhaps this rule is hard to put into words because it is not simply a description of a truth table; rather, it is a principle that we use to prove universally quantified statements.

Rules of Inference for Direct Proofs

We have seen the ingredients of a typical proof. What do we mean by a proof in general? A proof of a statement is a convincing argument that the statement is true. To be more precise, we can agree that a **direct proof** consists of a sequence of statements, each of which is either a hypothesis,[8] a generally accepted fact, or the result of one of the following rules of inference for compound statements.

Rules of Inference for Direct Proofs

1. From an example x that does not satisfy $p(x)$, we may conclude $\neg p(x)$.

2. From $p(x)$ and $q(x)$, we may conclude $p(x) \wedge q(x)$.

3. From either $p(x)$ or $q(x)$, we may conclude $p(x) \vee q(x)$.

4. From either $q(x)$ or $\neg p(x)$, we may conclude $p(x) \Rightarrow q(x)$.

5. From $p(x) \Rightarrow q(x)$ and $q(x) \Rightarrow p(x)$, we may conclude $p(x) \Leftrightarrow q(x)$.

6. From $p(x)$ and $p(x) \Rightarrow q(x)$, we may conclude $q(x)$.

7. From $p(x) \Rightarrow q(x)$ and $q(x) \Rightarrow r(x)$, we may conclude $p(x) \Rightarrow r(x)$.

8. If we can derive $q(x)$ from the hypothesis that x satisfies $p(x)$, then we may conclude $p(x) \Rightarrow q(x)$.

9. If we can derive $p(x)$ from the hypothesis that x is a (generic) member of our universe U, we may conclude $\forall x \in U(p(x))$.

10. From an example of an $x \in U$ satisfying $p(x)$, we may conclude $\exists x \in U(p(x))$.

The first rule is a statement of the principle of the excluded middle as it applies to statements about variables. The next four rules are, in effect, descriptions of the truth tables for "and," "or," "implies," and "if and only if." Rule 5 tells us what we must do to write a proof of an "if and only if" statement. Rule 6, exemplified in our earlier discussion, is the principle of direct inference, and it describes one row of the truth table for $p \Rightarrow q$. Rule 7 is the transitive law, a law that we could derive by analysis of truth tables. Rule 8, the principle of conditional proof, which is also exemplified earlier, may be regarded as yet another description of one row of the truth table of $p \Rightarrow q$. Rule 9 is the principle of universal generalization, discussed and exemplified earlier. Rule 10 specifies what we mean by the truth of an existentially quantified statement according to Principle 3.2.

Although some of our rules of inference are, strictly speaking, redundant, we include them because they allow us to express proofs more concisely. For example, we could have written a portion of our proof that the sum of even numbers is even as follows, without using Rule 8.

---

[8]If we are proving an implication $s \Rightarrow t$, we call s a hypothesis. If we make assumptions by saying, "Let ...," "Suppose ...," or something similar before we give the statement to be proved, then these assumptions are also hypotheses.

Let m and n be integers. If m is even, then there is a k with $m = 2k$. If n is even, then there is a j with $n = 2j$. Thus, if m is even and n is even, there are a k and j such that $m + n = 2k + 2j = 2(k + j)$. Thus, if m is even and n is even, there is an integer $h = k + j$ such that $m + n = 2h$. Thus, if m is even and n is even, $m + n$ is even.

Because this kind of argument could always be used to circumvent the use of Rule 8, that rule is not required as a rule of inference. However, because it permits us to avoid such unnecessarily complicated "silliness" in our proofs, we choose to include it. Rule 7, the transitive law, has a similar role.

Exercise 3.3-1 Prove that if m is even, then m^2 is even. Explain which steps of the proof use one of the ten rules of inference.

For Exercise 3.3-1, we can mimic the proof that the sum of even integers is even:

Let m be an integer. Suppose that m is even. If m is even, then there is a k with $m = 2k$. Thus, there is a k such that $m^2 = 4k^2$. Therefore, there is an integer $h = 2k^2$ such that $m^2 = 2h$. This tells us that if m is even, then m^2 is even. Therefore, for all integers m, if m is even, then m^2 is even.

Our first sentence sets us up to use Rule 9. The second sentence simply states an implicit hypothesis. The next two sentences use Rule 6, the principle of direct inference. When we say, "Therefore, there is an integer $h = 2k^2$ such that $m^2 = 2h$," we are simply stating an algebraic fact. The next sentences use Rule 8 and Rule 9. (You might have written the proof in a different way and used different rules of inference.)

Contrapositive Rule of Inference

Exercise 3.3-2 Show that "p implies q" is equivalent to "$\neg q$ implies $\neg p$."

Exercise 3.3-3 Is "p implies q" equivalent to "q implies p"?

To do Exercise 3.3-2, we construct the double truth table in Table 3.6. Because the columns under $p \Rightarrow q$ and under $\neg q \Rightarrow \neg p$ are exactly the same, we know the two statements are equivalent.

Table 3.6: *A double truth table for $p \Rightarrow q$ and $\neg q \Rightarrow \neg p$*

p	q	$p \Rightarrow q$	$\neg p$	$\neg q$	$\neg q \Rightarrow \neg p$
T	T	T	F	F	T
T	F	F	F	T	F
F	T	T	T	F	T
F	F	T	T	T	T

Exercise 3.3-2 tells us that if we know that $\neg q \Rightarrow \neg p$, then we can conclude that $p \Rightarrow q$. This is called the principle of **proof by contraposition**.

Principle 3.6   **(Proof by Contraposition)** The statements $p \Rightarrow q$ and $\neg q \Rightarrow \neg p$ are equivalent, and so a proof of one is a proof of the other.

The statement $\neg q \Rightarrow \neg p$ is called the **contrapositive** of the statement $p \Rightarrow q$. The proof of the following lemma demonstrates the utility of proof by contraposition.

Lemma 3.5   If n is a positive integer with $n^2 > 100$, then $n > 10$.

Proof: Suppose n is not greater than 10. (We now use the rule of algebra for inequalities that says if $x \leq y$ and $c \geq 0$, then $cx \leq cy$.) Then, because $1 \leq n \leq 10$,

$$n \cdot n \leq n \cdot 10 \leq 10 \cdot 10 = 100 .$$

Thus, n^2 is not greater than 100. Therefore, if n is not greater than 10, then n^2 is not greater than 100. By the principle of proof by contraposition, if $n^2 > 100$, then n must be greater than 10. ∎

We adopt Principle 3.6 as a rule of inference called the **contrapositive rule of inference**:

11. From $\neg q(x) \Rightarrow \neg p(x)$, we may conclude $p(x) \Rightarrow q(x)$.

In our proof of the Chinese remainder theorem (Theorem 2.24), we wanted to prove for a certain function f, that if x and y were different integers between 0 and $mn - 1$, then $f(x) \neq f(y)$. To prove this, we assumed that, in fact, $f(x) = f(y)$ and proved that x and y were not different integers between 0 and $mn - 1$. Had we known the principle of contrapositive inference, we could have concluded then and there that f was one-to-one. Instead, we used the more common principle of proof by contradiction, which is the major topic of the remainder of this section, to complete our proof. If you look back at the proof of the Chinese remainder theorem, you will see that we might have been able to use contrapositive inference to shorten it by a sentence.

For Exercise 3.3-3, a quick look at the double truth table for $p \Rightarrow q$ and $q \Rightarrow p$ in Table 3.7 demonstrates that these two statements are *not* equivalent. The statement $q \Rightarrow p$ is called the **converse** of $p \Rightarrow q$. Notice that $p \Leftrightarrow q$ is true exactly when $p \Rightarrow q$ and its converse are true. It is surprising how often people, even professional mathematicians, absentmindedly try to prove the converse of a statement when they mean to prove the statement itself. Try not to join this crowd!

Proof by Contradiction

Proof by contrapositive inference is an example of what we call **indirect proof**. We actually saw another example of indirect proof in the principle of proof by

Table 3.7: *A double truth table for $p \Rightarrow q$ and $q \Rightarrow p$*

p	q	$p \Rightarrow q$	$q \Rightarrow p$
T	T	T	T
T	F	F	T
F	T	T	F
F	F	T	T

contradiction. We introduced the principle of proof by contradiction (Principle 2.1) in our proof to Corollary 2.6, in which we were trying to prove:

> Suppose there is a b in Z_n such that the equation $a \cdot_n x = b$ does not have a solution. Then a does not have a multiplicative inverse in Z_n.

We assumed that the hypothesis that $a \cdot_n x = b$ does not have a solution was true. We also assumed that the conclusion that a does not have a multiplicative inverse was false. We showed that these two assumptions together led to a contradiction. Using the principle of the excluded middle (Principle 3.1), but without saying so, we concluded that if the hypothesis was in fact true, then the only possibility was that the conclusion was also true.

We used the principle of proof by contradiction again in our proof of Euclid's division theorem. Recall that in that proof, we began by assuming that there was an integer m for which there were no integers q and r with $m = qn + r$ and $0 \leq r < n$. We then chose the smallest integer m such that there was not a pair of integers q and r with $m = qn + r$ and $0 \leq r < n$. We then made some computations by which we proved that, in this case, there *are* integers q and r with $0 \leq r < n$ such that $m = qn + r$. In overview, we started out by assuming the theorem was false, and from that assumption, we drew a contradiction (to the assumption itself). Because all our reasoning, except for the assumption that the theorem was false, used accepted rules of inference, the only source of that contradiction was our assumption. Thus, by the principle of the excluded middle, our assumption had to be incorrect. We adopt the principle of proof by contradiction (also called the principle of **reduction to absurdity**) as our last rule of inference.

12. If from assuming $p(x)$ and $\neg q(x)$, we can derive both $r(x)$ and $\neg r(x)$ for some statement $r(x)$, then we may conclude $p(x) \Rightarrow q(x)$.

There can be many variations of proof by contradiction. These variations are all examples of what we call an "indirect proof." Each of the next three indirect proofs of the same statement gets a slightly different contradiction. In each case, p is the statement $x^2 + x - 2 = 0$, and s is the statement $x \neq 0$. In each case, we prove that p implies q.

1. We may assume p is true and q is false; from this, we derive the contradiction that p is false, as in the following example.

    Prove that if $x^2 + x - 2 = 0$, then $x \neq 0$.

    Proof: Suppose that $x^2 + x - 2 = 0$. Assume that $x = 0$. Substituting 0 for x in the polynomial gives $x^2 + x - 2 = 0 + 0 - 2 = -2$, which contradicts the assumption that $x^2 + x - 2 = 0$. Thus, by the principle of proof by contradiction, if $x^2 + x - 2 = 0$, then $x \neq 0$. ∎

    Here the statement r was identical to p, namely, $x^2 + x - 2 = 0$.

2. We may assume p is true and q is false and derive a contradiction of a known fact. Here is an example.

    Prove that if $x^2 + x - 2 = 0$, then $x \neq 0$.

    Proof: Suppose that $x^2 + x - 2 = 0$. Assume that $x = 0$. Then $x^2 + x - 2 = 0 + 0 - 2 = -2$. Thus, $0 = -2$, which is a contradiction. Thus, by the principle of proof by contradiction, if $x^2 + x - 2 = 0$, then $x \neq 0$. ∎

    Here the statement r is the known fact that $0 \neq -2$.

3. Sometimes the statement r that appears in the principle of proof by contradiction is simply a statement that arises naturally as we try to construct our proof, as in the following example.

    Prove that if $x^2 + x - 2 = 0$, then $x \neq 0$.

    Proof: Suppose that $x^2 + x - 2 = 0$. Then $x^2 + x = 2$. Assume that $x = 0$. Then $x^2 + x = 0 + 0 = 0$. But this is a contradiction to our observation that $x^2 + x = 2$. Thus, by the principle of proof by contradiction, if $x^2 + x - 2 = 0$, then $x \neq 0$. ∎

    Here the statement r is $x^2 + x = 2$.

4. Finally, if proof by contradiction seems to you not to be much different from proof by contraposition, you are right, as the following example shows.

    Prove that if $x^2 + x - 2 = 0$, then $x \neq 0$.

    Proof: Assume that $x = 0$. Then $x^2 + x - 2 = 0 + 0 - 2 = -2$, so that $x^2 + x - 2 \neq 0$. Thus, by the principle of proof by contraposition, if $x^2 + x - 2 = 0$, then $x \neq 0$. ∎

Any proof that uses one of the indirect methods of inference, either contradiction or contraposition, is called an **indirect proof**. The previous four examples illustrate the rich possibilities that indirect proof provides us. Of course, they also illustrate why indirect proof can be confusing. There is no set formula that we use in writing a proof by contradiction, so there is no rule we can memorize to formulate indirect proofs. Instead, we have to ask ourselves whether assuming the opposite of what we are trying to prove gives insight into why the assumption makes no sense. If it does, we have the basis of an indirect proof. The way in which we choose to write that proof is a matter of personal choice.

Without extracting square roots, prove that if n is a positive integer such that $n^2 < 9$, then $n < 3$. You may use rules of algebra for dealing with inequalities.

Prove that $\sqrt{5}$ is not rational.

To prove the statement in Exercise 3.3-4, we assume, for purposes of contradiction, that $n \geq 3$. Squaring both sides of this equation, we obtain

$$n^2 \geq 9,$$

which contradicts our hypothesis that $n^2 < 9$. Therefore, by the principle of proof by contradiction, $n < 3$.

To prove the statement in Exercise 3.3-5, we assume, for the purpose of contradiction, that $\sqrt{5}$ is rational. This means that it can be expressed as the fraction m/n, where m and n are integers. Squaring both sides of the equation $m/n = \sqrt{5}$, we obtain

$$\frac{m^2}{n^2} = 5,$$

or

$$m^2 = 5n^2.$$

Now, m^2 must have an even number of prime factors (counting each prime factor as many times as it occurs), as must n^2. But $5n^2$ has an odd number of prime factors. Thus, a product of an even number of prime factors is equal to a product of an odd number of prime factors. This is a contradiction, because each positive integer may be expressed uniquely as a product of (positive) prime numbers. Thus, by the principle of proof by contradiction, $\sqrt{5}$ is not rational.

Important Concepts, Formulas, and Theorems

1. *Principle of direct inference or modus ponens.* From p and $p \Rightarrow q$, we may conclude q.

2. *Principle of conditional proof.* If by assuming p we may prove q, then the statement $p \Rightarrow q$ is true.

3. *Principle of universal generalization.* If we can prove a statement about x by assuming x is a member of our universe, then we can conclude it is true for every member of our universe.

4. *Rules of inference.* The following 12 rules of inference appear in this chapter.

   1. From an example x that does not satisfy $p(x)$, we may conclude $\neg p(x)$.

   2. From $p(x)$ and $q(x)$, we may conclude $p(x) \wedge q(x)$.

   3. From either $p(x)$ or $q(x)$, we may conclude $p(x) \vee q(x)$.

   4. From either $q(x)$ or $\neg p(x)$, we may conclude $p(x) \Rightarrow q(x)$.

5. From $p(x) \Rightarrow q(x)$ and $q(x) \Rightarrow p(x)$, we may conclude $p(x) \Leftrightarrow q(x)$.

6. From $p(x)$ and $p(x) \Rightarrow q(x)$, we may conclude $q(x)$.

7. From $p(x) \Rightarrow q(x)$ and $q(x) \Rightarrow r(x)$, we may conclude $p(x) \Rightarrow r(x)$.

8. If we can derive $q(x)$ from the hypothesis that x satisfies $p(x)$, then we may conclude $p(x) \Rightarrow q(x)$.

9. If we can derive $p(x)$ from the hypothesis that x is a (generic) member of our universe U, we may conclude $\forall x \in U(p(x))$.

10. From an example of an $x \in U$ satisfying $p(x)$, we may conclude $\exists x \in U(p(x))$.

11. From $\neg q(x) \Rightarrow \neg p(x)$, we may conclude $p(x) \Rightarrow q(x)$.

12. If from assuming $p(x)$ and $\neg q(x)$, we can derive both $r(x)$ and $\neg r(x)$ for some statement r, then we may conclude $p(x) \Rightarrow q(x)$.

5. *Contrapositive of $p \Rightarrow q$.* The contrapositive of the statement $p \Rightarrow q$ is the statement $\neg q \Rightarrow \neg p$.

6. *Converse of $p \Rightarrow q$.* The converse of the statement $p \Rightarrow q$ is the statement $q \Rightarrow p$.

7. *Contrapositive rule of inference.* From $\neg q \Rightarrow \neg p$, we may conclude $p \Rightarrow q$.

8. *Principle of proof by contradiction.* If from assuming p and $\neg q$ we can derive both r and $\neg r$ for some statement r, then we may conclude $p \Rightarrow q$.

Problems

1. Write the converse and contrapositive of each statement.

 a. If the hose is 60 ft long, then the hose will reach the tomatoes. a/h

 b. George goes for a walk only if Mary goes for a walk. a/h

 c. Pamela recites a poem if Andre asked for a poem. a/h

2. Construct a proof that if m is odd, then m^2 is odd.

3. Construct a proof that for all integers m and n, if m is even and n is odd, then $m + n$ is odd.

4. What does it really mean to say, "Prove that if m is odd, and n is odd, then $m + n$ is even"? Prove this more precise statement. a/h

5. Prove that for all integers m and n, if m is odd and n is odd, then mn is odd.

6. Is the statement $p \Rightarrow q$ equivalent to the statement $\neg p \Rightarrow \neg q$? a/h

7. Construct a contrapositive proof that for all real numbers x, if $x^2 - 2x \neq -1$, then $x \neq 1$. a/h

8. Construct a proof by contradiction that for all real numbers x, if $x^2 - 2x \neq -1$, then $x \neq 1$.

9. Prove that if $x^3 > 8$, then $x > 2$. a/h

10. Prove that $\sqrt{3}$ is irrational.

11. Construct a proof that if m is an integer such that m^2 is even, then m is even. a/h

12. Prove or disprove the following statement: "For every positive integer n, if n is prime, then 12 and $n^3 - n^2 + n$ have a common factor greater than 1." a/h

13. Prove or disprove the following statement: "For all integers b, c, and d, if x is a rational number such that $x^2 + bx + c = d$, then x is an integer." (*Hints:* Are all the quantifiers given explicitly? It is okay, but not necessary, to use the quadratic formula.)

14. Prove that there is no largest prime number. a/h

15. Prove that if f, g, and h are functions from R^+ to R^+ such that $f(x) = O(g(x))$ and $g(x) = O(h(x))$, then $f(x) = O(h(x))$.

4 INDUCTION, RECURSION, AND RECURRENCES

4.1 MATHEMATICAL INDUCTION

Smallest Counterexamples

In Section 3.3, we demonstrated one way of proving statements about infinite universes: We considered a "generic" member of the universe and derived the desired statement about that member. When our universe is the universe of integers, or when it is in a one-to-one correspondence with the integers, there is a second technique we can use.

Recall our proof of Euclid's division theorem (Theorem 2.12), which says that when n is a positive integer, for each nonnegative integer m, there exist unique nonnegative integers q and r such that $m = nq + r$ and $0 \leq r < n$. For the purpose of a proof by contradiction, we assumed that there is a nonnegative integer m for which no such q and r exist. We chose a smallest such m and observed that $m - n$ is a nonnegative integer less than m. Then we said:

> Therefore, there exist integers q' and r' such that $m - n = nq' + r'$ with $0 \leq r < n$. But then $m = n(q' + 1) + r'$. So, by taking $q = q' + 1$ and $r = r'$, we obtain $m = qn + r$ with $0 \leq r < n$. This is a contradiction to the assumption that there are no integers q and r with $0 \leq r < n$ such that $m = qn + r$. Thus, by the principle of proof by contradiction, such integers q and r exist.

To analyze these sentences, let $p(m)$ denote the statement "There exist integers q' and r' such that $m - n = nq' + r'$ with $0 \leq r < n$." The first two sentences of the quotation provide a proof that $p(m - n) \Rightarrow p(m)$. This implication is the crux of the proof. Let us give an analysis of the proof that shows the pivotal role of this implication.

- We assumed that a counterexample with a smallest m existed.[1]

- Using the fact that $p(m')$ had to be true for every m' smaller than m, we chose $m' = m - n$ and observed that $p(m')$ had to be true.

- We used the implication $p(m - n) \Rightarrow p(m)$ to conclude the truth of $p(m)$.

- However, we had assumed that $p(m)$ was false, so this is the assumption we contradicted in the proof by contradiction.

Exercise 4.1-1

In Chapter 1, we learned Gauss's trick for showing that for all positive integers n,

$$1 + 2 + 3 + 4 + \cdots + n = \frac{n(n + 1)}{2} . \tag{4.1}$$

Use the technique of asserting that if there is a counterexample, then there is a smallest counterexample and deriving a contradiction to prove that the sum is $n(n + 1)/2$. What implication did you have to prove in the process?

Exercise 4.1-2

For what values of $n \geq 0$ do you think $2^{n+1} \geq n^2 + 2$? Use the technique of asserting that if there is a counterexample, then there is a smallest counterexample and deriving a contradiction to prove you are right. What implication did you have to prove in the process?

Exercise 4.1-3

For what values of $n \geq 0$ do you think $2^{n+1} \geq n^2 + 3$? Is it possible to use the technique of asserting that if there is a counterexample, then there is a smallest counterexample and deriving a contradiction to prove you are right? If so, do so and describe the implication you had to prove in the process? If not, why not?

In Exercise 4.1-1, suppose the formula for the sum is false. Then there must be a smallest n such that the formula does not hold for the sum of the first n positive integers. Thus, for any positive integer i smaller than n,

$$1 + 2 + \cdots + i = \frac{i(i + 1)}{2} . \tag{4.2}$$

Because $1 = 1 \cdot 2/2$, Equation 4.1 holds when $n = 1$. Therefore, the smallest counterexample is not $n = 1$. So, $n > 1$, and $n - 1$ is one of the positive integers i for which the formula holds. Substituting $n - 1$ for i in Equation 4.2 yields

$$1 + 2 + \cdots + n - 1 = \frac{(n - 1)n}{2} .$$

---

[1] The fact that every set of nonnegative integers has a smallest element is called the "well-ordering principle" for the natural numbers, and we say that the nonnegative integers are "well ordered."

Adding n to both sides gives

$$1 + 2 + \cdots + n - 1 + n = \frac{(n-1)n}{2} + n$$

$$= \frac{n^2 - n + 2n}{2}$$

$$= \frac{n(n+1)}{2} .$$

Thus, n is not a counterexample after all. Therefore, there is no counterexample to the formula. Hence, the formula holds for all positive integers n. Note that the crucial step was proving that $p(n-1) \Rightarrow p(n)$, where $p(n)$ is the formula

$$1 + 2 + \cdots + n = \frac{n(n+1)}{2} .$$

In Exercise 4.1-2, let $p(n)$ be the statement $2^{n+1} \geq n^2 + 2$. Some experimenting with small values of n leads us to believe this statement is true for all nonnegative integers. Thus, we want to prove $p(n)$ is true for all nonnegative integers n. To do so, we assume that the statement, "$p(n)$ is true for all nonnegative integers n" is false. When a "for all" statement is false, there must be some n for which it is false. Therefore, there is some smallest nonnegative integer n so that $2^{n+1} \not\geq n^2 + 2$. Assume now that n has this value. This means that $2^{i+1} \geq i^2 + 2$ for all nonnegative integers i with $i < n$. Because we know from our experimentation that $n \neq 0$, we know $n - 1$ is a nonnegative integer less than n. Thus, using $n - 1$ in place of i, we get

$$2^{(n-1)+1} \geq (n-1)^2 + 2 ,$$

or

$$2^n \geq n^2 - 2n + 1 + 2$$

$$= n^2 - 2n + 3 . \tag{4.3}$$

From this, we want to draw a contradiction—presumably, a contradiction to $2^{n+1} \not\geq n^2 + 2$.

To get the contradiction, we want to convert the left side of Equation 4.3 to 2^{n+1}. For this purpose, we multiply both sides by 2. Because $2^{n+1} = 2 \cdot 2^n$, we may use Equation 4.3 to write

$$2^{n+1} \geq 2 \cdot (n^2 - 2n + 3) , \text{ or}$$

$$2^{n+1} \geq 2n^2 - 4n + 6 . \tag{4.4}$$

You may get this far and wonder, "What next?" Because we want to obtain a contradiction, we want to convert the right side of Inequality 4.4 into something like $n^2 + 2$. More precisely, we will convert the right side into $n^2 + 2$ plus an additional term. If we can show that the additional term is nonnegative, the proof will be complete. Thus, we write

$$2^{n+1} \geq 2n^2 - 4n + 6$$
$$= (n^2 + 2) + (n^2 - 4n + 4)$$
$$= n^2 + 2 + (n - 2)^2$$
$$\geq n^2 + 2, \qquad\qquad\qquad (4.5)$$

where the last inequality holds because $(n - 2)^2 \geq 0$. This is a contradiction, so there must not have been a smallest counterexample. Thus, there must be no counterexample. Therefore, $2^n \geq n^2 + 2$ for all nonnegative integers n.

What implication did we prove? Let $p(n)$ stand for $2^{n+1} \geq n^2 + 2$. In Equations 4.3 and 4.5, we proved that $p(n - 1) \Rightarrow p(n)$. At one point in our proof, we had to note that we had considered the case with $n = 0$ already. Although we have given a proof by smallest counterexample, it is natural to ask whether it would make more sense to try to prove the statement directly.

Now that we have

$$p(n - 1) \Rightarrow p(n) \,,$$

we see that $p(0)$ implies $p(1)$, $p(1)$ implies $p(2)$, $p(2)$ implies $p(3)$, and so on. In this way, we have $p(k)$ for every k. Isn't this a more direct proof? We will address this question shortly.

First, let's consider Exercise 4.1-3. Notice that $2^{n+1} \not> n^2 + 3$ for $n = 0$ and $n = 1$, but $2^{n+1} > n^2 + 3$ for any larger n we look at. Let us try to prove that $2^{n+1} > n^2 + 3$ for $n \geq 2$. We now let $p'(n)$ be the statement $2^{n+1} > n^2 + 3$. We can easily prove $p'(2)$ as follows: $8 = 2^3 \geq 2^2 + 3 = 7$. Now, suppose that among the integers larger than 2, there is a counterexample m to $p'(n)$. That is, suppose there is an m such that $m > 2$ and $p'(m)$ is false. Then there is a smallest such m. Then $p'(k)$ is true for k between 2 and $m - 1$. If you look back at your proof that $p(n - 1) \Rightarrow p(n)$, you will see that when $n \geq 2$, essentially the same proof applies to p' as well. That is, with very similar computations, we can show that $p'(n - 1) \Rightarrow p'(n)$, so long as $n \geq 2$. Thus, because $p'(m - 1)$ is true, our implication tells us that $p'(m)$ is also true. This is a contradiction to our assumption that $p'(m)$ is false. Therefore, $p'(m)$ is true.

Again, we could conclude from $p'(2)$ and $p'(2) \Rightarrow p'(3)$ that $p'(3)$ is true (and similarly for $p'(4)$ and so on). Again, this seems to give a more direct proof. The implication we had to prove was $p'(n - 1) \Rightarrow p'(n)$.

The Principle of Mathematical Induction

It may seem clear that repeatedly using the implication $p(n - 1) \Rightarrow p(n)$ will prove $p(n)$ for all n (or all $n \geq 2$). This observation is the central idea of the principle of mathematical induction, which we are about to introduce. In a theoretical discussion of the integers, the principle of mathematical induction (or the equivalent **well-ordering principle**—every set of nonnegative integers has a smallest element, which allows us to use the "smallest counterexample" technique) is one of the first principles we assume. The principle of mathematical induction is usually described in two forms. The one we have talked about so far, called the "weak form," applies to statements about integers n.

Principle 4.1 **(The Weak Principle of Mathematical Induction)** If the statement $p(b)$ is true and the statement $p(n-1) \Rightarrow p(n)$ is true for all $n > b$, then $p(n)$ is true for all integers $n \geq b$.

Suppose, for example, we wish to give a direct inductive proof that $2^{n+1} > n^2 + 3$ for $n \geq 2$. We would proceed as follows. (The material in square brackets is not part of the proof; it is a running commentary on what is going on in the proof.)

> We will prove by induction that $2^{n+1} > n^2 + 3$ for $n \geq 2$. First, $2^{2+1} = 2^3 = 8$, while $2^2 + 3 = 7$. [We just proved $p(2)$. We will now proceed to prove $p(n-1) \Rightarrow p(n)$.] Suppose now that $n > 2$ and $2^n > (n-1)^2 + 3$. [We just made the hypothesis of $p(n-1)$ in order to use Rule 8 of our rules of inference.]
>
> Now we multiply both sides of this inequality by 2, giving
>
> $$2^{n+1} > 2(n^2 - 2n + 1) + 6 \,.$$
>
> But
>
> $$2(n^2 - 2n + 1) + 6 = n^2 + 3 + n^2 - 4n + 4 + 1$$
> $$= n^2 + 3 + (n-2)^2 + 1 \,.$$
>
> Therefore, $2^{n+1} > n^2 + 3 + (n-2)^2 + 1$.
>
> Because $(n-2)^2 + 1$ is positive, this proves $2^{n+1} > n^2 + 3$. [We just showed that from the hypothesis of $p(n-1)$, we can derive $p(n)$. Now we can apply Rule 8 to assert that $p(n-1) \Rightarrow p(n)$.] Therefore, $2^n > (n-1)^2 + 3 \Rightarrow 2^{n+1} > n^2 + 3$, and by the principle of mathematical induction, $2^{n+1} > n^2 + 3$ for $n \geq 2$.

In this proof, the sentence "First, $2^{2+1} = 2^3 = 8$, while $2^2 + 3 = 7$" is called the **base case**. It consists of directly proving that $p(b)$ is true, where, in this case, b is 2 and $p(n)$ is $2^{n+1} > n^2 + 3$. The sentence "Suppose now that $n > 2$ and $2^n > (n-1)^2 + 3$" is called the **inductive hypothesis**, which is the assumption that $p(n-1)$ is true. In inductive proofs, we always make such a hypothesis[2] to prove the implication $p(n-1) \Rightarrow p(n)$. The proof of the implication is called the **inductive step**. The final sentence of the proof is called the **inductive conclusion**.

---

[2] At times, it might be more convenient to assume that $p(n)$ is true and use this assumption to prove that $p(n+1)$ is true. This proves the implication $p(n) \Rightarrow p(n+1)$, which lets us reason in the same way.

Use mathematical induction to show that

$$1 + 3 + 5 + \cdots + (2k - 1) = k^2$$

for each positive integer k.

Exercise 4.1-5

For what values of n is $2^n > n^2$? Use mathematical induction to show that your answer is correct.

For Exercise 4.1-4, we note that the formula holds when $k = 1$. Assume inductively that the formula holds when $k = n - 1$, so that $1 + 3 + \cdots + (2n - 3) = (n - 1)^2$. Adding $2n - 1$ to both sides of this equation gives

$$1 + 3 + \cdots + (2n - 3) + (2n - 1) = n^2 - 2n + 1 + 2n - 1$$
$$= n^2 . \tag{4.6}$$

Thus, the formula holds when $k = n$, and so, by the principle of mathematical induction, the formula holds for all positive integers k.

Notice that in our discussion of Exercise 4.1-4, nowhere did we mention a statement $p(n)$. In fact, $p(n)$ is the statement we get by substituting n for k in the formula. In Equation 4.6, we were proving $p(n - 1) \Rightarrow p(n)$. Next, notice that we did not explicitly say we were going to give a proof by induction; instead, we indicated that we were making an inductive proof when we were making the inductive hypothesis by saying, "Assume inductively that ...." This convention makes the prose flow nicely but still tells the reader that he or she is reading a proof by induction. Notice also how the notation in the statement of the exercise helped us write the proof. If we state what we are trying to prove in terms of a variable other than n, such as k, then we can assume that our desired statement holds when this variable, k, is $n - 1$ and then prove that the statement holds when $k = n$. Without this notational device, we have to either mention our statement $p(n)$ explicitly or avoid any discussion of substituting values into the formula we are trying to prove. Our proof that $2^{n+1} > n^2 + 3$ demonstrates this last approach to writing an inductive proof in plain English. This is usually the "slickest" way of writing an inductive proof (though it is often the hardest to master). We will use this approach first for the next exercise.

For Exercise 4.1-5, we note that $2 = 2^1 > 1^2 = 1$, but then the inequality fails for $n = 2, 3, 4$. However, $32 > 25$. Now we assume inductively that for $n > 5$, we have $2^{n-1} > (n - 1)^2$. Multiplying by 2 gives us the following:

$$2^n > 2(n^2 - 2n + 1)$$
$$= n^2 + n^2 - 4n + 2$$
$$> n^2 + n^2 - n \cdot n$$
$$= n^2 ,$$

because $n > 5$ implies that $-4n > -n \cdot n$. (We also used the fact that $n^2 + n^2 - 4n + 2 > n^2 + n^2 - 4n$.) Thus, by the principle of mathematical induction, $2^n > n^2$ for all $n \geq 5$.

Alternatively, we could write the following: Let $p(n)$ denote the inequality $2^n > n^2$. Then $p(5)$ is true, because $32 > 25$. Assume that $n > 5$ and that $p(n-1)$ is true. This gives us $2^{n-1} > (n-1)^2$. Multiplying by 2 gives us the following:

$$2^n > 2(n^2 - 2n + 1)$$
$$= n^2 + n^2 - 4n + 2$$
$$> n^2 + n^2 - n \cdot n$$
$$= n^2 \, ,$$

because $n > 5$ implies that $-4n > -n \cdot n$. Therefore, $p(n-1) \Rightarrow p(n)$. Thus, by the principle of mathematical induction, $2^n > n^2$ for all $n \geq 5$.

Notice how the "slick" method simply assumes that the reader knows we are doing a proof by induction from our "Assume inductively. . . ." It also assumes the reader mentally supplies the appropriate $p(n)$ and observes that we have proved $p(n-1) \Rightarrow p(n)$ at the right moment.

Here is a slight variation of the technique of changing variables. To prove that $2^n > n^2$ when $n \geq 5$, we observe that the inequality holds when $n = 5$, because $32 > 25$. Assume inductively that the inequality holds when $n = k$, so that $2^k > k^2$. Now, when $k \geq 5$, multiplying both sides of this inequality by 2 yields the following sequence of inequalities (which are explained in the text that follows):

$$2^{k+1} > 2k^2$$
$$= k^2 + k^2$$
$$> k^2 + 5k$$
$$> k^2 + 2k + 1$$
$$= (k+1)^2 \, ,$$

because $k \geq 5$ implies that $k^2 \geq 5k$ and $5k = 2k + 3k > 2k + 1$. Thus, by the principle of mathematical induction, $2^n > n^2$ for all $n \geq 5$.

This last variation of the proof illustrates two ideas. First, there is no need to save the name n for the variable we use in applying mathematical induction. We used k as our inductive variable in this case. Second, as suggested in footnote 2, there is no need to restrict ourselves to proving the implication $p(n-1) \Rightarrow p(n)$. In this case, we proved the implication $p(k) \Rightarrow p(k+1)$. Clearly, these two implications are equivalent as n ranges over all integers larger than b and as k ranges over all integers larger than or equal to b.

Strong Induction

In our proof of Euclid's division theorem, we had a statement of the form $p(m)$, and, assuming that it was false, we chose a smallest m such that $p(m)$ is false for some n.

This meant we could assume that $p(m')$ is true for *all* nonnegative $m' < m$. We needed this assumption because we had to show that $p(m - n) \Rightarrow p(m)$ in order to get our contradiction. This situation differs from the examples we used to introduce mathematical induction, for in those we used an implication of the form $p(n - 1) \Rightarrow p(n)$. The essence of our method in proving Euclid's division theorem is the following:

1. We have a statement $q(k)$ that we want to prove for all k larger than some integer.

2. We suppose it is false; so, there must be a smallest k for which $q(k)$ is false.

3. This means we may assume $q(k')$ is true for *all* k' in the universe of q with $k' < k$.

4. We then use this assumption to derive a proof of $q(k)$, thus generating our contradiction.

Again, we can avoid the step of generating a contradiction in the following way. Suppose first we have a proof of $q(0)$. Suppose also we have a proof that

$$q(0) \wedge q(1) \wedge q(2) \wedge \cdots \wedge q(k - 1) \Rightarrow q(k)$$

for all k larger than 0. Then, from $q(0)$, we can prove $q(1)$; from $q(0) \wedge q(1)$, we can prove $q(2)$; from $q(0) \wedge q(1) \wedge q(2)$, we can prove $q(3)$; and so on. This gives us a proof of $q(n)$ for any n we desire. This is another form of the principle of mathematical induction. We use this when, as in Euclid's division theorem, we can get an implication of the form $q(k') \Rightarrow q(k)$ for *some* $k' < k$ or when we can get an implication of the form $q(0) \wedge q(1) \wedge q(2) \wedge \cdots \wedge q(k - 1) \Rightarrow q(k)$. (As is the case in Euclid's division theorem, we often don't really know what the k' is, so the first kind of situation is really just a special case of the second. It is for this reason that we do not treat the first of the two implications separately.) We have just described the method of proof known as the strong principle of mathematical induction.

Principle 4.2 **(The Strong Principle of Mathematical Induction)** If the statement $p(b)$ is true and the statement $p(b) \wedge p(b + 1) \wedge \cdots \wedge p(n - 1) \Rightarrow p(n)$ is true for all $n > b$, then $p(n)$ is true for all integers $n \geq b$.

Exercise 4.1-6 Prove that every positive integer is either a power of a prime number or the product of powers of prime numbers.

In Exercise 4.1-6, we observe that 1 is a power of a prime number; for example, $1 = 2^0$. Suppose now that we know that every number less than n is a power of a prime number or a product of powers of prime numbers. Then, if n is not a prime number, it is a product of two smaller numbers, each of which is, by our supposition, a power of a prime number or a product of powers of prime numbers. But multiplying two powers of primes or products of powers of primes gives a product of powers of primes. Therefore, n is a power of a prime number or a product of powers of prime numbers. Thus, by the strong principle of mathematical induction, every positive integer is a power of a prime number or a product of powers of prime numbers.

Note that there was no explicit mention of an implication of the form

$$p(b) \wedge p(b+1) \wedge \cdots \wedge p(n-1) \Rightarrow p(n) \, .$$

This is common with inductive proofs. Note also that we did not explicitly identify the base case or the inductive hypothesis in our proof. This is common, too. Readers of inductive proofs are expected to recognize when the base case is being given and when an implication of the form $p(n-1) \Rightarrow p(n)$ or $p(b) \wedge p(b+1) \wedge \cdots \wedge p(n-1) \Rightarrow p(n)$ is being proved.

Mathematical induction is used frequently in discrete math and computer science. Many quantities that we are interested in measuring, such as running time or space used in memory, typically are restricted to positive integers. Thus, mathematical induction is a natural way to prove facts about these quantities. We will use it frequently throughout this book. We typically will not distinguish between strong and weak induction; we just think of them both as induction. (Problems 13 and 14 ask you to derive each version of the principle from the other.)

Induction in General

We now summarize what we have said so far. A typical proof by mathematical induction showing that a statement $p(n)$ is true for all integers $n \geq b$ consists of three steps.

1. We show that $p(b)$ is true. This is called establishing a base case.

2. We either show that

$$p(n-1) \Rightarrow p(n)$$

   for all $n > b$ or show that

$$p(b) \wedge p(b+1) \wedge \cdots \wedge p(n-1) \Rightarrow p(n)$$

   for all $n > b$. For this purpose, we make either the inductive hypothesis $p(n-1)$ or the inductive hypothesis $p(b) \wedge p(b+1) \wedge \cdots \wedge p(n-1)$. Then we derive $p(n)$ to complete the proof of the implication we desire—either $p(n-1) \Rightarrow p(n)$ or $p(b) \wedge p(b+1) \wedge \cdots \wedge p(n-1) \Rightarrow p(n)$.

3. We conclude on the basis of the principle of mathematical induction that $p(n)$ is true for all integers n greater than or equal to b.

The second step is the core of an inductive proof. This is usually where we need the most insight into what we are trying to prove. Looking back on the examples of induction in this chapter, you may notice that in Example 4.1-5, we did not show that $p(n-1) \Rightarrow p(n)$; instead, we showed that $p(n) \Rightarrow p(n+1)$. Logically, in the context of an inductive proof, these statements are equivalent (simply substitute m for $n-1$).

For convenience, we now restate an alternate to condition 2:

2′. We either show that for all $n \geq b$, either

$$p(n) \Rightarrow p(n+1)$$

or

$$p(b) \wedge p(b+1) \wedge \cdots \wedge p(n) \Rightarrow p(n+1) .$$

For this purpose, we make either the inductive hypothesis $p(n)$ or the inductive hypothesis $p(b) \wedge p(b+1) \wedge \cdots \wedge p(n)$. Then we derive $p(n+1)$ to complete the proof of the implication we desire—either $p(n) \Rightarrow p(n+1)$ or $p(b) \wedge p(b+1) \wedge \cdots \wedge p(n) \Rightarrow p(n+1)$.

It is important to realize that induction arises in some circumstances that do not fit the typical description we just gave. These circumstances seem to arise often in computer science. First, instead of a single base case, we may need multiple base cases. Second, instead of needing to show just one implication that demonstrates that $p(n)$ is true given that $p(n')$ is true for some set of $n' < n$, we may need to show a set of such implications.

For example, consider the problem of proving the following statement:

$$\sum_{i=0}^{n} \left\lfloor \frac{i}{2} \right\rfloor = \begin{cases} \frac{n^2}{4} & \text{if } n \text{ is even} , \\ \frac{n^2-1}{4} & \text{if } n \text{ is odd} . \end{cases} \qquad (4.7)$$

To prove this, we must show that $p(0)$ is true, $p(1)$ is true, $p(n-2) \Rightarrow p(n)$ if n is odd, and $p(n-2) \Rightarrow p(n)$ if n is even. Putting all these together, we see that our formulas hold for all $n \geq 0$. We can view this as either two proofs by induction, one for even and one for odd numbers, or one proof in which we have two base cases and two methods of deriving results from previous ones. The second view is more profitable because it expands our idea of what induction means and makes it easier to find inductive proofs. In this proof of Equation 4.7, we have two base cases and two inductive implications. We could also find situations where we have just one implication to prove but several base cases to check (we will see one such situation shortly) or just one base case but several different implications to prove.

Logically speaking, we could rework the proof of Equation 4.7 above so that it fits the pattern of strong induction. For example, when we prove a second base case, then we have just proved that the first base case implies it, because a true statement implies a true statement. However, in the mathematics literature and especially in the computer science literature, inductive proofs are written with multiple base cases and multiple implications with no effort to reduce them to one of the standard forms of mathematical induction. So long as it is possible to cover all the cases under consideration with such a proof, it can be rewritten as a standard inductive proof. Because readers of such proofs are expected to know that this is possible, and because reworking such a proof as a standard inductive proof adds unnecessary verbiage, this is almost always left out.

A Recursive View of Induction

Those familiar with recursive programs might notice similarities between induction and recursion.[3] Both talk about base cases. Both may appear at first glance to be circular. In recursion, a function calls itself. When we prove the implication in the inductive step of an inductive proof, we prove a property for an instance of size n by assuming the property is true for other instances. In both cases, the same thing prevents circularity:

- The instances solved by recursion when the function calls itself are always smaller than the current instance and the recursion eventually gets down to base cases that are dealt with directly.

- The instances assumed in the inductive step of an inductive proof are always smaller than the current instance and the induction eventually gets down to base cases that are dealt with directly.

Students who have written a number of recursive programs come to understand that recursion works. As long as all the recursive calls are all to smaller-sized instances, the recursion is not circular and will terminate. As long as the base cases are handled correctly and larger instances are correctly solved by building on solutions to smaller instances, the recursion will compute the correct answer.

An inductive proof can be seen as a description of a recursive program that will print a complete, horribly detailed proof for any chosen instance of size n as long as n is bigger than some value. Because recursion works, we can call this program to print a proof for any n. Because there is a complete proof for any n, the property must be true for all n.

For our example, we again consider Euclid's division theorem (Theorem 2.12). We will prove that for each pair (m, n) of positive integers, there are nonnegative integers q and r such that $m = qn + r$ and $0 \leq r < n$. (To simplify the program, we do not include the proof of uniqueness.) The basic idea of an inductive proof of this theorem is the observation that if we know that a solution exists for the pair $(m - n, n)$, then it is easy to show that a solution exists for the pair (m, n). In particular, if $m - n = q'n + r'$ for $0 \leq r' < n$, then $m = (q' + 1)n + r'$, so we have a solution with $q = q' + 1$ and $r = r'$. Thus, if we can use either a base case or recursion to show that the theorem is true for the pair $(m - n, n)$, we can use this fact and recursion to show that the theorem is true for the pair (m, n).

What would a recursive program for writing such a proof for any pair (m, n) look like? We first ask what the base case of the induction is. The recursion works by solving a smaller problem of the same form as the problem we are trying to solve. We are to prove the theorem for any pair of positive integers. We can prove the theorem for a pair of positive integers (m, n) by recursively proving it for $(m - n, n)$, so long as $m - n$ and n are positive integers. This is true when $m - n > 0$, or $m > n$. Once $m \leq n$, we can no longer use recursion, which means that for $1 \leq m \leq n$ we must find a different way to solve the problem. The cases with $1 \leq m \leq n$ are the m base

---

[3]In this section, when we speak of recursion, we mean recursion in a computer program. We will simply say recursion to avoid being repetitive.

cases. For $1 \leq m < n$, the theorem is true with $q = 0$ and $r = m$. For $m = n$, the theorem is true with $q = 1$ and $r = 0$.

The following pseudocode shows how the arguments of the previous two paragraphs can be written as a program to generate a proof for any m and n. The pseudocode uses recursion as described above to calculate q and r for each problem and then uses these values in the proof that it writes. Note that the problem requires calculation of two numbers, both of which need to be returned. The pseudocode indicates this by returning an ordered pair (q, r). Thus, `Return (0, m)` means return an ordered pair of values, the first of which is 0 and the second of which is m. Similarly, `(q, r) = ProveEuclidDivision(m − n, n)` means that the first value of the ordered pair returned should be copied into q and the second value into r.

`ProveEuclidDivision(m,n)`

```
// Assume that m and n are positive integers.
// This is a recursive program that inputs m and n and, after some intermediate
// steps, returns the ordered pair (q,r) so that m = nq + r with 0 ≤ r < n.
// As it does the computation, it prints a proof that the pair (q,r) works.
(1)   Print "We wish to prove that:"
(2)   Print m, " = q*", n, " +r for some q ≥ 0 and 0 ≤ r < ", n
(3)   If (m < n)
(4)       Print "This is true because ", m, " = 0*", n, " + ", m
(5)       Return (0, m)
(6)   Else If (m == n)
(7)       Print "This is true because ", m, " = 1*", n, " + 0"
(8)       Return (1, 0)
(9)   Else
(10)      Print "We first prove that:"
(11)      Print m−n, " = q*", n, " +r for some q ≥ 0 and 0 ≤ r < ", n
(12)      (q, r) = ProveEuclidDivision(m−n, n)
(13)      Print "Adding ", n, " to both sides gives"
(14)      Print m, " = ", q + 1, "*", n, " + ", r
(15)      Print "This proves that"
(16)      Print m, " = q*", n, "+r for some q ≥ 0 and 0 ≤ r < ", n
(17)      Return (q + 1, r)
```

The print statements are messy, but the code is fairly straightforward. In particular, without the print statements, this program would be a recursively defined function that inputs the values m and n and returns the values q and r with $m = nq + r$ and $0 \leq r < n$. The output of the call `ProveEuclidDivision(10, 4)` is:

```
We wish to prove that
10 = q*4 + r for some q ≥ 0 and 0 ≤ r < 4
We first prove that
6 = q*4 + r for some q ≥ 0 and 0 ≤ r < 4
We wish to prove that
6 = q*4 + r for some q ≥ 0 and 0 ≤ r < 4
We first prove that
2 = q*4 + r for some q ≥ 0 and 0 ≤ r < 4
```

```
We wish to prove that
2 = q*4 + r for some q ≥ 0 and 0 ≤ r < 4
This is true because 2 = 0*4 + 2
Adding 4 to both sides gives
6 = 1*4 + 2
This proves that
6 = q*4 + r for some q ≥ 0 and 0 ≤ r < 4
Adding 4 to both sides gives
10 = 2*4 + 2
This proves that
10 = q*4 + r for some q ≥ 0 and 0 ≤ r < 4
```

We do not expect you to write such a program when you are asked to do a proof by induction. However, thinking recursively is the easiest way to discover an inductive proof. Given an instance for which you are trying to prove a property, start by figuring out how to break it down into one or more smaller instances of the same form. Because the instances are smaller, you can assume that the property is true for them; after all, if you had to, you could generate a proof for that case by writing a recursive program, as we just did. Then show how the fact that the property holds for these smaller instances implies that the property holds for the original instance. Finally, decide at what point the recursive decomposition stops giving problems of the same form. These problems, which cannot be recursively decomposed into smaller problems, are the base cases. You must check directly that the property holds for the smaller problems.

Note that this is the reverse of the way that a proof is actually written. The recursive decomposition is developed first. The decomposition then determines whether the induction is strong or weak and what base cases are needed. Proofs could be written in this form, but it is traditional to prove the base cases first and then show that the smaller cases imply the larger one.

When proving the validity of *formulas* via induction, it sometimes helps, as we did implicitly in Exercise 4.1-4, to think of how to "grow" a smaller case into a larger one. However, it is usually more profitable to think of decomposing a larger case into smaller ones than to think of building a smaller case to create a larger one. (Building a smaller case to a larger one is just one way of seeing how to decompose the larger case into smaller ones.) As we will see in Sections 6.1 and 6.2 (especially in Exercises 6.2-6 and 6.2-7), there are times when this way of thinking is clearly the best way to get a valid proof. Such examples occur throughout computer science. Therefore, it is good to get in the habit of doing induction by starting with a larger instance and recursively decomposing it to get smaller instances. There is another advantage to doing this: There is no question about what our base case or base cases should be. The base cases are the ones where the recursive decomposition no longer works. This answers a question students often ask, namely, "How do I choose my base case or cases?"

To demonstrate this idea, we reconsider the proof that every positive integer is a prime or the product of powers of primes. The recursive decomposition is to factor a number into two smaller factors, which is always possible unless the number is a prime or is 1.

Thus, our base cases are $1 = 2^0$ and all the primes. (As you may recall, our base case in our first solution of Exercise 4.1-6 was simply the case where the number is 1.) In all these cases, our number is either a prime or a power of a prime. For any other number n, we assume that the property holds for all $k < n$. Because the number is not a prime or 1, we can factor it into two smaller numbers, and, by our inductive hypothesis, each is either prime or the product of powers of primes. Multiplying two products of powers of primes gives another product of powers of primes. Thus, our number is the product of powers of primes. By the strong principle of mathematical induction, every positive integer is a prime or a product of powers of primes.

It may seem strange to talk about an infinite number of base cases (all the primes), but these would be the base cases of a recursive program to factor a number into a product of powers of primes. If you reread the original solution to Exercise 4.1-6, then you will see that primes are a special case handled without using the inductive hypothesis. They are base cases in the recursive sense, and whether we choose to call them base cases in the inductive sense or to view them as inductive cases that don't require the inductive hypothesis to prove them is a matter of taste.[4] Inductive proofs are often cleaner if we define a base case to be any case that does not use the inductive hypothesis in its proof.

Important Concepts, Formulas, and Theorems

1. *Weak principle of mathematical induction.* The *weak principle of mathematical induction* states that if the statement $p(b)$ is true and the statement $p(n-1) \Rightarrow p(n)$ is true for all $n > b$, then $p(n)$ is true for all integers $n \geq b$.

2. *Strong principle of mathematical induction.* The *strong principle of mathematical induction* states that if the statement $p(b)$ is true and the statement $p(b) \wedge p(b+1) \wedge \cdots \wedge p(n-1) \Rightarrow p(n)$ is true for all $n > b$, then $p(n)$ is true for all integers $n \geq b$.

3. *Base case.* Every proof by mathematical induction, strong or weak, begins with a *base case,* which establishes the result being proved for at least one value of the variable on which we are inducting. This base case should prove the result for the smallest value of the variable for which we are asserting the result. In a proof with multiple base cases, the base cases should cover all values of the variable that are not covered by the inductive step of the proof.

4. *Inductive hypothesis.* Every proof by induction includes an inductive hypothesis in which we assume that the result $p(n)$ we are trying to prove is true when $n = k - 1$ or when $n < k$ (or in which we assume an equivalent statement).

---

[4]Recall that in the truth table for $p \Rightarrow q$, in each row in which q is true, the statement $p \Rightarrow q$ is true as well. Thus, one thing that would prove $p \Rightarrow q$ to be true is a proof of q that does not make any assumption about p. This is what we were doing when we wrote, "Suppose now we know that every number less than n is a power of a prime number or a product of powers of prime numbers. Then, if n is not a prime number, it is a product of two smaller numbers." We were treating the cases where n is a prime number as special cases in which our conclusion could be shown to be true without using the hypothesis.

5. *Inductive step.* Every proof by induction includes an inductive step in which we prove the implication that $p(k-1) \Rightarrow p(k)$ or the implication that $p(b) \wedge p(b+1) \wedge \cdots \wedge p(k-1) \Rightarrow p(k)$, or some equivalent implication.

6. *Inductive conclusion.* A proof by mathematical induction should include, at least implicitly, a concluding statement of the form "Thus, by the principle of mathematical induction ...," which asserts that by the principle of mathematical induction, the result $p(n)$ that we are trying to prove is true for all values of n, including and beyond the base case(s).

Problems

1. This problem explores ways to prove that

$$\frac{2}{3} + \frac{2}{9} + \cdots + \frac{2}{3^n} = 1 - \left(\frac{1}{3}\right)^n$$

for all positive integers n.

a. First, we explore how to prove the formula by contradiction. In other words, assume that there is some integer n that makes the formula false. In this case, there must be some smallest n that makes the formula false.

i. Can this smallest n be 1?

ii. What do you know about

$$\frac{2}{3} + \frac{2}{9} + \cdots + \frac{2}{3^i}$$

when i is a positive integer smaller than this smallest n?

iii. Is $n - 1$ a positive integer for this smallest n?

iv. What do you know about

$$\frac{2}{3} + \frac{2}{9} + \cdots + \frac{2}{3^{n-1}}$$

for this smallest n?

v. Write the answer to part iv as an equation, add $2/3^n$ to both sides, and simplify the right side.

vi. What does the equation that results from part v say about your assumption that the formula is false?

vii. What can you conclude about the truth of the formula?

viii. If $p(k)$ is the statement

$$\frac{2}{3} + \frac{2}{9} + \cdots + \frac{2}{3^k} = 1 - \left(\frac{1}{3}\right)^k,$$

what implication did you prove in the process of deriving your contradiction? a/h

b.   i. What is the base case in a proof by mathematical induction that

$$\frac{2}{3} + \frac{2}{9} + \cdots + \frac{2}{3^n} = 1 - \left(\frac{1}{3}\right)^n$$

     for all positive integers n?

    ii. What would you assume as an inductive hypothesis?

   iii. What would you prove in the inductive step of a proof of this formula by induction?

   iv. Prove it.

    v. What does the principle of mathematical induction allow you to conclude?

   vi. If $p(k)$ is the statement

$$\frac{2}{3} + \frac{2}{9} + \cdots + \frac{2}{3^k} = 1 - \left(\frac{1}{3}\right)^k ,$$

     what implication did you prove in the process of doing your proof by induction? a/h

2. Use contradiction to prove $1 \cdot 2 + 2 \cdot 3 + \cdots + n(n+1) = n(n+1)(n+2)/3$.

3. Use induction to prove that $1 \cdot 2 + 2 \cdot 3 + \cdots + n(n+1) = n(n+1)(n+2)/3$. a/h

4. Prove that $1^3 + 2^3 + 3^3 + \cdots + n^3 = n^2(n+1)^2/4$.

5. Use strong induction to write a careful proof of Euclid's division theorem. a/h

6. Prove that $\sum_{i=j}^{n} \binom{i}{j} = \binom{n+1}{j+1}$. In addition to an inductive proof, there is a nice "story" proof of this formula. It is well worth trying to figure out both proofs.

7. Prove that every number greater than 7 is a sum of a nonnegative integer multiple of 3 and a nonnegative integer multiple of 5. a/h

8. We can define the nonnegative powers of a number a by the rules $a^0 = 1$ and $a^{n+1} = a^n \cdot a$. Explain why this defines a^n for all nonnegative integers n. From this definition, prove the rule of exponents $a^{m+n} = a^m a^n$ for nonnegative integers m and n.

9. Our arguments in favor of the sum principle were quite intuitive. In fact, the sum principle for n sets follows from the sum principle for two sets. Use induction to prove the sum principle for a union of n sets from the sum principle for a union of two sets. a/h

10. We have proved that every positive integer is a power of a prime number or a product of powers of prime numbers. Show that this factorization is unique in the following sense: If you have two factorizations of a positive integer, both factorizations use exactly the same primes, and each prime occurs to the same

power in both factorizations. For this purpose, it is helpful to know that if a prime divides a product of integers, then it divides one of the integers in the product. (Another way to say this is that if a prime is a factor of a product of integers, then it is a factor of one of the integers in the product.)

11. Find the error in the following "proof" that all positive integers n are equal: Let $p(n)$ be the statement that all numbers in an n-element set of positive integers are equal. Then $p(1)$ is true. Now assume $p(n-1)$ is true, and let N be the set of the first n integers. Let N' be the set of the first $n-1$ integers, and let N'' be the set of the last $n-1$ integers. By $p(n-1)$, all members of N' are equal, and all members of N'' are equal. Thus, the first $n-1$ elements of N are equal and the last $n-1$ elements of N are equal, and so all elements of N are equal. Therefore, all positive integers are equal.

12. Prove by induction that the number of subsets of an n-element set is 2^n. a/h

13. Prove that the strong principle of mathematical induction implies the weak principle of mathematical induction.

14. Prove that the weak principle of mathematical induction implies the strong principle of mathematical induction. a/h

15. Prove Statement 4.7.

4.2 RECURSION, RECURRENCES, AND INDUCTION

Recursion

Exercise 4.2-1

Describe how you have used recursion when writing programs. Include as many uses as you can.

Exercise 4.2-2

A standard problem for computer science students who are learning about recursion is the Tower of Hanoi problem. In this problem, we have three pegs numbered 1, 2, and 3. One peg has a stack of n disks, each smaller in diameter than the one below it, as in Figure 4.1. An allowable move consists of removing a disk from one peg and sliding it onto another peg so that it is not above another disk of smaller size. We are to determine how many allowable moves are needed to move the disks from one peg to another. Describe the strategy you have used or would use in a recursive program to solve this problem.

Figure 4.1: *The Tower of Hanoi*

For the Tower of Hanoi problem, to solve the problem with no disks, do nothing. To solve the problem of moving all n disks to Peg 3, do the following:

1. Recursively solve the problem of moving the top $n - 1$ disks from Peg 1 to Peg 2.

2. Move Disk n to Peg 3.

3. Recursively solve the problem of moving the $n - 1$ disks on Peg 2 to Peg 3.

Thus, if $M(n)$ is the number of moves needed to move n disks from Peg i to Peg j, we have

$$M(n) = 2M(n - 1) + 1 .$$

This is an example of a **recurrence equation** or **recurrence**. A recurrence equation for a function defined on the set of integers greater than or equal to some number b is one that tells us how to compute the nth value of a function from the $(n - 1)$st value or how to compute the nth value from some or all the first $n - 1$ values. To specify completely a function on the basis of a recurrence, we have to give enough information about the function to get started. This information is called the **initial condition** (or the initial conditions) (which we also call the *base case*) for the recurrence. In this case, we have said that $M(0) = 0$. Using this, we get from the recurrence that $M(1) = 1$, $M(2) = 3$, $M(3) = 7$, $M(4) = 15$, and $M(5) = 31$. We are led to guess that $M(n) = 2^n - 1$.

Formally, we write our recurrence and initial condition together as

$$M(n) = \begin{cases} 0 & \text{if } n = 0 , \\ 2M(n - 1) + 1 & \text{otherwise} . \end{cases} \tag{4.8}$$

Now we give an inductive proof that our guess is correct. The base case is trivial, because we have defined $M(0) = 0$, and $0 = 2^0 - 1$. For the inductive step, we assume that $n > 0$ and $M(n - 1) = 2^{n-1} - 1$. From the recurrence, $M(n) = 2M(n - 1) + 1$. But, by the inductive hypothesis, $M(n - 1) = 2^{n-1} - 1$; so, we get that

$$M(n) = 2M(n - 1) + 1$$
$$= 2(2^{n-1} - 1) + 1$$
$$= 2^n - 1 .$$

Thus, by the principle of mathematical induction, $M(n) = 2^n - 1$ for all nonnegative integers n.

The ease with which we solved this recurrence and proved our solution correct is no accident. Recursion, recurrences, and induction are all intimately related. The relationship between recursion and recurrences is reasonably transparent—recurrences give a natural way of analyzing recursive algorithms. Both recursion and recurrences specify the solution to an instance of a problem in terms of solutions to one or more smaller instances. Induction also falls naturally into this paradigm in that we are deriving a statement $p(n)$ from statements $p(n')$ for $n' < n$. In fact, we saw at the end of

Section 4.1 that proof by induction can be thought of as proof by recursion. Thus, we really have three variations on the same theme.

We also observe, more concretely, that the mathematical correctness of solutions to recurrences is naturally proved via induction. Also, the correctness of a recurrence that describes the number of steps needed to solve a recursive problem is also naturally proved by induction. The recurrence or recursive structure of the problem makes setting up the inductive proof straightforward.

Examples of First-Order Linear Recurrences

Exercise 4.2-3

The empty set ($\emptyset$) is a set with no elements. How many subsets does it have? How many subsets does the one-element set $\{1\}$ have? How many subsets does the two-element set $\{1, 2\}$ have? How many of these subsets contain 2? How many subsets does $\{1, 2, 3\}$ have? How many contain 3? Give a recurrence for the number $S(n)$ of subsets of an n-element set, and prove that your recurrence is correct.

Exercise 4.2-4

When paying off a loan with initial amount A and monthly payment M at an interest rate of p percent, the total amount $T(n)$ of the loan after n months is computed by adding $p/12$ percent to the amount due after $n - 1$ months and then subtracting the monthly payment M. Convert this description into a recurrence for the amount owed after n months.

Exercise 4.2-5

Given the recurrence

$$T(n) = rT(n - 1) + a ,$$

where r and a are constants, find a recurrence that expresses $T(n)$ in terms of $T(n - 2)$ instead of $T(n - 1)$. Now find a recurrence that expresses $T(n)$ in terms of $T(n - 3)$ instead of $T(n - 2)$ or $T(n - 1)$. Now find a recurrence that expresses $T(n)$ in terms of $T(n - 4)$ rather than $T(n - 1)$, $T(n - 2)$, or $T(n - 3)$. Based on your work so far, find a general formula for the solution to the recurrence

$$T(n) = rT(n - 1) + a ,$$

with $T(0) = b$ and where r and a are constants.

If we construct small examples for Exercise 4.2-3, we see that $\emptyset$ has only one subset, $\{1\}$ has two subsets, $\{1, 2\}$ has four subsets, and $\{1, 2, 3\}$ has eight subsets. This gives us a good guess as to what the general formula is, but to prove it, we will need to think recursively. Consider the subsets of $\{1, 2, 3\}$:

$$\emptyset \quad \{1\} \quad \{2\} \quad \{1, 2\}$$
$$\{3\} \quad \{1, 3\} \quad \{2, 3\} \quad \{1, 2, 3\}$$

The first four subsets do not contain 3, but the second four do. Further, the first four subsets are exactly the subsets of $\{1, 2\}$, while the second four are the four subsets of $\{1, 2\}$ with 3 added into each one. So, we get a subset of $\{1, 2, 3\}$ either by taking a subset of $\{1, 2\}$ or by adjoining 3 to a subset of $\{1, 2\}$. This suggests that the recurrence for the number of subsets of an n-element set (which we may assume is $\{1, 2, \ldots, n\}$) is

$$S(n) = \begin{cases} 2S(n-1) & \text{if } n \geq 1 , \\ 1 & \text{if } n = 0 . \end{cases} \tag{4.9}$$

To prove that this recurrence is correct, we note that the subsets of $\{1, 2, \ldots, n\}$ can be partitioned according to whether they contain element n. The subsets of $\{1, 2, \ldots, n\}$ containing element n can be constructed by adjoining the element n to the subsets not containing element n. So, the number of subsets containing element n is the same as the number of subsets not containing element n. The number of subsets not containing element n is simply the number of subsets of an $(n-1)$-element set. Therefore, each block of our partition has size equal to the number of subsets of an $(n-1)$-element set. Thus, by the sum principle, the number of subsets of $\{1, 2, \ldots, n\}$ is twice the number of subsets of $\{1, 2, \ldots, n-1\}$. This proves that $S(n) = 2S(n-1)$ if $n > 0$. We already observed that $\emptyset$ has only one subset (itself), so we have proved the correctness of Recurrence 4.9.

For Exercise 4.2-4, we can algebraically describe what the problem said in words by

$$T(n) = \left(1 + \frac{0.01p}{12}\right) \cdot T(n-1) - M ,$$

with $T(0) = A$. Note that we add $0.01p/12$ times the principal to the amount due each month, because $p/12$ percent of a number is $0.01p/12$ times the number.

Iterating a Recurrence

Turning to Exercise 4.2-5, we can substitute the right side of the equation $T(n-1) = rT(n-2) + a$ for $T(n-1)$ in our recurrence and then substitute the similar equations for $T(n-2)$ and $T(n-3)$:

$$\begin{aligned} T(n) &= r\left(rT(n-2) + a\right) + a \\ &= r^2 T(n-2) + ra + a \\ &= r^2\left(rT(n-3) + a\right) + ra + a \\ &= r^3 T(n-3) + r^2 a + ra + a \\ &= r^3\left(rT(n-4) + a\right) + r^2 a + ra + a \\ &= r^4 T(n-4) + r^3 a + r^2 a + ra + a . \end{aligned}$$

From this, we can guess that

$$T(n) = r^n T(0) + a \sum_{i=0}^{n-1} r^i$$

$$= r^n b + a \sum_{i=0}^{n-1} r^i . \tag{4.10}$$

The method we used to guess the solution is called **iterating the recurrence** because we repeatedly use the recurrence with smaller and smaller values in place of n. We could instead have written

$$T(0) = b$$

$$T(1) = r T(0) + a$$
$$= rb + a$$

$$T(2) = r T(1) + a$$
$$= r(rb + a) + a$$
$$= r^2 b + ra + a$$

$$T(3) = r T(2) + a$$
$$= r^3 b + r^2 a + ra + a ,$$

which leads us to the same guess. Why, then, have we introduced two methods? Having different approaches to solving a problem often yields insights we would not get with just one approach. For example, when we study recursion trees, we will see how to visualize the process of iterating certain kinds of recurrences to simplify the algebra involved in solving them.

Geometric Series

You may recognize the sum $\sum_{i=0}^{n-1} r^i$ in Equation 4.10. It is called a **finite geometric series with common ratio r**. The sum $\sum_{i=0}^{n-1} ar^i$ is called a **finite geometric series with common ratio r and initial value a**. Recall from algebra the factorizations

$$(1 - x)(1 + x) = 1 - x^2$$
$$(1 - x)(1 + x + x^2) = 1 - x^3$$
$$(1 - x)(1 + x + x^2 + x^3) = 1 - x^4 .$$

These factorizations are easy to verify, and they suggest that $(1 - r)(1 + r + r^2 + \cdots + r^{n-1}) = 1 - r^n$, or

$$\sum_{i=0}^{n-1} r^i = \frac{1 - r^n}{1 - r} . \tag{4.11}$$

In fact, this formula is true and lets us rewrite the formula for $T(n)$ in a very nice form.

Theorem 4.1 If $T(n) = rT(n-1) + a$, $T(0) = b$, and $r \neq 1$, then

$$T(n) = r^n b + a\frac{1 - r^n}{1 - r} \qquad (4.12)$$

for all nonnegative integers n.

Proof: We prove our formula by induction. Notice that the formula gives

$$T(0) = r^0 b + a\frac{1 - r^0}{1 - r},$$

which is b. So, the formula is true when $n = 0$. Now assume that $n > 0$ and

$$T(n-1) = r^{n-1} b + a\frac{1 - r^{n-1}}{1 - r}.$$

Then we have

$$\begin{aligned}
T(n) &= rT(n-1) + a \\
&= r\left(r^{n-1}b + a\frac{1 - r^{n-1}}{1 - r}\right) + a \\
&= r^n b + \frac{ar - ar^n}{1 - r} + a \\
&= r^n b + \frac{ar - ar^n + a - ar}{1 - r} \\
&= r^n b + a\frac{1 - r^n}{1 - r}.
\end{aligned}$$

Therefore, by the principle of mathematical induction, our formula holds for all integers n greater than or equal to 0. ∎

One possible value for r in Theorem 4.1 is 0. With $r = 0$, our recurrence gives us $T(0) = b$, so we expect Equation 4.12 to give b as well. It is standard for many reasons to define 0^0 to be 1, which is exactly what we need to make Equation 4.12 correct in this special case. We did not prove Equation 4.11. However, it is easy to use Theorem 4.1 to prove it.

Corollary 4.2 The formula for the sum of a geometric series with $r \neq 1$ is

$$\sum_{i=0}^{n-1} r^i = \frac{1 - r^n}{1 - r}. \qquad (4.13)$$

Proof: Define $T(n) = \sum_{i=0}^{n-1} r^i$ for $n > 0$ and $T(0) = 0$. Then $T(n) = r T(n-1) + 1$. Applying Theorem 4.1, with $b = 0$ and $a = 1$, gives us

$$T(n) = \frac{1 - r^n}{1 - r} \ . \ \blacksquare$$

Often, when we see a geometric series, we will only be concerned with expressing the sum in big O notation. In this case, we can show that the sum of a geometric series is at most the largest term times a constant factor, where the constant factor depends on r but not on n. For example, if $|r| < 1$, then the largest term in the sum is 1 and the numerator of $(1 - r^n)/(1 - r)$ is less than 1; so, the quotient is no more than the constant $1/(1 - r)$. Thus, the sum of the series is no more than the constant $1/(1 - r)$ times 1. In other words, the sum of the series is $O(1)$.

Lemma 4.3   Let r be a quantity whose value is independent of n and not equal to 1. Let $t(n)$ be the largest term of the geometric series

$$\sum_{i=0}^{n-1} r^i \ .$$

Then the value of the geometric series is $O\big(t(n)\big)$.

Proof: It is straightforward to see that we may limit ourselves to proving the lemma for $r > 0$. We consider two cases, depending on whether $r > 1$ or $r < 1$. If $r > 1$, then

$$\sum_{i=0}^{n-1} r^i = \frac{1 - r^n}{1 - r}$$

$$= \frac{r^n - 1}{r - 1}$$

$$\leq \frac{r^n}{r - 1}$$

$$= r^{n-1} \frac{r}{r - 1}$$

$$= O(r^{n-1}) \ .$$

On the other hand, if $r < 1$, then the largest term is $r^0 = 1$, and the sum has value

$$\frac{1 - r^n}{1 - r} < \frac{1}{1 - r} \ .$$

Thus, the sum is $O(1)$, and because $t(n) = 1$, the sum is $O\big(t(n)\big)$. $\blacksquare$

In fact, when r is nonnegative, an even stronger statement is true. Recall that we said that for two functions f and g from the real numbers to the real numbers, $f = \Theta(g)$ if $f = O(g)$ and $g = O(f)$.

| Theorem 4.4 | Let r be a nonnegative quantity whose value is independent of n and not equal to 1. Let $t(n)$ be the largest term of the geometric series |

$$\sum_{i=0}^{n-1} r^i .$$

Then the value of the geometric series is $\Theta(t(n))$.

Proof: By Lemma 4.3, we need only show that

$$t(n) = O\left(\frac{r^n - 1}{r - 1}\right).$$

Because all r^i are nonnegative, the sum $\sum_{i=0}^{n-1} r^i$ is at least as large as any of its summands. But $t(n)$ is one of these summands, so

$$t(n) = O\left(\frac{r^n - 1}{r - 1}\right). \blacksquare$$

Note from the proofs that $t(n)$ and the constant in the big O and big Θ upper bounds depend on r. We will use this theorem in subsequent sections.

First-Order Linear Recurrences

A recurrence of the form $T(n) = f(n)T(n-1) + g(n)$ is called a **first-order linear recurrence**. When $f(n)$ is a constant, such as r, the general solution is almost as easy to write as in Theorem 4.1. Iterating the recurrence gives us

$$
\begin{aligned}
T(n) &= rT(n-1) + g(n) \\
&= r\left(rT(n-2) + g(n-1)\right) + g(n) \\
&= r^2 T(n-2) + rg(n-1) + g(n) \\
&= r^2\left(rT(n-3) + g(n-2)\right) + rg(n-1) + g(n) \\
&= r^3 T(n-3) + r^2 g(n-2) + rg(n-1) + g(n) \\
&= r^3\left(rT(n-4) + g(n-3)\right) + r^2 g(n-2) + rg(n-1) + g(n) \\
&= r^4 T(n-4) + r^3 g(n-3) + r^2 g(n-2) + rg(n-1) + g(n) \\
&\;\;\vdots \\
&= r^n T(0) + \sum_{i=0}^{n-1} r^i g(n-i) .
\end{aligned}
$$

This suggests our next theorem.

| **Theorem 4.5** | For any positive constants a and r and any function g defined on the nonnegative integers, the solution to the first-order linear recurrence |

$$T(n) = \begin{cases} rT(n-1) + g(n) & \text{if } n > 0, \\ a & \text{if } n = 0, \end{cases}$$

is

$$T(n) = r^n a + \sum_{i=1}^{n} r^{n-i} g(i). \tag{4.14}$$

Proof: Let's prove this by induction.

Because the sum $\sum_{i=1}^{n} r^{n-i} g(i)$ in Equation 4.14 has no terms when $n = 0$, the formula gives $T(0) = 0$ and, so, is valid[5] when $n = 0$. We now assume that n is positive and $T(n-1) = r^{n-1}a + \sum_{i=1}^{n-1} r^{(n-1)-i} g(i)$. Using the definition of the recurrence and the inductive hypothesis, we get that

$$T(n) = rT(n-1) + g(n)$$

$$= r\left(r^{n-1}a + \sum_{i=1}^{n-1} r^{(n-1)-i} g(i)\right) + g(n)$$

$$= r^n a + \sum_{i=1}^{n-1} r^{(n-1)+1-i} g(i) + g(n)$$

$$= r^n a + \sum_{i=1}^{n-1} r^{n-i} g(i) + g(n)$$

$$= r^n a + \sum_{i=1}^{n} r^{n-i} g(i).$$

Therefore, by the principle of mathematical induction, the solution to

$$T(n) = \begin{cases} rT(n-1) + g(n) & \text{if } n > 0, \\ a & \text{if } n = 0, \end{cases}$$

is given by Equation 4.14 for all nonnegative integers n. ∎

The formula in Theorem 4.5 is a little less easy to use than that in Theorem 4.1 because it gives us a sum to compute. Fortunately, for a number of commonly occurring functions g, the sum $\sum_{i=1}^{n} r^{n-i} g(i)$ is not too hard to compute.

---

[5]Part of the definition of *summation notation* is that we assign 0 to a summation that has no terms because the value of the summation index below the summation sign is larger than the value of the summation index above the sign.

Solve the recurrence $T(n) = 4T(n-1) + 2^n$, with $T(0) = 6$.

Solve the recurrence $T(n) = 3T(n-1) + n$, with $T(0) = 10$.

For Exercise 4.2-6, we can use Equation 4.14 to write

$$T(n) = 6 \cdot 4^n + \sum_{i=1}^{n} 4^{n-i} \cdot 2^i$$

$$= 6 \cdot 4^n + 4^n \sum_{i=1}^{n} 4^{-i} \cdot 2^i$$

$$= 6 \cdot 4^n + 4^n \sum_{i=1}^{n} \left(\frac{1}{2}\right)^i$$

$$= 6 \cdot 4^n + 4^n \cdot \frac{1}{2} \cdot \sum_{i=0}^{n-1} \left(\frac{1}{2}\right)^i$$

$$= 6 \cdot 4^n + \left(1 - \left(\frac{1}{2}\right)^n\right) \cdot 4^n$$

$$= 7 \cdot 4^n - 2^n \ .$$

For Exercise 4.2-7, we begin in the same way and quickly face a bit of a surprise. Using Equation 4.14, we write

$$T(n) = 10 \cdot 3^n + \sum_{i=1}^{n} 3^{n-i} \cdot i$$

$$= 10 \cdot 3^n + 3^n \sum_{i=1}^{n} i 3^{-i}$$

$$= 10 \cdot 3^n + 3^n \sum_{i=1}^{n} i \left(\frac{1}{3}\right)^i . \tag{4.15}$$

Now we are faced with a sum that you may not recognize, a sum that has the form

$$\sum_{i=1}^{n} i x^i = x \sum_{i=1}^{n} i x^{i-1} \ ,$$

with $x = 1/3$. However, by writing it in this form, we can use calculus to recognize it as x times a derivative. In particular, using the fact that $0x^0 = 0$, we can write

$$\sum_{i=1}^{n} i x^i = x \sum_{i=0}^{n} i x^{i-1} = x \frac{d}{dx} \sum_{i=0}^{n} x^i = x \frac{d}{dx} \left(\frac{1 - x^{n+1}}{1 - x}\right) .$$

Using the formula from calculus for the derivative of a quotient, we may write

$$x \frac{d}{dx}\left(\frac{1-x^{n+1}}{1-x}\right) = x \frac{(1-x)\left(-(n+1)x^n\right) - (1-x^{n+1})(-1)}{(1-x)^2}$$

$$= \frac{nx^{n+2} - (n+1)x^{n+1} + x}{(1-x)^2}.$$

Connecting our first and last equations, we get

$$\sum_{i=1}^{n} ix^i = \frac{nx^{n+2} - (n+1)x^{n+1} + x}{(1-x)^2}. \tag{4.16}$$

Substituting $x = 1/3$ and simplifying gives us

$$\sum_{i=1}^{n} i\left(\frac{1}{3}\right)^i = -\frac{3}{2}(n+1)\left(\frac{1}{3}\right)^{n+1} - \frac{3}{4}\left(\frac{1}{3}\right)^{n+1} + \frac{3}{4}.$$

Substituting this into Equation 4.15 gives us

$$T(n) = 10 \cdot 3^n + 3^n \left(-\frac{3}{2}(n+1)\left(\frac{1}{3}\right)^{n+1} - \frac{3}{4}\left(\frac{1}{3}\right)^{n+1} + \frac{3}{4}\right)$$

$$= 10 \cdot 3^n - \frac{n+1}{2} - \frac{1}{4} + \frac{3^{n+1}}{4}$$

$$= \frac{43}{4}3^n - \frac{n+1}{2} - \frac{1}{4}.$$

The sum that arises in this exercise occurs so often that we give its formula as a theorem. Because the formula is so complicated, we prefer deriving it when we need it rather than memorizing it.[6]

Theorem 4.6	For any real number $x \neq 1$,

$$\sum_{i=1}^{n} ix^i = \frac{nx^{n+2} - (n+1)x^{n+1} + x}{(1-x)^2}.$$

Proof: The proof for this theorem was given before the statement of the theorem. ∎

---

[6]The derivation consists of recognizing the left side of the formula as x times a derivative of a geometric series, using the quotient rule for this derivative, and substituting.

Important Concepts, Formulas, and Theorems

1. *Recurrence equation or recurrence.* A *recurrence equation* for a function defined on the set of integers greater than or equal to some number b is one that tells us how to compute the nth value of a function from the $(n-1)$st value or how to compute the nth value from some or all the first $n-1$ values.

2. *Initial condition.* To specify completely a function on the basis of a recurrence, we have to give enough information about the function to get started. This information is called the *initial condition* (or the initial conditions) for the recurrence.

3. *First-order linear recurrence.* A recurrence $T(n) = f(n)T(n-1) + g(n)$ is called a *first-order linear recurrence.*

4. *Constant coefficient recurrence.* A recurrence in which $T(n)$ is expressed in terms of a sum of constant multiples of $T(k)$ for certain values $k < n$ (and perhaps another function of n) is called a *constant coefficient recurrence.*

5. *Solution to a first-order constant coefficient linear recurrence.* If $T(n) = rT(n-1) + a$, $T(0) = b$, and $r \neq 1$, then

$$T(n) = r^n b + a\frac{1 - r^n}{1 - r}$$

   for all nonnegative integers n.

6. *Finite geometric series.* A *finite geometric series with common ratio r* is a sum of the form $\sum_{i=0}^{n-1} r^i$. The formula for the sum of a geometric series with $r \neq 1$ is

$$\sum_{i=0}^{n-1} r^i = \frac{1 - r^n}{1 - r} \, .$$

7. *Big Θ bounds on the sum of a geometric series.* Let r be a nonnegative quantity whose value is independent of n and not equal to 1. Let $t(n)$ be the largest term of the geometric series

$$\sum_{i=0}^{n-1} r^i.$$

   Then the value of the geometric series is $\Theta(t(n))$.

8. *Solution to a first-order linear recurrence.* For any positive constants a and r and any function g defined on the nonnegative integers, the solution to the first-order linear recurrence

$$T(n) = \begin{cases} rT(n-1) + g(n) & \text{if } n > 0, \\ a & \text{if } n = 0, \end{cases}$$

is

$$T(n) = r^n a + \sum_{i=1}^{n} r^{n-i} g(i) \, .$$

9. *Iterating a recurrence.* We are *iterating a recurrence* when we guess its solution by

   a. using the equation that expresses $T(n)$ in terms of $T(k)$ for k smaller than n to reexpress $T(n)$ in terms of $T(k)$ for k smaller than $n - 1$,

   b. reexpressing $T(n)$ in terms of $T(k)$ for k smaller than $n - 2$, and

   c. repeating this procedure until we can guess the formula for the sum.

10. *An important sum.* For any real number $x \neq 1$,

$$\sum_{i=1}^{n} ix^i = \frac{nx^{n+2} - (n+1)x^{n+1} + x}{(1-x)^2} \, .$$

The derivation of this formula consists of recognizing the left side of the formula as x times a derivative of a geometric series, using the quotient rule for this derivative, and substituting.

Problems

1. Prove Equation 4.13 directly by induction. (Recall that $r \neq 1$.)

2. Prove Equation 4.16 directly by induction. (Assume $x \neq 1$.)

3. Solve the recurrence $M(n) = 2M(n - 1) + 2$, with a base case of $M(1) = 1$. How does it differ from the solution to Recurrence 4.8? a/h

4. Solve the recurrence $M(n) = 3M(n - 1) + 1$, with a base case of $M(1) = 1$. How does it differ from the solution to Recurrence 4.8? a/h

5. Solve the recurrence $M(n) = M(n - 1) + 2$, with a base case of $M(1) = 1$. How does it differ from the solution to Recurrence 4.8? a/h

6. There are m functions from a one-element set to the set $\{1, 2, \ldots, m\}$. How many functions are there from a two-element set to $\{1, 2, \ldots, m\}$? From a three-element set? Give a recurrence for the number $T(n)$ of functions from an n-element set to $\{1, 2, \ldots, m\}$. Solve the recurrence. a/h

7. Solve the recurrence derived in Exercise 4.2-4.

8. At the end of each year, a state fish hatchery puts 2000 fish into a lake. The number of fish in the lake at the beginning of the year doubles by the end of the year due to reproduction. Give a recurrence for the number of fish in the lake after n years, and solve the recurrence. a/h

9. Consider the recurrence $T(n) = 3T(n-1) + 1$, with the initial condition $T(0) = 2$. You could write the solution from Theorem 4.1. Instead of using the theorem, try to guess the solution from the first four values of $T(n)$ and then try to guess the solution by iterating the recurrence four times.

10. What sort of big Θ bound can you give on the value of a geometric series $1 + r + r^2 + \cdots + r^n$, with common ratio $r = 1$? a/h

11. Solve the recurrence $T(n) = 2T(n-1) + n2^n$, with the initial condition $T(0) = 1$.

12. Solve the recurrence $T(n) = 2T(n-1) + n^3 2^n$, with the initial condition $T(0) = 2$. a/h

13. Solve the recurrence $T(n) = 2T(n-1) + 3^n$, with $T(0) = 1$.

14. Solve the recurrence $T(n) = rT(n-1) + r^n$, with $T(0) = 1$. a/h

15. Solve the recurrence $T(n) = rT(n-1) + r^{2n}$, with $T(0) = 1$. (Assume that $r \neq 1$.)

16. Solve the recurrence $T(n) = rT(n-1) + s^n$, with $T(0) = 1$. (Assume that $r \neq s$.) a/h

17. Solve the recurrence $T(n) = rT(n-1) + n$, with $T(0) = 1$. (Assume that $r \neq 1$.)

18. The Fibonacci numbers are defined by the recurrence

$$T(n) = \begin{cases} T(n-1) + T(n-2) & \text{if } n > 0, \\ 1 & \text{if } n = 0 \text{ or } n = 1. \end{cases}$$

a. Write the first ten Fibonacci numbers, starting with $T(0)$.

b. Show that $((1 + \sqrt{5})/2)^n$ and $((1 - \sqrt{5})/2)^n$ are solutions to the equation $F(n) = F(n-1) + F(n-2)$.

c. Why is

$$c_1 \left(\frac{1 + \sqrt{5}}{2} \right)^n + c_2 \left(\frac{1 - \sqrt{5}}{2} \right)^n$$

a solution to the equation $F(n) = F(n-1) + F(n-2)$ for any real numbers c_1 and c_2?

d. Find constants c_1 and c_2 such that the Fibonacci numbers are given by

$$F(n) = c_1 \left(\frac{1 + \sqrt{5}}{2} \right)^n + c_2 \left(\frac{1 - \sqrt{5}}{2} \right)^n.$$

4.3 GROWTH RATES OF SOLUTIONS TO RECURRENCES

Divide and Conquer Algorithms

One of the most basic and powerful algorithmic techniques is **divide and conquer**. Consider, for example, the binary search algorithm, which we will describe in the context of guessing a number between 1 and 100. Suppose someone picks a number between 1 and 100 and allows you to ask questions of the form "Is the number greater than k?" or "Is the number equal to k?" where k is an integer you choose. Your goal is to ask as few questions as possible to get a "yes" to a question of the form "Is the number equal to k?" Why should your first question be, "Is the number greater than 50?" After asking if the number is bigger than 50, you have learned either that the number is between 1 and 50 or that the number is between 51 and 100. In either case, you have reduced your problem to one in which the range is only half as big. Thus, you have *divided* the problem into a problem that is only half as big, and you can now (recursively) *conquer* this remaining problem. (If you ask any other question, the size of one of the possible ranges of values you could end up with would be more than half the size of the original problem.) If you continue in this fashion, always cutting the problem size in half, you will reduce the problem size to 1 fairly quickly, and then you will know what the number is. Of course, if we started with a number in the range from 1 to 128, it would be easier to cut the problem size exactly in half each time, but the question doesn't sound quite so plausible then. Thus, to analyze the problem, we will assume someone asks you to figure out a number between 0 and n, where n is a power of 2.

Exercise 4.3-1   Let $T(n)$ be the number of questions in a binary search on the range of numbers between 1 and n. Assuming that n is a power of 2, give a recurrence for $T(n)$.

For Exercise 4.3-1, we get

$$T(n) = \begin{cases} T(n/2) + 1 & \text{if } n \geq 2 \\ 1 & \text{if } n = 1 . \end{cases} \tag{4.17}$$

That is, the number of questions needed to carry out binary search on n items is equal to one step (the first question) plus the time to perform binary search on the remaining $n/2$ items. Note that the base case is $T(1) = 1$ because we have to ask a question of the form "Is the number k?" when we have reduced the range of possible values to 1.

What we are really interested in is how much time it takes to use binary search in a computer program that looks for an item in an ordered list. While the number of questions gives us a feel for the amount of time, processing each question may take several steps in our computer program. The exact amount of time these steps take might depend on some factors over which we have little control, such as where portions of the list are stored. Also, we may have to deal with lists with lengths that are not a power of 2. Thus, a more realistic description of the time needed would be

$$T(n) \leq \begin{cases} T(\lceil n/2 \rceil) + C_1 & \text{if } n \geq 2, \\ C_2 & \text{if } n = 1, \end{cases} \tag{4.18}$$

where C_1 and C_2 are constants.

Note that $\lceil x \rceil$ stands for the smallest integer larger than or equal to x, whereas $\lfloor x \rfloor$ stands for the largest integer less than or equal to x. It turns out that the solution to Recurrences 4.17 and 4.18 are roughly the same, in a sense that should become clear later. For now, let's not worry about floors and ceilings and the distinction between things that take 1 unit of time and things that take no more than some constant amount of time.

Instead, let's turn to **merge sort**, another example of a divide-and-conquer algorithm. In this algorithm, we wish to sort a list of n items. Assume that the data are stored in Positions 1 through n of an array A and that n is a power of 2. If the list has only one element, we don't need to do anything to sort it. Otherwise, to sort the list, we divide A into the portions from 1 to $n/2$ and from $n/2 + 1$ to n. We recursively sort the first half, we recursively sort the second half, and then we merge the two sorted "half lists" into one sorted list. (We saw examples of one way to merge two lists in the beginning of Section 3.1.) Merge sort can be described in pseudocode as follows:

```
MergeSort(A,low,high)

// This algorithm sorts the portion of the list A between two locations given
// by the values of the variables low and high.
    if (low == high)
        return
    else
        mid = ⌊(low + high)/2⌋
        MergeSort(A,low,mid)
        MergeSort(A,mid+1,high)
        Merge the sorted lists from the previous two steps
        return
```

More details on merge sort can be found in almost any algorithms textbook. The base case (`low == high`) takes one step. The other case executes one step, makes two recursive calls on problems of size $n/2$, and then executes the merge instruction, which can be done in n steps.

Thus, we obtain the following recurrence for the running time of merge sort:

$$T(n) = \begin{cases} 2T(n/2) + n & \text{if } n > 1, \\ 1 & \text{if } n = 1. \end{cases} \tag{4.19}$$

Recurrences such as this one can be understood via the idea of a recursion tree, which we introduce next. This concept allows us to analyze recurrences that arise in divide-and-conquer algorithms, as well as those that arise in other recursive situations, such as the Tower of Hanoi.

Recursion Trees

A recursion tree for a recurrence is a visual and conceptual representation of the process of iterating the recurrence. We use several examples to introduce the idea of a recursion tree. To understand recursion trees, it is helpful to have an "algorithmic" interpretation of a recurrence. For example, ignoring for a moment the base case, we can interpret the recurrence

$$T(n) = 2T\left(\frac{n}{2}\right) + n \qquad (4.20)$$

as, "To solve a problem of size n, we must solve two problems of size $n/2$ and do n units of additional work." Similarly, we can interpret

$$T(n) = T\left(\frac{n}{4}\right) + n^2$$

as, "To solve a problem of size n, we must solve one problem of size $n/4$ and do n^2 units of additional work." We can also interpret the recurrence

$$T(n) = 3T(n - 1) + n$$

as, "To solve a problem of size n, we must solve three subproblems of size $n - 1$ and do n additional units of work."

In Figure 4.2, we draw the beginning of the recursion tree diagram for Recurrence 4.20. For now, assume n is a power of 2. We draw the diagram in levels, each level representing a level of recursion. Equivalently, each level of the diagram represents a level of iteration of the recurrence. A level of a recursion tree diagram has five parts: two on the left, one in the middle, and two on the right. On the left, we keep track of the problem size and the number of problems; in the middle, we draw the tree; and on the right, we keep track of the work done per problem and the total amount of work done on the current level. So, to begin the recursion tree diagram for Recurrence 4.20, we show, in Level 0 on the left, that we have one problem of size n. Then, by drawing a root vertex with two edges leaving it, we show in the middle that we are splitting our problem into two problems. We note on the right that we do n units of work in addition to whatever is done on the two new problems we created. Because there is only one problem on this level, the total work done on this level is n units of work. In the next level, we draw two vertices in the middle, representing the two problems into which we split our main problem, and we show on the left that we have two problems of size $n/2$.

Figure 4.2: *The initial stage of drawing a recursion tree diagram*

Number of problems	Problem size		Work per problem	Work per level
1	n		n	n
2	$n/2$			

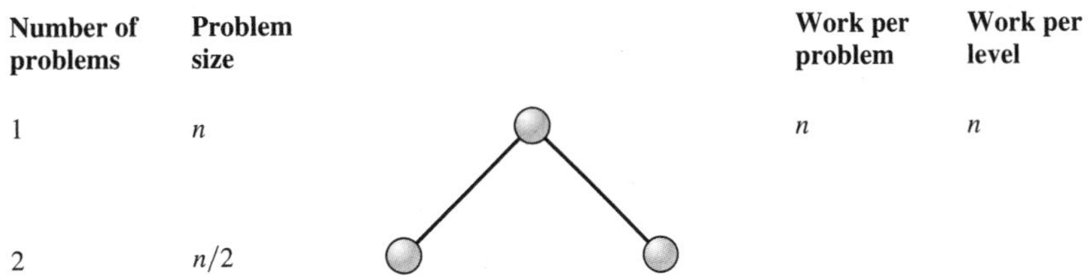

Notice how the recurrence is reflected in Levels 0 and 1 of the recursion tree. The top vertex of the tree represents $T(n)$. On the next level, we have two problems of size $n/2$, representing the recursive term $2T(n/2)$ of our recurrence. After we solve these two problems, we return to Level 0 of the tree and do n additional units of work for the nonrecursive term of the recurrence.

Now we continue to draw the tree in the same manner. Filling in the rest of Level 1 (which is the second level because the first is Level 0) and adding a few more levels, we get Figure 4.3.

Figure 4.3: *Four levels of a recursion tree diagram*

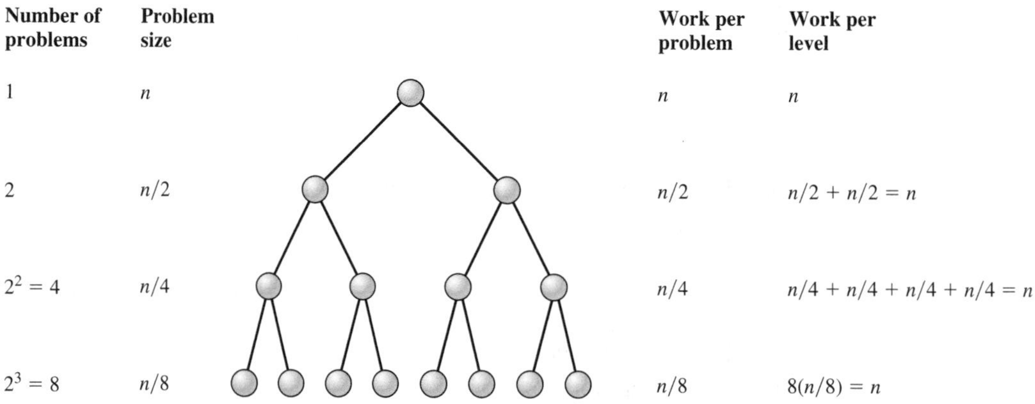

Number of problems	Problem size		Work per problem	Work per level
1	n		n	n
2	$n/2$		$n/2$	$n/2 + n/2 = n$
$2^2 = 4$	$n/4$		$n/4$	$n/4 + n/4 + n/4 + n/4 = n$
$2^3 = 8$	$n/8$		$n/8$	$8(n/8) = n$

Let us summarize what the diagram tells us so far. At Level 0 (the top level), n units of work are done. We see that at each succeeding level, we halve the problem size and double the number of subproblems. We also see that at Level 1, each of the two subproblems requires $n/2$ units of additional work; thus, a total of n units of additional work are done. Similarly, Level 2 has four subproblems of size $n/4$; thus, $4(n/4) = n$ units of additional work are done. Notice that to compute the total work done on a level, we add the amount of work done on each subproblem. When the problems all have the same size, as they do here, this is equivalent to multiplying the number of subproblems by the amount of additional work per subproblem.

To see how iteration of the recurrence is reflected in the diagram, we iterate the recurrence once to obtain

$$T(n) = 2T\left(\frac{n}{2}\right) + n$$

$$= 2\left(2T\left(\frac{n}{4}\right) + \frac{n}{2}\right) + n$$

$$= 4T\left(\frac{n}{4}\right) + n + n$$

$$= 4T\left(\frac{n}{4}\right) + 2n \ .$$

If we examine Levels 0, 1, and 2 in Figure 4.3, we see that at Level 2 we have four vertices, which represent four problems, each of size $n/4$. This corresponds to the recursive term that we obtained after iterating the recurrence. However, after we solve these problems, we return to Level 1, where we do $n/2$ additional units of work twice, and to Level 0, where we do another n additional units of work. In this way, each time we add a level to the tree, we are showing the result of one more iteration of the recurrence.

We now have enough information to describe the recursion tree diagram in general. To do this, we need to determine four things for each level:

- the number of subproblems
- the size of each subproblem
- the amount of work done per subproblem
- the total work done at that level

Once we know, for each level, the total work done at that level, we can sum over all levels to obtain the total overall work. For this purpose, we also need to figure out how many levels there are in the recursion tree.

We see that for this problem, at Level i, we have 2^i subproblems of size $n/2^i$. Further, because a problem of size 2^i requires 2^i units of additional work, there are $(2^i)(n/(2^i)) = n$ units of work done per level. To figure out how many levels there are in the tree, we notice that at each level, the problem size is cut in half, and the tree stops when the problem size is 1. Therefore, there are $\log_2 n + 1$ levels of the tree, because we start with the top level and cut the problem size in half $\log_2 n$ times.[7] We can thus visualize the whole tree in Figure 4.4.

Figure 4.4: *A finished recursion tree diagram*

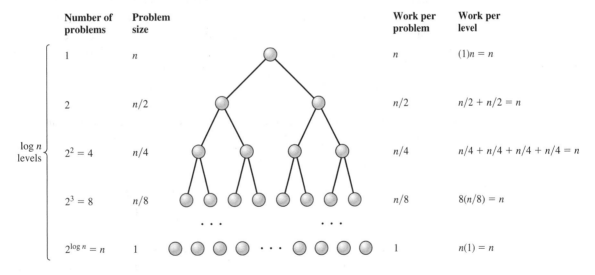

---

[7]To simplify notation for the remainder of the book, if we omit the base of a logarithm, it should be assumed to be base 2.

The computation of the work done at the bottom level is different from the other levels. In the other levels, the work is described by the recursive equation of the recurrence. At the bottom level, the work comes from the base case. Thus, we must compute the number of problems of size 1 (for this recurrence, the base case is $n = 1$) and then multiply this value by $T(1) = 1$. In the recursion tree in Figure 4.4, the number of nodes at the bottom level is $2^{\log_2 n} = n$. Because $T(1) = 1$, we do n units of work at the bottom level of the tree. But if we had chosen to say that $T(1)$ was some constant c other than 1, the work done at the bottom level would have been cn. We emphasize that the correct value of work per problem at the bottom level always comes from the base case.

The bottom level of the tree represents the final stage of iterating the recurrence. We have seen that at this level, we have n problems, each requiring work $T(1) = 1$, giving us total work n for the level. After we solve the problems represented by the bottom level, we have to do all the additional work from all the earlier levels. For this reason, we sum the work done at all the levels of the tree to get the total work done. *Iteration of the recurrence shows that the solution to the recurrence is the sum of all the work done at all the levels of the recursion tree.*

The important thing is that we now know how much work is done at each level. Once we know this, we can sum the total amount of work done over all the levels, giving us the solution to our recurrence. In this case, there are $\log_2 n + 1$ levels; at each level, the amount of work we do is n units. Thus, we conclude that the total amount of work done to solve the problem described by Recurrence 4.20 is $n(\log_2 n + 1)$.

Because 1 unit of time will vary from computer to computer, and because some kinds of work might take longer than other kinds, we are usually interested in the big Θ behavior of $T(n)$. For example, we can consider a recurrence that is identical to Recurrence 4.19, except that $T(1) = a$ for some constant a. In this case, $T(n) = an + n \log n$, because an units of work are done at Level 1, and n additional units of work are done at each of the remaining $\log n$ levels. It is still true that $T(n) = \Theta(n \log n)$, because the different base case did not change the solution to the recurrence by more than a constant factor.[8] Although recursion trees can give the exact solutions (such as $T(n) = an + n \log n$) to recurrences, our interest in the big Θ behavior of solutions will usually lead us to use a recursion tree to determine the big Θ or, in complicated cases, the big O behavior of the actual solution to the recurrence. Problem 18 explores whether the value of $T(1)$ actually influences the big Θ behavior of the solution to a recurrence that arises from a divide-and-conquer algorithm.

Let's look at one more recurrence:

$$T(n) = \begin{cases} T(n/2) + n & \text{if } n > 1, \\ 1 & \text{if } n = 1. \end{cases} \tag{4.21}$$

Again, assume n is a power of 2. We can interpret this as follows: To solve a problem of size n, we must solve one problem of size $n/2$ and do n units of additional work. Figure 4.5 shows the recursion tree diagram for this problem. We see that the problem

---

[8]More precisely, $n \log n < an + n \log n < (a + 1)n \log n$ for any $a > 0$.

Figure 4.5: *A recursion tree diagram for Recurrence 4.21*

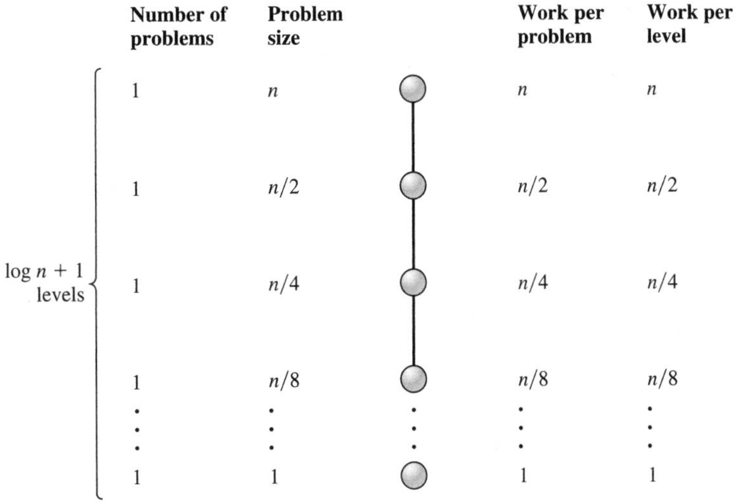

	Number of problems	Problem size		Work per problem	Work per level
	1	n		n	n
	1	$n/2$		$n/2$	$n/2$
$\log n + 1$ levels	1	$n/4$		$n/4$	$n/4$
	1	$n/8$		$n/8$	$n/8$
	⋮	⋮	⋮	⋮	⋮
	1	1		1	1

sizes are the same as in the previous tree. The remainder, however, is different. The number of subproblems does not double, rather it remains at 1 on each level. Consequently, the amount of work halves at each level. Note that there are still $\log n + 1$ levels, because the number of levels is determined by how the problem size changes, not by how many subproblems there are. So, on Level i, we have one problem of size $n/2^i$, for total work of $n/2^i$ units.

We now wish to compute how much work is done in solving a problem that gives this recurrence. Note that the additional work done is different on each level, so we have that the total amount of work is

$$n + \frac{n}{2} + \frac{n}{4} + \cdots + 2 + 1 = n \left(1 + \frac{1}{2} + \frac{1}{4} + \cdots + \left(\frac{1}{2}\right)^{\log n} \right),$$

which is n times a geometric series. By Theorem 4.4, the value of a geometric series in which the largest term is 1 is $\Theta(1)$. This implies that the work done is described by $T(n) = \Theta(n)$.

We emphasize that there is exactly one solution to Recurrence 4.21; it is the one we get by using the recurrence to compute $T(2)$ from $T(1)$, then to compute $T(4)$ from $T(2)$, and so on. Here, we have shown that $T(n) = \Theta(n)$. In fact, for the kinds of recurrences we have been examining, once we know $T(1)$, we can compute $T(n)$ for any relevant n by repeatedly using the recurrence. Thus, there is no question that solutions do exist and can, in principle, be computed for any value of n. In most applications, we are not interested in the exact form of the solution; rather, we are interested in a big O upper bound or a big Θ bound on the solution.

Use a recursion tree to find a big Θ bound for the solution to the recurrence

$$T(n) = \begin{cases} 3T(n/3) + n & \text{if } n \geq 3, \\ 1 & \text{if } n < 3. \end{cases}$$

Assume that n is a power of 3.

Use a recursion tree to solve the recurrence

$$T(n) = \begin{cases} 4T(n/2) + n & \text{if } n \geq 2, \\ 1 & \text{if } n = 1. \end{cases}$$

Assume that n is a power of 2. Convert your solution to a big Θ statement about the behavior of the solution.

Can you give a general big Θ bound for solutions to recurrences of the form $T(n) = aT(n/2) + n$ when n is a power of 2? You may have different answers for different values of a.

The recurrence in Exercise 4.3-2 is similar to the merge sort recurrence. One difference is that at each step, we divide into three problems of size $n/3$ rather than two problems of size $n/2$. Thus, we get the picture in Figure 4.6. Another difference is that the number of levels, instead of being $\log_2 n + 1$, is now $\log_3 n + 1$, so that the total work is still $\Theta(n \log n)$ units. (Note that $\log_b n = \Theta(\log_2 n)$ for any $b > 1$.)

Figure 4.6: *The recursion tree diagram for the recurrence in Exercise 4.3-2*

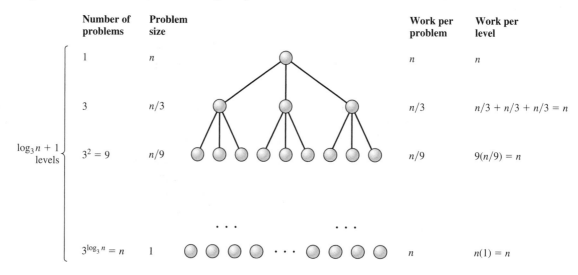

Now let's look at the recursion tree for Exercise 4.3-3. A node of size n has four children of size $n/2$, and we get Figure 4.7. Just as in the merge sort tree, there are

Figure 4.7: *The Recursion tree for Exercise 4.3-3*

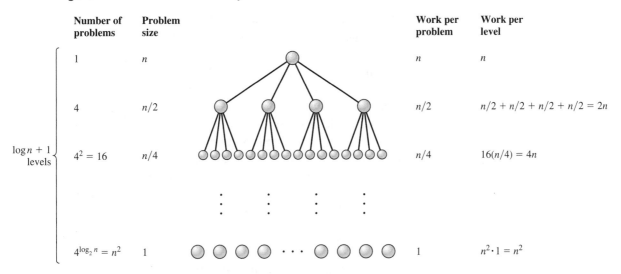

There are many ways to simplify this expression. For example, from our formula for the sum of a geometric series, we get

$$T(n) = n \sum_{i=0}^{\log n} 2^i$$

$$= n \frac{1 - 2^{(\log n)+1}}{1 - 2}$$

$$= n \frac{1 - 2n}{-1}$$

$$= 2n^2 - n$$

$$= \Theta(n^2).$$

More simply, by Theorem 4.4, we have that $T(n) = n \cdot \Theta(2^{\log n}) = \Theta(n^2)$.

Three Different Behaviors

Let's compare the recursion tree diagrams for the recurrences $T(n) = 2T(n/2) + n$, $T(n) = T(n/2) + n$, and $T(n) = 4T(n/2) + n$. Note that all three trees have depth $1 + \log_2 n$, as this is determined by the size of the subproblems relative to the parent problem, and that in each case, the size of each subproblem is half the size of the parent problem. The trees differ, however, in the amount of work done per level. For the first recurrence, the amount of work on each level is the same. In the second, the amount of work done on a level decreases as we go down the tree, with the most work being at the top level. In fact, it decreases geometrically; by Theorem 4.4, the total work done is bounded above and below by a constant multiplied by the work done at the root node. In the third recurrence, the number of nodes per level is growing at a faster rate than the problem size is decreasing, and the level with the largest amount of work is the bottom one. Again, we have a geometric series; and so, by Theorem 4.4, the total work is bounded above and below by a constant multiplied by the amount of work done at the last level.

If you understand these three cases and the differences among them, then you understand the great majority of the recursion trees that arise in algorithms.

So, to answer Exercise 4.3-4, which asks for a general big Θ bound for the solutions to recurrences of the form $T(n) = aT(n/2) + n$, we can conclude the following:

Lemma 4.7	Suppose that we have a recurrence of the form

$$T(n) = aT\left(\frac{n}{2}\right) + n ,$$

where a is a positive integer and $T(1)$ is nonnegative. Then we have the following big Θ bounds on the solution:

1. If $a < 2$, then $T(n) = \Theta(n)$.

2. If $a = 2$, then $T(n) = \Theta(n \log n)$.

3. If $a > 2$, then $T(n) = \Theta(n^{\log_2 a})$.

Proof: Cases 1 and 2 follow immediately from our earlier observations. We can verify Case 3 as follows: At Level i, we have a^i nodes, each corresponding to a problem of size $n/2^i$. Thus, at Level i, the total amount of work is $a^i(n/2^i) = n(a/2)^i$ units. Summing over the $\log_2 n$ levels, we get

$$a^{\log_2 n} T(1) + n \sum_{i=0}^{(\log n)-1} \left(\frac{a}{2}\right)^i .$$

The sum given by the summation sign is a geometric series. Therefore, because $a/2 \neq 1$, the sum will be big Θ of the largest term (see Theorem 4.4). Because $a > 2$, the largest term in this case is clearly the last one, namely, $n(a/2)^{(\log n)-1}$. Applying rules of exponents and logarithms, we get that n times the largest term is

$$n \left(\frac{a}{2} \right)^{(\log_2 n)-1} = \frac{2}{a} \cdot \frac{n \cdot a^{\log n}}{2^{\log n}}$$

$$= \frac{2}{a} \cdot \frac{n \cdot a^{\log n}}{n}$$

$$= \frac{2}{a} \cdot a^{\log n}$$

$$= \frac{2}{a} (2^{\log a})^{\log n}$$

$$= \frac{2}{a} (2^{\log n})^{\log a}$$

$$= \frac{2}{a} \cdot n^{\log a} . \tag{4.22}$$

Thus, $T(1)a^{\log n} = T(1)n^{\log a}$. Because $2/a$ and $T(1)$ are both nonnegative, the total work done is $\Theta(n^{\log_2 a})$. ∎

In fact, Lemma 4.7 holds for all positive real numbers a; we can iterate the recurrence to see this. Because a recursion tree diagram is a way to visualize iterating the recurrence when a is an integer, iteration is the natural thing to try when a is not an integer.

Notice that in the last two equalities of the computation made in Equation 4.22, we showed that $a^{\log n} = n^{\log a}$. This is a useful and, perhaps, surprising fact, so we state it (in slightly more generality) as a corollary to the proof.

Corollary 4.8   For any base b, we have $a^{\log_b n} = n^{\log_b a}$.

Important Concepts, Formulas, and Theorems

1. *Divide-and-conquer algorithm.* A *divide-and-conquer algorithm* is one that solves a problem by dividing the problem into "subproblems" that are smaller than, but otherwise of the same type as, the original one; recursively solving these subproblems; and then assembling the solution of these subproblems into a solution of the original one. Although not all problems can be solved by such a strategy, a great many problems of interest in computer science can be.

2. *Merge sort.* In *merge sort,* we sort a list of items that have some underlying order by dividing the list in half, sorting the first half (by recursively using merge sort), sorting the second half (by recursively using merge sort), and then merging the two sorted lists. For a list of length 1, merge sort returns the same list.

3. *Recursion tree diagram.* We draw a *recursion tree diagram* for a recurrence by levels, with each level representing a level of recursion. A level of a recursion

tree diagram has five parts: two on the left, one in the middle, and two on the right. On the left, we keep track of the problem size and the number of problems; in the middle, we draw the tree; and on the right, we keep track of the work done per problem and the total amount of work done on the current level. The tree has a vertex representing the initial problem and one representing each subproblem to be solved. The work done per problem at each level, other than the bottom, is given by the "additional work" part of the recurrence. The work done at the bottom level is determined by the base case of the recurrence, as is the size of a problem at the bottom level. The solution to the recurrence is the sum of the total work done at each level of the recursion tree.

4. *The base level of a recursion tree.* The amount of work done on the lowest level in a recursion tree is the number of nodes times the value given by the initial condition; it is not determined by attempting to make a computation of "additional work" done at the lowest level.

5. *Bases for logarithms.* We use $\log n$ as an alternate notation for $\log_2 n$. A fundamental fact about logarithms is that $\log_b n = \Theta(\log_2 n)$ for any real number $b > 1$.

6. *An important fact about logarithms.* For any $b > 0$, we have $a^{\log_b n} = n^{\log_b a}$.

7. *Three behaviors of solutions.* The solution to a recurrence of the form $T(n) = aT(n/2) + n$ behaves in one of the following ways:

   a. If $a < 2$, then $T(n) = \Theta(n)$.

   b. If $a = 2$, then $T(n) = \Theta(n \log n)$.

   c. If $a > 2$, then $T(n) = \Theta(n^{\log_2 a})$.

Problems

1. Draw a recursion tree diagram for

$$T(n) = \begin{cases} 4T(n/4) + n & \text{if } n \geq 2, \\ 1 & \text{if } n = 1. \end{cases}$$

Use it to find the exact solution to the recurrence. Assume n is a power of 4.

2. Draw a recursion tree diagram for

$$T(n) = \begin{cases} 2T(n/2) + 2n & \text{if } n \geq 2, \\ 2 & \text{if } n = 1. \end{cases}$$

Use it to find the exact solution to the recurrence. Assume n is a power of 2. a/h

3. Draw a recursion tree diagram for

$$T(n) = \begin{cases} 9T(n/3) + n & \text{if } n > 1, \\ 1 & \text{if } n = 1. \end{cases}$$

Use it to find a big Θ bound on the solution to the recurrence. Assume n is a power of 3. a/h

4. Draw a recursion tree diagram for

$$T(n) = \begin{cases} T(n/4) + n & \text{if } n \geq 2, \\ 1 & \text{if } n = 1. \end{cases}$$

Use it to find a big Θ bound to the solution to the recurrence. Assume n is a power of 4.

5. Draw a recursion tree diagram for

$$T(n) = \begin{cases} 2T(n/4) + n & \text{if } n \geq 2, \\ 1 & \text{if } n = 1. \end{cases}$$

Use it to find a big Θ bound on the solution to the recurrence. Assume n is a power of 4. a/h

6. Draw a recursion tree diagram for

$$T(n) = \begin{cases} 4T(n/2) + n^2 & \text{if } n \geq 2, \\ 3 & \text{if } n = 1. \end{cases}$$

Use it to find the exact solution to the recurrence. Assume n is a power of 2.

7. Draw a recursion tree diagram for

$$T(n) = \begin{cases} 3T(n/3) + 1 & \text{if } n \geq 2, \\ 2 & \text{if } n = 1. \end{cases}$$

Use it to find the exact solution to the recurrence. Assume n is a power of 3. a/h

8. Draw a recursion tree diagram for $T(n) = T(n/3) + 1$, with $T(1) = 3$. Use it to find an exact solution to the recurrence.

9. Draw recursion trees, and use them to find big Θ bounds on the solutions to the following recurrences. For each, assume that $T(1) = 1$ and that n is a power of the appropriate integer.

   a. $T(n) = 8T(n/2) + n$ a/h
   b. $T(n) = 8T(n/2) + n^3$ a/h
   c. $T(n) = 3T(n/2) + n$
   d. $T(n) = T(n/4) + 1$ a/h
   e. $T(n) = 3T(n/3) + n^2$

10. Draw recursion trees and find exact solutions to the following recurrences. For each, assume that $T(1) = 1$ and that n is a power of the appropriate integer.

   a. $T(n) = 8T(n/2) + n$ `a/h`

   b. $T(n) = 8T(n/2) + n^3$ `a/h`

   c. $T(n) = 3T(n/2) + n$

   d. $T(n) = T(n/4) + 1$ `a/h`

   e. $T(n) = 3T(n/3) + n^2$

11. Find the exact solution to Recurrence 4.21.

12. Show that $\log_b n = \Theta(\log_2 n)$ for any constant $b > 1$. `a/h`

13. Prove Corollary 4.8 by showing that $a^{\log_b n} = n^{\log_b a}$ for any $b > 0$.

14. Recursion trees work, even if the problems do not break up geometrically or if the work per level is not n^c units. Draw recursion trees and find the best big O bounds you can for solutions to the following recurrences. For each, assume that $T(1) = 1$.

   a. $T(n) = T(n-1) + n$ `a/h`

   b. $T(n) = 2T(n-1) + n$

   c. $T(n) = T(\lfloor \sqrt{n} \rfloor) + 1$ (Assume n has the form $n = 2^{2^i}$.) `a/h`

   d. $T(n) = 2T(n/2) + n \log n$ (Assume n is a power of 2.) `a/h`

15. In each case in Problem 14, is the big O bound you found a big Θ bound? `a/h`

16. If $S(n) = aS(n-1) + g(n)$ and $g(n) < c^n$ with $1 \le c < a$, how fast does $S(n)$ grow (in big Θ terms)?

17. If $S(n) = aS(n-1) + g(n)$ and $g(n) = c^n$ with $0 < a < c$, how fast does $S(n)$ grow in big Θ terms? `a/h`

18. Suppose you are given recurrences of the form $T(n) = aT(n/b) + g(n)$, with $T(1) = d > 0$ and $g(n) > 0$ for all n, and $S(n) = aS(n/b) + g(n)$, with $S(1) = 0$ (and the same a, b, and $g(n)$). Is there any difference in the big Θ behavior of the solutions to the two recurrences? What does this say about the influence of the initial condition on the big Θ behavior of such recurrences?

4.4 THE MASTER THEOREM

Master Theorem

In Section 4.3, we saw three different kinds of behavior for recurrences of the form

$$T(n) = \begin{cases} aT(n/2) + n & \text{if } n > 1, \\ d & \text{if } n = 1. \end{cases}$$

These behaviors depend upon whether $a < 2$, $a = 2$, or $a > 2$. Remember that a is the number of subproblems into which our problem is divided. Dividing by 2 cuts our problem size in half each time. The n term says that after we complete our recursive work, we have n additional units of work to do for a problem of size n. There is no reason that the amount of additional work required by each subproblem needs to be the size of the subproblem. In many applications, it will be something else. In the master theorem, which follows, we consider a more general case. Similarly, the sizes of the subproblems don't have to be half the size of the parent problem. We get the following theorem, our first version of the **master theorem**. (Later in this chapter, we will develop some stronger forms of this theorem.)

Theorem 4.9

(Master Theorem, Preliminary Version) Let a be an integer greater than or equal to 1, and let b be a real number greater than 1. Let c be a positive real number, and d, a nonnegative real number. Given a recurrence of the form

$$T(n) = \begin{cases} aT(n/b) + n^c & \text{if } n > 1, \\ d & \text{if } n = 1, \end{cases}$$

in which n is restricted to be a power of b, we get the following:

1. If $\log_b a < c$, then $T(n) = \Theta(n^c)$.

2. If $\log_b a = c$, then $T(n) = \Theta(n^c \log n)$.

3. If $\log_b a > c$, then $T(n) = \Theta(n^{\log_b a})$.

Proof: We will prove the special case $d = 1$; the case for general d is not much more difficult and is dealt with in Problem 6.

Let's think about the recursion tree for this recurrence. There will be $1 + \log_b n$ levels. At each level, the number of subproblems will be multiplied by a; so, the number of subproblems at Level i will be a^i. Each subproblem at Level i is a problem of size n/b^i. A subproblem of size n/b^i requires $(n/b^i)^c$ additional work, and because there are a^i problems on Level i, the total number of units of work on Level i is

$$a^i \left(\frac{n}{b^i} \right)^c = n^c \left(\frac{a^i}{b^{ci}} \right) = n^c \left(\frac{a}{b^c} \right)^i. \tag{4.23}$$

At the bottom level, $n/b^i = 1$ and there are a^i subproblems, each requiring 1 unit of work, so Equation 4.23 gives the work for the bottom level as well. In Lemma 4.7, the different cases for $c = 1$ occurred when the work per level was decreasing, constant, or increasing. The same analysis applies here. From our formula for work on Level i, we see that the work per level is decreasing, constant, or increasing exactly when $(a/b^c)^i$ is decreasing, constant, or increasing, respectively. These three cases depend on whether (a/b^c) is less than 1, equal to 1, or greater than 1, respectively. Now observe that

$$\left(\tfrac{a}{b^c}\right) = 1$$

$$\Leftrightarrow \quad a = b^c$$

$$\Leftrightarrow \log_b a = c \log_b b$$

$$\Leftrightarrow \quad \log_b a = c \,.$$

This shows us where the three cases in the statement of the theorem come from. Now we need to show the bound on $T(n)$ in the different cases. In the next few paragraphs, we will use the following facts (whose proofs are a straightforward application of the definition of logarithms and rules of exponents).

- For any x, y, and z, each greater than 1, we have $x^{\log_y z} = z^{\log_y x}$. (See Corollary 4.8, Problem 13 from Section 4.3, and Problem 7 at the end of this section.)

- For any $y > 0$ and any real number $x > 1$, we have $\log_x y = \Theta(\log_2 y)$. (See Problem 12 from Section 4.3.)

In general, we compute the total work done by summing the expression, given in Equation 4.23, for the work per level over all the levels. This gives

$$\sum_{i=0}^{\log_b n} n^c \left(\frac{a}{b^c}\right)^i = n^c \sum_{i=0}^{\log_b n} \left(\frac{a}{b^c}\right)^i \,.$$

In Case 1 (part 1 in the statement of the theorem), this is n^c times a geometric series with a common ratio less than 1. We now complete the proof in Case 1 and leave Cases 2 and 3 as exercises. Theorem 4.4 tells us that

$$n^c \sum_{i=0}^{\log_b n} \left(\frac{a}{b^c}\right)^i = \Theta(n^c) \,.$$

This concludes the proof of Case 1. ∎

Exercise 4.4-1   Prove Case 2 of the master theorem.

Exercise 4.4-2   Prove Case 3 of the master theorem.

In Case 2, we have that $a/b^c = 1$, and so

$$n^c \sum_{i=0}^{\log_b n} \left(\frac{a}{b^c}\right)^i = n^c \sum_{i=0}^{\log_b n} 1^i$$

$$= n^c (1 + \log_b n)$$

$$= \Theta(n^c \log n) \,.$$

In Case 3, we have that $a/b^c > 1$. So, in the series

$$\sum_{i=0}^{\log_b n} n^c \left(\frac{a}{b^c}\right)^i = n^c \sum_{i=0}^{\log_b n} \left(\frac{a}{b^c}\right)^i ,$$

the largest term is the last one. Then by Theorem 4.4, the sum is $\Theta\left(n^c(a/b^c)^{\log_b n}\right)$. But

$$n^c \left(\frac{a}{b^c}\right)^{\log_b n} = n^c \cdot \frac{a^{\log_b n}}{(b^c)^{\log_b n}}$$

$$= n^c \cdot \frac{n^{\log_b a}}{n^{\log_b b^c}}$$

$$= n^c \cdot \frac{n^{\log_b a}}{n^c}$$

$$= n^{\log_b a} .$$

Thus, the solution is $\Theta(n^{\log_b a})$.

Note that we may assume that a is a real number with $a > 1$ and give a somewhat similar proof of the master theorem (replacing the recursion tree with an iteration of the recurrence), but we do not give the details here.

Solving More General Kinds of Recurrences

Exercise 4.4-3 What can we say about the big Θ behavior of the solution to

$$T(n) = \begin{cases} 2T(n/3) + 4n^{3/2} & \text{if } n > 1 , \\ d & \text{if } n = 1 , \end{cases}$$

where n can be any nonnegative power of 3?

Exercise 4.4-4 If $f(n) = n\sqrt{n+1}$, what can we say about the big Θ behavior of solutions to

$$S(n) = \begin{cases} 2S(n/3) + f(n) & \text{if } n > 1 , \\ d & \text{if } n = 1 , \end{cases}$$

where n can be any nonnegative power of 3?

For Exercise 4.4-3, the work done at each level of the tree, except for the bottom level, will be four times the work done by the recurrence

$$T'(n) = \begin{cases} 2T'(n/3) + n^{3/2} & \text{if } n > 1 , \\ d & \text{if } n = 1 . \end{cases}$$

Thus, the work done by T will be no more than four times the work done by T' but will be larger than the work done by T'. Therefore, $T(n) = \Theta(T'(n))$. By the master theorem, because $\log_3 2 < 1 < 3/2$, we have that $T(n) = \Theta(n^{3/2})$.

For Exercise 4.4-4, because $n\sqrt{n+1} > n\sqrt{n} = n^{3/2}$, we have that $S(n)$ is at least as big as the solution to the recurrence

$$T'(n) = \begin{cases} 2T'(n/3) + n^{3/2} & \text{if } n > 1 , \\ d & \text{if } n = 1 , \end{cases}$$

where n can be any nonnegative power of 3. But the solution to the recurrence for S will be no more than the solution to the recurrence in Exercise 4.4-3 for T, because $n\sqrt{n+1} \leq 4n^{3/2}$ for $n \geq 0$. Because $T(n) = \Theta(T'(n))$, we have that $S(n) = \Theta(T'(n))$ as well. Thus $S(n) = \Theta(n^{3/2})$.

Extending the Master Theorem

As Exercises 4.4-3 and 4.4-4 suggest, there is a whole range of interesting recurrences that do not fit the preliminary version of the master theorem but are closely related to recurrences that do. These recurrences have the same kind of behavior predicted by our original version of the master theorem. However, the original version of the theorem does not apply to them, just as it does not apply to the recurrences of Exercises 4.4-3 and 4.4-4.

We now state a second version of the master theorem that covers these cases. A still stronger version of the theorem may be found in *Introduction to Algorithms* by Cormen, et al.[13]; the version here captures much of the interesting behavior of recurrences that arise from the analysis of algorithms.

| Theorem 4.10 | **(Master Theorem)** Let a and b be positive real numbers, with $a \geq 1$ and $b > 1$. Let $T(n)$ be defined for integers n that are powers of b by |

$$T(n) = \begin{cases} aT(n/b) + f(n) & \text{if } n > 1 , \\ d & \text{if } n = 1 . \end{cases}$$

Then we have the following:

1. If $f(n) = \Theta(n^c)$, where $\log_b a < c$, then $T(n) = \Theta(n^c) = \Theta(f(n))$.

2. If $f(n) = \Theta(n^c)$, where $\log_b a = c$, then $T(n) = \Theta(n^c \log n) = \Theta(f(n) \log n)$.

3. If $f(n) = \Theta(n^c)$, where $\log_b a > c$, then $T(n) = \Theta(n^{\log_b a})$.

Proof: We construct a recursion tree or iterate the recurrence. Because we have assumed that $f(n) = \Theta(n^c)$, there are constants c_1 and c_2, independent of the level, so that the work at each level is between $c_1 n^c (a/b^c)^i$ and $c_2 n^c (a/b^c)^i$. From this point on, the proof is largely a translation of the original proof. ∎

What does the master theorem tell us about the solutions to the recurrence

$$T(n) = \begin{cases} 3T(n/2) + n\sqrt{n+1} & \text{if } n > 1, \\ 1 & \text{if } n = 1? \end{cases}$$

Our solution to Exercise 4.4-4 showed us that $x\sqrt{x+1} = \Theta(x^{3/2})$. Because $2^{3/2} = \sqrt{2^3} = \sqrt{8} < 3$, we have that $\log_2 3 > 3/2$. Then, by the third conclusion of the master theorem, $T(n) = \Theta(n^{\log_2 3})$.

The remainder of this section is devoted to careful analysis of divide-and-conquer recurrences in which n is not a power of b and $T(n/b)$ is replaced by $T(\lceil n/b \rceil)$. Although the details are somewhat technical, the end result is that the big Θ behavior of such recurrences is the same as the corresponding recurrences for functions defined on powers of b. In particular, the following theorem is a consequence of what we prove.

Theorem 4.11

Let a and b be positive real numbers, with $a \geq 1$ and $b \geq 2$. Let $T(n)$ satisfy the recurrence

$$T(n) = \begin{cases} aT(\lceil n/b \rceil) + f(n) & \text{if } n > 1, \\ d & \text{if } n = 1. \end{cases}$$

Then we have the following:

1. If $f(n) = \Theta(n^c)$, where $\log_b a < c$, then $T(n) = \Theta(n^c) = \Theta(f(n))$.

2. If $f(n) = \Theta(n^c)$, where $\log_b a = c$, then $T(n) = \Theta(n^c \log n) = \Theta(f(n) \log n)$.

3. If $f(n) = \Theta(n^c)$, where $\log_b a > c$, then $T(n) = \Theta(n^{\log_b a})$.

(The condition that $b \geq 2$ can be changed to $b > 1$ with an appropriate change in the base case of the recurrence, but the base case will then depend on b. We do not prove this here.)

More General Recurrences*

So far, we have considered divide-and-conquer recurrences for functions $T(n)$ defined on integers n that are powers of b. To consider a more general recurrence in the master theorem, namely,

$$T(n) = \begin{cases} aT(\lceil n/b \rceil) + n^c & \text{if } n > 1, \\ d & \text{if } n = 1, \end{cases}$$

---

*The reader can skip this subsection without loss of continuity.

or

$$T(n) = \begin{cases} aT(\lfloor n/b \rfloor) + n^c & \text{if } n \geq 1, \\ d & \text{if } n = 0, \end{cases}$$

or even

$$T(n) = \begin{cases} a'T(\lceil n/b \rceil) + (a - a')T(\lfloor n/b \rfloor) + n^c & \text{if } n > 1, \\ d & \text{if } n = 1, \end{cases}$$

it is easiest first to extend the domain for our recurrences to a much bigger set than the nonnegative integers, either the positive real or the positive rational numbers, and then to work backward.

For example, we can write a recurrence of the form

$$t(x) = \begin{cases} f(x)t(x/b) + g(x) & \text{if } x \geq b, \\ k(x) & \text{if } 1 \leq x < b, \end{cases}$$

for two (known) functions f and g, defined on the real (or rational) numbers greater than 1, and one (known) function k defined on the real (or rational) numbers x, with $1 \leq x < b$. Then, as long as $b > 1$, it is possible to prove that there is a unique function t defined on the real (or rational) numbers greater than or equal to 1 that satisfies the recurrence. We use the lowercase t in this situation as a signal that we are considering a recurrence whose domain is the real or rational numbers greater than or equal to 1.

<table>
<tr><td>Exercise 4.4-6</td><td>

How would we compute $t(x)$ in the recurrence

$$t(x) = \begin{cases} 3t(x/2) + x^2 & \text{if } x \geq 2, \\ 5x & \text{if } 1 \leq x < 2, \end{cases}$$

if x were 7? How would we show that there is one and only one function t that satisfies the recurrence?

</td></tr>
<tr><td>Exercise 4.4-7</td><td>

Is it the case that there is one and only one solution to the recurrence

$$T(n) = \begin{cases} f(n)T(\lceil n/b \rceil) + g(n) & \text{if } n \geq 1, \\ k & \text{if } n = 1, \end{cases}$$

when f and g are (known) functions defined on the positive integers, and k and b are (known) constants with b an integer larger than or equal to 2?

</td></tr>
</table>

To compute $t(7)$ in Exercise 4.4-6, we need to know $t(7/2)$. To compute $t(7/2)$, we need to know $t(7/4)$. Because $1 < 7/4 < 2$, we know that $t(7/4) = 35/4$. Then we may write

$$t\left(\frac{7}{2}\right) = 3 \cdot \frac{35}{4} + \frac{49}{4} = \frac{154}{4} = \frac{77}{2} \, .$$

Next, we may write

$$t(7) = 3t\left(\frac{7}{2}\right) + 7^2$$

$$= 3 \cdot \frac{77}{2} + 49$$

$$= \frac{329}{2} \, .$$

Clearly we can compute $t(x)$ in this way for any x, though we are unlikely to enjoy the arithmetic. On the other hand, suppose that all we need to do is show that there is a unique value of $t(x)$ determined by the recurrence for all real numbers $x \geq 1$. If $1 \leq x < 2$, then $t(x) = 5x$, which uniquely determines $t(x)$. Given a number $x \geq 2$, there is a smallest integer i such that $x/2^i < 2$, and for this i, we have $1 \leq x/2^i$. We can now prove by induction on i that $t(x)$ is uniquely determined by the recurrence relation.

In Exercise 4.4-7, there is one and only one solution. Why? Clearly $T(1)$ is determined by the recurrence. Now assume inductively that $n > 1$ and that $T(m)$ is uniquely determined for positive integers $m < n$. We know that $n \geq 2$, so that $n/2 \leq n - 1$ (one could show this quickly by induction). Because $b \geq 2$, we know that $n/2 \geq n/b$, so that $n/b \leq n - 1$. Therefore, $\lceil n/b \rceil < n$, so that we know by the inductive hypothesis that $T(\lceil n/b \rceil)$ is uniquely determined by the recurrence. Then by the recurrence, we have that

$$T(n) = f(n)T\left(\left\lceil \frac{n}{b} \right\rceil\right) + g(n) \, ,$$

which uniquely determines $T(n)$. Thus, by the principle of mathematical induction, $T(n)$ is determined for all positive integers n.

For every realistic kind of recurrence we have dealt with, there is similarly one and only one solution. Because we know solutions exist, we don't find formulas for solutions to demonstrate that solutions exist; rather we do so to understand properties of the solutions. In this section and Section 4.3, for example, we were interested in how fast the solutions grew as n grew large. This is why we were finding big O and big Θ bounds for our solutions.

Recurrences for General n^*

We will now show how recurrences for arbitrary real numbers relate to recurrences involving floors and ceilings. We begin by showing that the conclusions of the master theorem apply to recurrences for arbitrary real numbers when we replace the real numbers with "nearby" powers of b.

---

*The reader can skip this subsection without loss of continuity.

Theorem 4.12	Let a and b be positive real numbers, with $b > 1$, and let c and d be real numbers. Let $t(x)$ be the solution to the recurrence

$$t(x) = \begin{cases} at(x/b) + x^c & \text{if } x \geq b\,, \\ d & \text{if } 1 \leq x < b\,. \end{cases}$$

Let $T(n)$ be the solution to the recurrence

$$T(n) = \begin{cases} aT(n/b) + n^c & \text{if } n \geq 0\,, \\ d & \text{if } n = 1\,, \end{cases}$$

defined when n is a nonnegative integer power of b. Let $m(x)$ be the largest integer power of b less than or equal to x. Then $t(x) = \Theta\big(T(m(x))\big)$.

Proof: If we iterate (or, in the case that a is an integer, draw recursion trees for) the two recurrences, we can see that the results of the iterations are nearly identical. This means the solutions to the recurrences have the same big Θ behavior. See the Proofs of Theorems, later in this section, for details. ∎

Removing Floors and Ceilings*

We have pointed out that a more realistic master theorem would apply to recurrences of the form $T(n) = aT(\lfloor n/b \rfloor) + n^c$, $T(n) = aT(\lceil n/b \rceil) + n^c$, or even $T(n) = a'T(\lceil n/b \rceil) + (a - a')T(\lfloor n/b \rfloor) + n^c$. For example, if we are applying merge sort to an array of size 101, we really break it into pieces of size 50 and 51. Thus, the recurrence we want is not really $T(n) = 2T(n/2) + n$ but rather $T(n) = T(\lfloor n/2 \rfloor) + T(\lceil n/2 \rceil) + n$.

We can show, however, that we can essentially ignore the floors and ceilings in typical divide-and-conquer recurrences. If we remove the floors and ceilings from a recurrence relation, we convert it from a recurrence relation defined on the integers to one defined on the rational numbers. However, we have already seen that such recurrences are not difficult to handle.

Our next theorem says that in recurrences covered by the master theorem, if we remove ceilings, our recurrences still have the same big Θ bounds on their solutions. A similar proof shows that we may remove floors and still get the same big Θ bounds. Without too much more work, we can see that we can remove floors and ceilings simultaneously without changing the big Θ bounds on our solutions. Because we may remove either floors or ceilings, we may deal with recurrences of the form $T(n) = a'T(\lceil n/b \rceil) + (a - a')T(\lfloor n/b \rfloor) + n^c$. We can replace the condition $b > 2$ with $b > 1$, but the base case for the recurrence will depend on b.

---

*The reader can skip this subsection without loss of continuity.

| **Theorem 4.13** | Let a and b be positive real numbers, with $b \geq 2$, and let c and d be real numbers. Let $T(n)$ be the function defined on the integers by the recurrence |

$$T(n) = \begin{cases} aT(\lceil n/b \rceil) + n^c & \text{if } n > 1, \\ d & \text{if } n = 1, \end{cases}$$

and let $t(x)$ be the function on the real numbers defined by the recurrence

$$t(x) = \begin{cases} at(x/b) + x^c & \text{if } x \geq b, \\ d & \text{if } 1 \leq x < b. \end{cases}$$

Then $T(n) = \Theta(t(n))$. The same statement applies with ceilings replaced by floors.

Proof: As in Theorem 4.12, we can consider iterating the two recurrences. Although dealing with the notation is difficult, it is straightforward to show that for a given value of n, the iteration for computing $T(n)$ has, at most, two more levels than the iteration for computing $t(n)$. The work per level also has the same big Θ bounds at each level, and the work for the two additional levels of the iteration for $T(n)$ has the same big Θ bounds as the work at the bottom level of the recursion tree for $t(n)$. For details, see the Proofs of Theorems at the end of this section. ∎

Theorems 4.12 and 4.13 tell us that the big Θ behavior of solutions to our more realistic recurrences

$$T(n) = \begin{cases} aT(\lceil n/b \rceil) + n^c & \text{if } n > 1, \\ d & n = 1, \end{cases}$$

is determined by their big Θ behavior on powers of the base b.

Floors and Ceilings in the Stronger Version of the Master Theorem*

We just proved Theorem 4.11. This means that to analyze the recurrence of Theorem 4.11, we can ignore the ceilings and treat n as if it were a power of b. In fact we can ignore floors and ceilings in circumstances where the function that tells us the work done at each level of our recursion tree is $\Theta(x^c)$ for some positive real number c. This lets us apply the second version of the master theorem to recurrences of the form $T(n) = aT(\lceil n/b \rceil) + f(n)$.

―――――――――――――

*The reader can skip this subsection without loss of continuity.

Theorem 4.14	Theorems 4.12 and 4.13 apply to recurrences in which the x^c or n^c term is replaced by $f(x)$ or $f(n)$ for a function f with $f(x) = \Theta(x^c)$.

Proof: We iterate the recurrences, or construct recursion trees, in the same way as in the proofs of the original theorems. We find that the condition $f(x) = \Theta(x^c)$ gives enough information to bound the solution above and below with multiples of the solution of the recurrence with x^c. The details are similar to those in the original proofs. ∎

Proofs of Theorems*

For convenience, we repeat the statements of the earlier theorems whose proofs we merely outlined.

Theorem 4.12	Let a and b be positive real numbers, with $b > 1$, and let c and d be real numbers. Let $t(x)$ be the solution to the recurrence

$$t(x) = \begin{cases} at(x/b) + x^c & \text{if } x \geq b, \\ d & \text{if } 1 \leq x < b. \end{cases}$$

Let $T(n)$ be the solution to the recurrence

$$T(n) = \begin{cases} aT(n/b) + n^c & \text{if } n \geq 0, \\ d & \text{if } n = 1, \end{cases}$$

defined for n, a nonnegative integer power of b. Let $m(x)$ be the largest integer power of b less than or equal to x. Then $t(x) = \Theta\big(T(m(x))\big)$.

Proof: By iterating each recursion four times (or using a four-level recursion tree in the case that a is an integer), we see that

$$t(x) = a^4 t\left(\frac{x}{b^4}\right) + \left(\frac{a}{b^c}\right)^3 x^c + \left(\frac{a}{b^c}\right)^2 x^c + \frac{a}{b^c} x^c$$

and

$$T(n) = a^4 T\left(\frac{n}{b^4}\right) + \left(\frac{a}{b^c}\right)^3 n^c + \left(\frac{a}{b^c}\right)^2 n^c + \frac{a}{b^c} n^c.$$

Continuing until we have a solution, in both cases, we get a solution that starts with a raised to an exponent, which we will denote as $e(x)$ or $e(n)$ when we

---

*The reader can skip this subsection without loss of continuity.

want to distinguish between them and e when it is unnecessary to distinguish. The solution for t will be

$$a^e t\left(\frac{x}{b^e}\right) + x^c \sum_{i=0}^{e-1}\left(\frac{a}{b^c}\right)^i .$$

The solution for T will be

$$a^e d + n^c \sum_{i=0}^{e-1}\left(\frac{a}{b^c}\right)^i .$$

In both cases, $t(x/b^e)$ (or $T(n/b^e)$) will be d. In both cases, the geometric series will be $\Theta(1)$, $\Theta(e)$, or $\Theta(a/b^c)^e$, depending on whether a/b^c is less than 1, equal to 1, or greater than 1. Clearly, $e(n) = \log_b n$. Suppose we want to divide x by b an integer number of times and have the result be in the range from 1 to b. Then this number of times must be greater than $\log_b(x) - 1$. Therefore, if m is the largest integer power of b less than or equal to x, then $0 \le e(x) - e(m) < 1$. If we use r to stand for the real number a/b^c, then we have $r^0 \le r^{e(x)-e(m)} < r$, or $r^{e(m)} \le r^{e(x)} \le r \cdot r^{e(m)}$. Then we have $r^{e(x)} = \Theta(r^{e(m)})$. Finally, $m^c \le x^c \le b^c m^c$, and so $x^c = \Theta(m^c)$. Thus, every term of $t(x)$ is Θ of the corresponding term of $T(m)$. Further, there are only a fixed number of different constants involved in our big Θ bounds. Therefore, because $t(x)$ is composed of sums and products of these terms, we have proved that $t(x) = \Theta(T(m))$. ∎

Theorem 4.13 Let a and b be positive real numbers, with $b \ge 2$, and let c and d be real numbers. Let $T(n)$ be the function defined on the integers by the recurrence

$$T(n) = \begin{cases} aT(\lceil n/b \rceil) + n^c & \text{if } n > 1 , \\ d & \text{if } n = 1 , \end{cases}$$

and let $t(x)$ be the function on the real numbers defined by the recurrence

$$t(x) = \begin{cases} at(x/b) + x^c & \text{if } x \ge b , \\ d & \text{if } 1 \le x < b . \end{cases}$$

Then $T(n) = \Theta(t(n))$.

Proof: As in the previous proof, we can iterate both recurrences. Let us compare the results of iterating the recurrence for $t(n)$ and the recurrence for $T(n)$ the same number of times. Note that

$$\left\lceil \frac{n}{b} \right\rceil < \frac{n}{b} + 1$$

$$\left\lceil \frac{\left\lceil \frac{n}{b} \right\rceil}{b} \right\rceil < \left\lceil \frac{n}{b^2} + \frac{1}{b} \right\rceil < \frac{n}{b^2} + \frac{1}{b} + 1$$

$$\left\lceil \frac{\left\lceil \frac{\left\lceil \frac{n}{b} \right\rceil}{b} \right\rceil}{b} \right\rceil < \left\lceil \frac{n}{b^3} + \frac{1}{b^2} + \frac{1}{b} \right\rceil < \frac{n}{b^3} + \frac{1}{b^2} + \frac{1}{b} + 1 .$$

As this suggests, if we define $n_0 = n$ and $n_i = \lceil n_{i-1}/b \rceil$, then using $b \geq 2$, it is straightforward to prove by induction, or with the formula for the sum of a geometric series, that $n_i < n/b^i + 2$. The number n_i is the argument of T in the ith iteration of the recurrence for T. We have just seen that n_i differs from the argument of t in the ith iteration of t by at most 2. In particular, to reach the base case, we might have to iterate the recurrence for T twice more than we iterate the recurrence for t. When we iterate the recurrence for t, we get the same solution we got in the previous theorem, with n substituted for x. When we iterate the recurrence for T, we get that

$$T(n) = a^j d + \sum_{i=0}^{j-1} a^i n_i^c ,$$

for some integer j, with $n/b^i \leq n_i \leq n/b^i + 2$. But, so long as $n/b^i \geq 2$, we have $n/b^i + 2 \leq n/b^{i-1}$. Because the number of iterations of T is at most two more than the number of iterations of t, and because the number of iterations of t is $\lfloor \log_b n \rfloor$, we have that j is at most $\lfloor \log_b n \rfloor + 2$. Therefore, all but perhaps the last three values of n_i are less than or equal to n/b^{i-1}. These last three values are at most b^2, b, and 1. Putting all these bounds together and using $n_0 = n$ gives us

$$\sum_{i=0}^{j-1} a^i \left(\frac{n}{b^i} \right)^c \leq \sum_{i=0}^{j-1} a^i n_i^c$$

$$\leq n^c + \sum_{i=1}^{j-4} a^i \left(\frac{n}{b^{i-1}} \right)^c + a^{j-2}(b^2)^c + a^{j-1}b^c + a^j 1^c ,$$

or

$$\sum_{i=0}^{j-1} a^i \left(\frac{n}{b^i} \right)^c \leq \sum_{i=0}^{j-1} a^i n_i^c$$

$$\leq n^c + b \sum_{i=1}^{j-4} a^i \left(\frac{n}{b^i} \right)^c + a^{j-2} \left(\frac{b^j}{b^{j-2}} \right)^c + a^{j-1} \left(\frac{b^j}{b^{j-1}} \right)^c + a^j \left(\frac{b^j}{b^j} \right)^c .$$

As we shall see, these last three "extra" terms and the b in front of the summation sign do not change the big Θ behavior of the right side.

As in the proof of the master theorem, the big Θ behavior of the left side depends on whether a/b^c is less than 1, in which case it is $\Theta(n^c)$; equal to 1, in which case it is $\Theta(n^c \log_b n)$; or greater than 1, in which case it is $\Theta(n^{\log_b a})$. But this is exactly the big Θ behavior of the right side, because $n < b^j < nb^2$. Then $b^j = \Theta(n)$, which means that $(b^j/b^i)^c = \Theta((n/b^i)^c)$. The b in front of the summation sign does not change the big Θ behavior. Adding $a^j d$ to the middle term of the inequality to get $T(n)$ does not change this behavior. But this modified middle term is exactly $T(n)$. Because the left and right sides have the same big Θ behavior as $t(n)$, we have $T(n) = \Theta(t(n))$. ∎

Important Concepts, Formulas, and Theorems

1. *Master theorem, preliminary version.* This simplified version of the *master theorem* states: Let a be an integer greater than or equal to 1 and b be a real number greater than 1. Let c be a positive real number and d a nonnegative real number. Given a recurrence of the form

$$T(n) = \begin{cases} aT(n/b) + n^c & \text{if } n > 1, \\ d & \text{if } n = 1, \end{cases}$$

for n a power of b, we have the following:

    a. If $\log_b a < c$, then $T(n) = \Theta(n^c)$.

    b. If $\log_b a = c$, then $T(n) = \Theta(n^c \log n)$.

    c. If $\log_b a > c$, then $T(n) = \Theta(n^{\log_b a})$.

2. *Properties of logarithms.* For any x, y, and z, each greater than 1, we have that $x^{\log_y z} = z^{\log_y x}$. Also, $\log_x y = \Theta(\log_2 y)$ if x is a constant.

3. *Master theorem, final version.* Let a and b be positive real numbers, with $a \geq 1$ and $b \geq 2$. Let $T(n)$ be defined for integers n that are powers of b by

$$T(n) = \begin{cases} aT(n/b) + f(n) & \text{if } n > 1, \\ d & \text{if } n = 1. \end{cases}$$

Then we have the following:

    a. If $f(n) = \Theta(n^c)$, where $\log_b a < c$, then $T(n) = \Theta(n^c) = \Theta(f(n))$.

    b. If $f(n) = \Theta(n^c)$, where $\log_b a = c$, then $T(n) = \Theta(n^c \log n) = \Theta(f(n) \log n)$.

    c. If $f(n) = \Theta(n^c)$, where $\log_b a > c$, then $T(n) = \Theta(n^{\log_b a})$.

A similar result with a base case that depends on b holds when $1 < b < 2$.

4. *Important recurrences have unique solutions (optional).* The recurrence

$$T(n) = \begin{cases} f(n)T(\lceil n/b \rceil) + g(n) & \text{if } n > 1, \\ k & \text{if } n = 1, \end{cases}$$

has a unique solution when f and g are (known) functions defined on the positive integers and k and b are (known) constants with b an integer greater than or equal to 2.

5. *Recurrences defined on the positive real numbers and recurrences defined on the positive integers (optional).* Let a and b be positive real numbers with $b > 1$. Let c and d be real numbers. Let $t(x)$ be the solution to the recurrence

$$t(x) = \begin{cases} at(x/b) + x^c & \text{if } x \geq b, \\ d & \text{if } 1 \leq x < b. \end{cases}$$

Let $T(n)$ be the solution to the recurrence

$$T(n) = \begin{cases} aT(n/b) + n^c & \text{if } n \geq 0, \\ d & \text{if } n = 1, \end{cases}$$

where n is a nonnegative integer power of b. Let $m(x)$ be the largest integer power of b less than or equal to x. Then $t(x) = \Theta\big(T(m(x))\big)$.

6. *Removing floors and ceilings from recurrences (optional).* Let a and b be positive real numbers with $b \geq 2$, and let c and d be real numbers. Let $T(n)$ be the function defined on the integers by the recurrence

$$T(n) = \begin{cases} aT(\lceil n/b \rceil) + n^c & \text{if } n > 1, \\ d & \text{if } n = 1, \end{cases}$$

and let $t(x)$ be the function on the real numbers defined by the recurrence

$$t(x) = \begin{cases} at(x/b) + x^c & \text{if } x \geq b, \\ d & \text{if } 1 \leq x < b. \end{cases}$$

Then $T(n) = \Theta(t(n))$. The same statement applies with ceilings replaced by floors.

7. *Extending Theorems 4.12 and 4.13 (optional).* In Theorems 4.12 and 4.13, summarized in 5 and 6 above, the n^c or x^c term may be replaced by a function f with $f(x) = \Theta(x^c)$.

8. *Solutions to realistic recurrences.* Theorems 4.12 and 4.13, summarized in 5, 6, and 7 above, tell us that the big Θ behavior of solutions to our more realistic recurrences

$$T(n) = \begin{cases} aT(\lceil n/b \rceil) + f(n) & \text{if } n > 1, \\ d & \text{if } n = 1, \end{cases}$$

where $f(n) = \Theta(n^c)$, is determined by their big Θ behavior on powers of the base b and with $f(n) = n^c$.

9. *A more general master theorem.* Let a and b be positive real numbers with $a > 1$ and $b > 2$. Let $T(n)$ satisfy the recurrence

$$T(n) = \begin{cases} aT(\lceil n/b \rceil) + f(n) & \text{if } n > 1, \\ d & \text{if } n = 1. \end{cases}$$

Then we have the following:

a. If $f(n) = \Theta(n^c)$, where $\log_b a < c$, then $T(n) = \Theta(n^c) = \Theta(f(n))$.

b. If $f(n) = \Theta(n^c)$, where $\log_b a = c$, then $T(n) = \Theta(n^c \log n) = \Theta(f(n) \log n)$.

c. If $f(n) = \Theta(n^c)$, where $\log_b a > c$, then $T(n) = \Theta(n^{\log_b a})$.

Problems

1. Use the master theorem to give big Θ bounds on the solutions to the following recurrences. For each, assume that $T(1) = 1$ and that n is a power of the appropriate integer.

a. $T(n) = 8T(n/2) + n$ a/h

b. $T(n) = 8T(n/2) + n^3$ a/h

c. $T(n) = 3T(n/2) + n$

d. $T(n) = T(n/4) + 1$ a/h

e. $T(n) = 3T(n/3) + n^2$

2. Give a big Θ bound on the solution to the recurrence

$$T(n) = \begin{cases} 3T(\lceil n/2 \rceil) + \sqrt{n+3} & \text{if } n > 1, \\ d & \text{if } n = 1. \end{cases}$$

3. Give a big Θ bound on the solution to the recurrence

$$T(n) = \begin{cases} 3T(\lceil n/2 \rceil) + \sqrt{n^3+3} & \text{if } n > 1, \\ d & \text{if } n = 1. \text{ a/h} \end{cases}$$

4. Give a big Θ bound on the solution to the recurrence

$$T(n) = \begin{cases} 3T(\lceil n/2 \rceil) + \sqrt{n^4+3} & \text{if } n > 1, \\ d & \text{if } n = 1. \end{cases}$$

5. Give a big Θ bound on the solution to the recurrence

$$T(n) = \begin{cases} 2T(\lceil n/2 \rceil) + \sqrt{n^2 + 3} & \text{if } n > 1, \\ d & \text{if } n = 1. \text{ a/h} \end{cases}$$

6. Extend the proof of the preliminary version of the master theorem (Theorem 4.9) to the case $T(1) = d$.

7. Prove Corollary 4.8 by showing that for any x, y, and z, each greater than 1, $x^{\log_y z} = z^{\log_y x}$. a/h

*8. Show that for each real number $x \geq 0$, there is one and only one value of $t(x)$ given by the recurrence

$$t(x) = \begin{cases} 7xt(x-1) + 1 & \text{if } x \geq 1, \\ 1 & \text{if } 0 \leq x < 1. \end{cases}$$

*9. Show that for each real number $x \geq 1$, there is one and only one value of $t(x)$ given by the recurrence

$$t(x) = \begin{cases} 3xt(x/2) + x^2 & \text{if } x \geq 2, \\ 1 & \text{if } 1 \leq x < 2. \text{ a/h} \end{cases}$$

*10. How many solutions are there to the recurrence

$$T(n) = \begin{cases} f(n)T(\lceil n/b \rceil) + g(n) & \text{if } n > 1, \\ k & \text{if } n = 1, \end{cases}$$

if $b < 2$? If $b = 10/9$, with what would you replace the conditions that $n > 1$ and $T(n) = k$ if $n = 1$ to get a unique solution?

*11. Explain why Theorem 4.11 is a consequence of Theorems 4.12 and 4.13.

4.5 MORE GENERAL KINDS OF RECURRENCES

Recurrence Inequalities

The recurrences we have been working with arise from idealized descriptions of important processes in computer science. For example, in merge sort on a list of n items, we divide the list into two parts of equal size, sort each part, and then merge the two sorted parts. The time it takes to do this is the time it takes to divide the list into two parts, plus the time it takes to sort each part, plus the time it takes to merge the

*This problem depends on material from an optional subsection.

two sorted lists. We don't specify how we are dividing the list or how we are doing the merging. We assume the sorting of smaller lists is done by applying the same method to the smaller lists, unless they have size 1, in which case we do nothing. What we know is that any sensible way of dividing the list into two parts takes no more than some constant multiple of n time units (and might take no more than constant time if we do it by leaving the list in place and manipulating pointers) and that any sensible algorithm for merging two lists will take no more than some (other) constant multiple of n time units. Thus, we know that if $T(n)$ is the time it takes to apply merge sort to n data items, then there is a constant c (the sum of the two constant multiples we mentioned) such that

$$T(n) \leq 2T\left(\frac{n}{2}\right) + cn . \tag{4.24}$$

Thus, rather than leading to recurrence equations, real-world problems often lead us to **recurrence inequalities**, which are inequalities that state that $T(n)$ is less than or equal to some expression involving values of $T(m)$ for $m < n$. (We could also include inequalities with a greater than or equal to sign, but they do not arise in the applications we are studying.) A **solution** to a recurrence inequality is a function T that satisfies the inequality. For simplicity, we will expand what we mean by the word "recurrence" to include either recurrence inequalities or recurrence equations.

In Recurrence 4.24, we are implicitly assuming that T is defined only on positive integer values, and because we said we divided the list into two equal parts each time, our analysis only makes sense if we assume that n is a power of 2.

Note that there are actually infinitely many solutions to Recurrence 4.24. (For example, for any $c' < c$, the unique solution to

$$T(n) = \begin{cases} 2T(n/2) + c'n & \text{if } n \geq 2 , \\ k & \text{if } n = 1 , \end{cases} \tag{4.25}$$

satisfies Recurrence 4.24 for any constant k.) The idea that Recurrence 4.24 has infinitely many solutions while Recurrence 4.25 has exactly one solution is analogous to the idea that $x - 3 \leq 0$ has infinitely many solutions while $x - 3 = 0$ has one solution. There are several ways to show that all the solutions to Recurrence 4.24 satisfy $T(n) = O(n \log_2 n)$. In other words, no matter how we sensibly implement merge sort, we have a $O(n \log_2 n)$ time bound on how long the merge sort process takes.

The Master Theorem for Inequalities

We commented that the unique solution to Recurrence 4.24 is also a solution to Recurrence 4.25. The largest solution to $x - 3 < 0$ is 3, which is the unique solution to $x - 3 = 0$. We have a similar phenomenon with recurrences.

Theorem 4.15	Let a and b be real numbers with $a > 0$ and $b > 1$, and let f be a function from nonnegative integer powers of b to the real numbers. Suppose that T is the unique solution to the recurrence

$$T(n) = \begin{cases} aT(n/b) + f(n) & \text{if } n \geq 1, \\ k & \text{if } n = 1, \end{cases}$$

defined on nonnegative integral powers n of b, and that S is a solution to

$$S(n) \leq \begin{cases} aS(n/b) + f(n) & \text{if } n > 1, \\ k & \text{if } n = 1. \end{cases}$$

Then $S(n) \leq T(n)$ for all $n \geq 1$.

Proof: We are given that $S(1) \leq k = T(1)$. Suppose that for $j < m$ for both powers of b, we have $S(j) \leq T(j)$. Then

$$S(m) \leq S\left(\frac{m}{b}\right) + f(m) \leq T\left(\frac{m}{b}\right) + f(m) = T(m) .$$

Thus, by the principle of mathematical induction, $S(n) \leq T(n)$ for all nonnegative integral powers n of b. ∎

Corollary 4.16	**(Master Theorem for Recurrence Inequalities)** Let a and b be real numbers with $a \geq 1$ and $b > 1$, and let S be a function from nonnegative integer powers of b to the real numbers. If

$$S(n) \leq \begin{cases} aS(n/b) + f(n) & \text{if } n > 1, \\ k & \text{if } n = 1, \end{cases}$$

then the conclusions of the master theorem (Theorem 4.10) hold for S with Θ replaced by O.

Proof: Define T by replacing $\leq$ with $=$ and S with T. Then T satisfies the conclusions of the master theorem, and, by Theorem 4.15, $S(n) \leq T(n)$. ∎

This tells us immediately that all solutions to Recurrence 4.24 are $O(n \log n)$. Thus, in situations where the function $f(n)$ that tells us the additional work for a problem of size n in a divide-and-conquer algorithm satisfies one of the three cases of the master theorem, we can analyze recurrence inequalities as easily as we analyze recurrence equations. However, not all realistic recurrences satisfy the hypotheses of the master theorem. For example, if $f(n) = n \log n$, none of the three conditions of the master theorem are satisfied. We can analyze recurrence inequalities via a recursion tree diagram. The process is virtually identical to our previous use of recursion trees; however, we must keep in mind that on each level we are really computing an upper bound on the work done on that level. We can also use a variant of the method that we used in solving Exercise 4.2-2—guessing an answer (in this case an upper bound) and verifying by induction. There are some technical aspects of induction that sometimes arise in inductive proofs in this context. Because it is possible to illustrate them more easily by using familiar recurrences, we shall do that.

A Wrinkle with Induction

Exercise 4.5-1 Carefully prove by induction that for any function T defined on the nonnegative integral powers of 2, if

$$T(n) \leq 2T\left(\frac{n}{2}\right) + cn$$

for some constant c, then $T(n) = O(n \log n)$.

We wish to show that $T(n) = O(n \log n)$. From the definition of big O, we can see that we wish to show $T(n) \leq kn \log n$ for some positive constant k (so long as n is larger than some value n_0).

We will now do something that may seem rather curious: We will consider the possibility that we have a value of k for which the inequality holds. Then, in analyzing the consequences of this possibility, we will discover that there are assumptions we need to make about k in order for such a k to exist. What we will really be doing is experimenting to see how to choose k to make an inductive proof work.

We are given that $T(n) \leq 2T(n/2) + cn$ for all positive integers n that are powers of 2. We want to prove there is another positive real number $k > 0$ and an $n_0 > 0$ such that $T(n) \leq kn \log n$ for $n > n_0$. We cannot expect to have the inequality $T(n) \leq kn \log n$ hold for $n = 1$, because $\log 1 = 0$. To have $T(2) \leq k \cdot 2 \log 2 = k \cdot 2$, we must choose $k \geq T(2)/2$. This is the first assumption we must make about k. Our inductive hypothesis is that if n is a power of 2 and m is a power of 2, with $2 \leq m < n$, then $T(m) \leq km \log m$. Now $n/2 < n$, and because n is a power of 2 greater than 2, we have that $n/2 \geq 2$. By the inductive hypothesis, $T(n/2) \leq k(n/2) \log n/2$. But then

$$T(n) \leq 2T\left(\frac{n}{2}\right) + cn \leq 2k\frac{n}{2}\log\frac{n}{2} + cn$$

$$= kn \log \frac{n}{2} + cn$$

$$= kn \log n - kn \log 2 + cn$$

$$= kn \log n - kn + cn \ .$$

Recall that we are trying to show that $T(n) \leq kn \log n$; but that is not quite what the preceding inequality tells us. Rather, the inequality shows that we need to make another assumption about k—namely, $-kn + cn \leq 0$, or $k \geq c$. If both of our assumptions about k are satisfied, we will have $T(n) < kn \log n$, and we can conclude, by the principle of mathematical induction, that for all $n > 1$ (so our n_0 is 2), $T(n) \leq kn \log n$; thus, $T(n) = O(n \log n)$.

A full inductive proof that $T(n) = O(n \log n)$ is actually embedded in the preceding discussion. However, because it might not appear to everyone to be a proof, in the next paragraph we summarize our observations in a more traditional-looking proof. Be aware that some authors and teachers prefer to write their proofs in a style that shows why they make certain choices about k. You should learn how to read discussions like the one above as proofs.

We want to show that if $T(n) \leq T(n/2) + cn$, then $T(n) = O(n \log n)$. We are given a real number $c > 0$ such that $T(n) \leq 2T(n/2) + cn$ for all $n > 1$. Choose k to be larger than or equal to $T(2)/2$ and larger than or equal to c. Then

$$T(2) \leq k \cdot 2 \log 2 \,,$$

because $k \geq T(n_0)/2$ and $\log 2 = 1$. Now assume that $n > 2$ and that for m with $2 \leq m < n$, we have $T(m) \leq km \log m$. Because n is a power of 2, we have $n \geq 4$, so that $n/2$ is an m with $2 \leq m < n$. Thus, by the inductive hypothesis,

$$T\left(\frac{n}{2}\right) \leq k\frac{n}{2} \log \frac{n}{2} \,.$$

Then by the recurrence,

$$\begin{aligned}
T(n) &\leq 2k\frac{n}{2} \log \frac{n}{2} + cn \\
&= kn(\log n - 1) + cn \\
&= kn \log n + cn - kn \\
&\leq kn \log n \,,
\end{aligned}$$

because $k \geq c$. Thus, by the principle of mathematical induction, $T(n) \leq kn \log n$ for all $n > 2$, and therefore, $T(n) = O(n \log n)$.

There are three things to note about this proof. First, without the preceding discussion, the choice of k seems arbitrary. Second, without the preceding discussion, the implicit choice of 2 for the n_0 in the big O statement also seems arbitrary. Third, the constant k is chosen in terms of the previous constant c. Because c was given to us by the recurrence, we may use it in choosing the constant that we use to prove a big O statement about solutions to the recurrence. If you compare the formal proof we just gave with the informal discussion that preceded it, you will find that each step of the formal proof actually corresponds to something we said in the informal discussion. Because the informal discussion explained why we were making the choices we did, it is natural that some people prefer the informal explanation to the formal proof.

Further Wrinkles in Induction Proofs

Exercise 4.5-2 Suppose that c is a real number greater than 0. Show by induction that any solution $T(n)$ to the recurrence

$$T(n) \le T\left(\frac{n}{3}\right) + cn ,$$

with n restricted to integer powers of 3, has $T(n) = O(n)$.

Exercise 4.5-3 Suppose that c is a real number greater than 0. Show by induction that any solution $T(n)$ to the recurrence

$$T(n) \le 4T\left(\frac{n}{2}\right) + cn ,$$

with n restricted to integer powers of 2, has $T(n) = O(n^2)$.

In Exercise 4.5-2, we are given a constant c such that $T(n) \le T(n/3) + cn$ if $n > 1$. Because we want to show that $T(n) = O(n)$, we want to find two more constants n_0 and k such that $T(n) \le kn$ whenever $n > n_0$.

We will choose $n_0 = 1$ here. (This was not an arbitrary choice; it is based on observing that the condition $T(n) \le kn$ is not impossible to satisfy when $n = 1$.) To have $T(n) \le kn$ for $n = 1$, we must assume $k \ge T(1)$. Assuming inductively that $T(m) \le km$ when $1 \le m < n$, we can write

$$T(n) \le T\left(\frac{n}{3}\right) + cn$$

$$\le k\left(\frac{n}{3}\right) + cn$$

$$= kn + \left(c - \frac{2k}{3}\right)n .$$

(Note that we used $kn/3 = kn - 2kn/3$ because we wanted to compare $T(n)$ with kn.) Thus, as long as $c - 2k/3 \le 0$, that is, $k \ge (3/2)c$, we may conclude, by mathematical induction, that $T(n) \le kn$ for all $n \ge 1$. Again, the elements of an inductive proof are in the preceding discussion; you should try to learn how to read the argument we just finished as a valid inductive proof. However, we now present something that looks more like an inductive proof.

We choose k to be the maximum of $T(1)$ and $3c/2$, and we choose $n_0 = 1$. To prove by induction that $T(x) \le kx$, we begin by observing that $T(1) \le k \cdot 1$. Next we assume that $n > 1$, and we assume inductively that for m with $1 \le m < n$, we have $T(m) \le km$. Now we may write

$$T(n) \le T\left(\frac{n}{3}\right) + cn$$

$$\le \frac{kn}{3} + cn$$

$$= kn + \left(c - \frac{2k}{3}\right)n$$

$$\le kn \ ,$$

because we chose k to be at least as large as $3c/2$, making $c - 2k/3$ negative or 0. Thus, by the principle of mathematical induction, we have $T(n) \le kn$ for all $n \ge 1$, and so $T(n) = O(n)$.

Now let's analyze Exercise 4.5-3. We won't dot all the i's and cross all the t's here because there is only one major difference between this exercise and the previous one. We wish to prove that there are an n_0 and a k such that $T(n) \le kn^2$ for $n > n_0$. Assuming we have chosen n_0 and k so that the base case holds, we can bound $T(n)$ inductively by assuming that $T(m) \le km^2$ for $m < n$ and reasoning as follows:

$$T(n) \le 4T\left(\frac{n}{2}\right) + cn$$

$$\le 4\left(k\left(\frac{n}{2}\right)^2\right) + cn$$

$$= 4\left(\frac{kn^2}{4}\right) + cn$$

$$= kn^2 + cn \ .$$

To proceed as before, we would like to choose a value of k so that $cn \le 0$. But we have a problem because both c and n are always positive! We have a statement that we know is true, by the master theorem, for example, and we have a proof method (induction) that worked nicely for similar problems. So, what went wrong?

The usual way to describe the problem we are facing is that although the statement is true, it is too weak to be proved by induction. To make the inductive proof work, we have to make an inductive hypothesis that puts some sort of negative quantity, such as a term like $-kn$, into the last line of our inequality. Let's see if we can prove something that is actually stronger than we were originally trying to prove—namely, $T(n) \le k_1 n^2 - k_2 n$ for some positive constants k_1 and k_2. Proceeding as before, we get

$$T(n) \leq 4T\left(\frac{n}{2}\right) + cn$$

$$\leq 4\left(k_1\left(\frac{n}{2}\right)^2 - k_2\left(\frac{n}{2}\right)\right) + cn$$

$$= 4\left(\frac{k_1 n^2}{4} - k_2\left(\frac{n}{2}\right)\right) + cn$$

$$= k_1 n^2 - 2k_2 n + cn$$

$$= k_1 n^2 - k_2 n + (c - k_2)n \ .$$

Now we have to make $(c - k_2)n \leq 0$ for the last line to be at most $k_1 n^2 - k_2 n$. So, we choose $k_2 \geq c$. Once we pick a value of k_2, we can then choose k_1 large enough to make the base case work. Thus, we have proved inductively that $T(n) \leq k_1 n^2 - k_2 n$ for some constants k_1 and k_2; so, $T(n) = O(n^2)$.

At first glance, this approach seems paradoxical: Why is it easier to prove a stronger statement than it is to prove a weaker one? The answer is related to the nature of induction, in which the proof of $p(n)$ depends on the proof of $p(m)$ for $m < n$. Therefore, if your statement is too weak, the base case may be easier to prove, but the weakness will hinder your ability to prove the statement for larger values of n. In other words, when you want to prove something about $p(n)$, you are using $p(1) \wedge \cdots \wedge p(n-1)$. Thus, if these are stronger, they will be of greater help in proving $p(n)$. In the case above, the problem was that the statements $p(1), \ldots, p(n-1)$ were too weak, and thus we were not able to use them to prove $p(n)$. By using a stronger $p(1), \ldots, p(n-1)$, however, we were able to prove a stronger $p(n)$, one that implied the original $p(n)$ we wanted. When we give an induction proof in this way, we are using a **stronger inductive hypothesis**.

Dealing with Functions Other Than n^c

Our statement of the master theorem involved a recursive term plus an added term that was $\Theta(n^c)$. Sometimes algorithmic problems lead us to consider other kinds of functions for the added term. The most common such example is when that added function involves logarithms. For example, consider the recurrence

$$T(n) = \begin{cases} 2T(n/2) + n \log n & \text{if } n > 1, \\ 1 & \text{if } n = 1, \end{cases} \qquad (4.26)$$

where n is a power of 2. Just as before, we can draw a recursion tree; the whole methodology works, but our sums may be a little more complicated. The tree for this recurrence is shown in Figure 4.8.

This is similar to the tree for $T(n) = 2T(n/2) + n$, except that the work on Level i is $n \log(n/2^i)$ for $i \geq 2$, and, for the bottom level, it is n (the number of subproblems)

Figure 4.8: *The recursion tree for Recurrence 4.26*

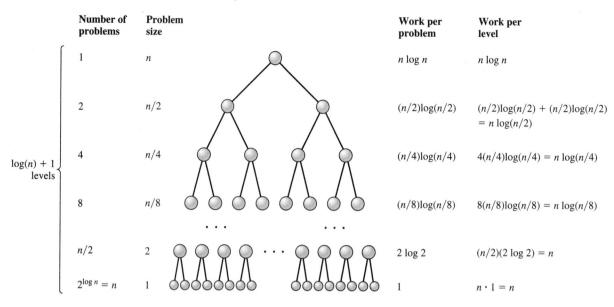

times 1. Thus, if we sum the work per level, we get

$$
\sum_{i=0}^{\log(n)-1} n \log\left(\frac{n}{2^i}\right) + n = n \left(\sum_{i=0}^{\log(n)-1} \log\left(\frac{n}{2^i}\right) + 1 \right)
$$

$$
= n \left(\sum_{i=0}^{\log(n)-1} (\log n - \log 2^i) + 1 \right)
$$

$$
= n \left(\sum_{i=0}^{\log(n)-1} \log n - \sum_{i=0}^{\log n-1} i \right) + n
$$

$$
= n \left((\log n)(\log n) - \frac{(\log n)(\log(n)-1)}{2} \right) + n
$$

$$
= O(n \log^2 n) .
$$

Notice that in the second-to-last line, there are two places where we multiplied $\log n$ by itself. Because of the 2 in the denominator, the second product will not cancel out the first (and the other terms we get by carrying out the indicated multiplications are smaller than $n \log^2 n$). Thus, our solution is in fact $\Theta(n \log^2 n)$.

Exercise 4.5-4   Find the best big O bound you can on the solution to the recurrence

$$
T(n) = \begin{cases} T(n/2) + n \log n & \text{if } n > 1 , \\ 1 & \text{if } n = 1 , \end{cases} \tag{4.27}
$$

assuming n is a power of 2. Is this bound a big Θ bound?

The tree for this recurrence is in Figure 4.9.

Figure 4.9: *The recursion tree for Recurrence 4.27*

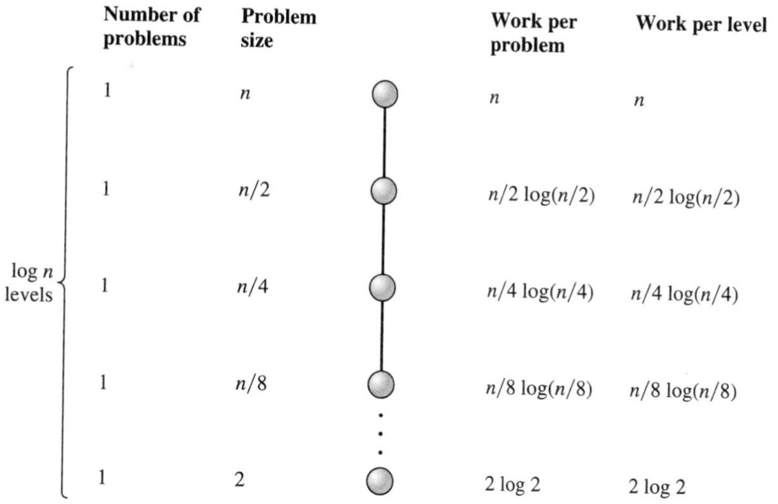

	Number of problems	Problem size		Work per problem	Work per level
	1	n	◯	n	n
log n levels	1	$n/2$	◯	$n/2 \log(n/2)$	$n/2 \log(n/2)$
	1	$n/4$	◯	$n/4 \log(n/4)$	$n/4 \log(n/4)$
	1	$n/8$	◯	$n/8 \log(n/8)$	$n/8 \log(n/8)$
			⋮		
	1	2	◯	$2 \log 2$	$2 \log 2$

Notice that the work done at the bottom nodes of the tree is determined by the statement $T(1) = 1$ in our recurrence; it is not $1 \log 1$. Summing the work, we get

$$1 + \sum_{i=0}^{\log(n)-1} \frac{n}{2^i} \log \frac{n}{2^i} = 1 + n \left(\sum_{i=0}^{\log(n)-1} \frac{1}{2^i} (\log n - \log 2^i) \right)$$

$$= 1 + n \left(\sum_{i=0}^{\log(n)-1} \left(\frac{1}{2}\right)^i (\log(n) - i) \right)$$

$$\leq 1 + n \left(\log n \sum_{i=0}^{\log(n)-1} \left(\frac{1}{2}\right)^i \right)$$

$$\leq 1 + n(\log n)(2)$$

$$= O(n \log n) .$$

Note that the largest term in the sum in our third-to-last line of equations and inequalities is $\log(n)$ and that none of the terms in the sum are negative. This means that n times the sum is at least $n \log n$. Therefore, we have that $T(n) = \Theta(n \log n)$.

Removing Ceilings and Treating Variables as Powers of b^*

In both versions of the master theorem, we showed that we could ignore ceilings and treat our variables as if they were powers of b. It might appear that the two theorems we used to show this do not apply to the more general functions we have studied in this section any more than the master theorem does. However, these two theorems actually only depend on properties of the powers n^c and not on the three different kinds of cases. Thus, we can extend them.

---

*The reader can skip this subsection without loss of continuity.

Notice that $(xb)^c = b^c x^c$ and that this proportionality holds for all values of x with constant of proportionality b^c. Putting this just a bit less precisely, we can write $(xb)^c = O(x^c)$, which suggests that we might be able to obtain big Θ bounds on $T(n)$ when T satisfies a recurrence of the form

$$T(n) = aT\left(\frac{n}{b}\right) + f(n),$$

with $f(nb) = \Theta(f(n))$. We also might be able to obtain big O bounds on T when T satisfies a recurrence of the form

$$T(n) \leq aT\left(\frac{n}{b}\right) + f(n),$$

with $f(nb) = O(f(n))$. But are these conditions satisfied by any functions of practical interest? Yes. For example, if $f(x) = \log x$, then

$$f(bx) = \log b + \log x = \Theta(\log x).$$

| Exercise 4.5-5 | Show that if $f(x) = x^2 \log x$, then $f(bx) = \Theta(f(x))$. |

| Exercise 4.5-6 | If $f(x) = 3^x$ and $b = 2$, then is $f(bx) = \Theta(f(x))$? Is $f(b(x)) = O(f(x))$? |

For Exercise 4.5-5, if $f(x) = x^2 \log x$, then

$$f(bx) = (bx)^2 \log(bx) = b^2 x^2 (\log b + \log x) = \Theta(x^2 \log x).$$

However, if $f(x) = 3^x$, then

$$f(2x) = 3^{2x} = (3^x)^2 = 3^x \cdot 3^x.$$

Because $3^x \cdot 3^x$ cannot be less than or equal to a constant multiple of 3^x, it is neither $\Theta(3^x)$ nor $O(3^x)$. Our exercises suggest that the kinds of functions that satisfy the condition $f(bx) = O(f(x))$ might include at least some of the kinds of functions of x that arise in the study of algorithms. They certainly include the power functions and, thus, polynomial functions and root functions.

There was one other property of power functions n^c that we used implicitly in both our discussion of removing floors and ceilings and our discussion of assuming that our variables were powers of b—namely, if $x > y$ (and $c \geq 0$), then $x^c \geq y^c$. A function f from the real numbers to the real numbers is called **(weakly) increasing** if whenever $x > y$, then $f(x) \geq f(y)$. Functions like $f(x) = \log x$ and $f(x) = x \log x$ are increasing functions. On the other hand, the function defined by

$$f(x) = \begin{cases} x & \text{if } x \text{ is a power of } b, \\ x^2 & \text{otherwise}, \end{cases}$$

is not increasing, even though it does satisfy the condition $f(bx) = \Theta(f(x))$.

| **Theorem 4.17** | Theorems 4.12 and 4.13 apply to recurrences in which the x^c term is replaced by an increasing function f for which $f(bx) = \Theta(f(x))$. |

Proof: We iterate the recurrences in the same way as in the proofs of the original theorems. The condition $f(bx) = \Theta(f(x))$, when applied to an increasing function, gives us enough information to give an upper and lower bound to the solution to either kind of recurrence such that each bound is a multiple of the solution of the other kind. The details are similar to those in the original proofs, so we omit them. ∎

In fact, there are versions of Theorems 4.12 and 4.13 for recurrence inequalities. Because the proofs involve a similar analysis of iterated recurrences or recursion trees, we omit them.

| **Theorem 4.18** | Let a and b be positive real numbers, with $b \geq 2$, and let $f : R^+ \to R^+$ be an increasing function such that $f(bx) = O(f(x))$. Then every solution $t(x)$ to the recurrence |

$$t(x) \leq \begin{cases} at(x/b) + f(x) & \text{if } x \geq b, \\ c & \text{if } 1 \leq x < b, \end{cases}$$

where a, b, and c are constants, satisfies $t(x) = O(h(x))$ if and only if every solution $T(n)$ to the recurrence

$$T(n) \leq \begin{cases} aT(n/b) + f(n) & \text{if } n > 1, \\ d & \text{if } n = 1, \end{cases}$$

where n is restricted to powers of b, satisfies $T(n) = O(h(n))$.

| **Theorem 4.19** | Let a and b be positive real numbers, with $b \geq 2$, and let $f : R^+ \to R^+$ be an increasing function such that $f(bx) = O(f(x))$. Then every solution $T(n)$ to the recurrence |

$$T(n) \leq \begin{cases} at(\lceil n/b \rceil) + f(n) & \text{if } n > 1, \\ d & \text{if } n = 1, \end{cases}$$

satisfies $T(n) = O(h(n))$ if and only if every solution $t(x)$ to the recurrence

$$t(x) \leq \begin{cases} aT(x/b) + f(x) & \text{if } x \geq b, \\ d & \text{if } 1 \leq x < b, \end{cases}$$

satisfies $t(x) = O(h(x))$.

Important Concepts, Formulas, and Theorems

1. *Recurrence inequality. Recurrence inequalities* state that $T(n)$ is less than or equal to some expression involving values of $T(m)$ for $m < n$. A *solution* to a recurrence inequality is a function T that satisfies the inequality.

2. *Recursion trees for recurrence inequalities.* We can analyze recurrence inequalities via a recursion tree. The process is virtually identical to our previous use of recursion trees. We must, however, keep in mind that on each level, we are really computing an upper bound on the work done on that level.

3. *Discovering necessary assumptions for an inductive proof.* Suppose we are trying to prove a statement that there is a value k such that an inequality of the form $f(n) \le kg(n)$ is true or that some other statement that involves the parameter k is true. We may start an inductive proof without knowing a value for k and determine conditions on k that make the proof valid by analyzing the assumptions that we need to make in order for the inductive proof to work. When written properly, such an explanation is a valid proof.

4. *Making a stronger inductive hypothesis.* If we are trying to prove by induction a statement of the form $p(n) \Rightarrow q(n)$ and we have a statement $s(n)$ such that $s(n) \Rightarrow q(n)$, it is sometimes useful to try to prove the statement $p(n) \Rightarrow s(n)$. This process is known as proving a *stronger* statement or making a *stronger* inductive hypothesis. It sometimes works because it gives an inductive hypothesis that suffices to prove the stronger statement, even though our original statement $q(n)$ did not give an inductive hypothesis sufficient to prove the original statement. However, we must be careful in our choice of $s(n)$, because we have to be able to succeed in proving $p(n) \Rightarrow s(n)$.

5. *When the master theorem does not apply.* To deal with recurrences of the form

$$T(n) = \begin{cases} aT(\lceil n/b \rceil) + f(n) & \text{if } n > 1, \\ d & \text{if } n = 1, \end{cases}$$

where $f(n)$ is not $\Theta(n^c)$, recursion trees and iterating the recurrence are appropriate tools even though the master theorem does not apply. The same holds for recurrence inequalities.

6. *Increasing function (optional).* A function $f : R \to R$ is said to be *(weakly) increasing* if $f(x) \ge f(y)$ whenever $x > y$.

7. *Removing floors and ceilings when the master theorem does not apply (optional).* To deal with big Θ bounds with recurrences of the form

$$T(n) = \begin{cases} aT(\lceil n/b \rceil) + f(n) & \text{if } n > 1, \\ d & \text{if } n = 1, \end{cases}$$

where $f(n)$ is not $\Theta(n^c)$, we may remove floors and ceilings and replace n with powers of b if f is increasing and f satisfies the condition

$f(nb) = \Theta(f(n))$. To deal with big O bounds for a similar recurrence inequality, we may remove floors and ceilings if f is increasing and if f satisfies the condition that $f(nb) = O(f(n))$.

Problems

1. Suppose that c is a real number greater than 0. Show by induction that any solution $T(n)$ to the recurrence

$$T(n) \leq T\left(\frac{n}{4}\right) + cn ,$$

   with n restricted to integer powers of 4, has $T(n) = O(n)$. a/h

2. Prove by induction that if $T(n) \leq 4T(n/2) + n^2$, then $T(n) = O(n^2 \log n)$ (assuming n is a power of 2).

3. Show by induction that any solution to a recurrence of the form

$$T(n) \leq 2T\left(\frac{n}{3}\right) + c \log_3 n$$

   is $O(n \log_3 n)$. What happens if you replace 2 with 3? Explain why. Would it make a difference if you used a different base for the logarithm (only an intuitive explanation is needed here)? a/h

4. What happens if you replace the 2 in Problem 3 with 4? Do you still get the same big O upper bound? If not, what do you get? (*Hint:* One way to attack this is with recursion trees. It might also be helpful to ask what happens if you replace the $\log_3 n$ with 1 and then with n.)

5. Is the big O upper bound in Problem 3 actually a big Θ bound? a/h

*6. Does the conclusion of Problem 2 hold if you have the recurrence $T(n) \leq 4T(\lceil n/2 \rceil) + n^2$ and do not require that n is a power of 2?

7. a. Find the best big O upper bound you can to any solution to the recurrence

$$T(n) = \begin{cases} 4T(n/2) + n \log n & \text{if } n > 1 , \\ 1 & \text{if } n = 1 . \end{cases}$$ a/h

   b. Assuming that you were able to guess the result you got in part a, prove by induction that your answer is correct. a/h

8. Is the big O upper bound in Problem 7 actually a big Θ bound? a/h

---

*This problem depends on material from an optional subsection marked with an asterisk.

9. Show by induction that

$$T(n) = \begin{cases} 8T(n/2) + n \log n & \text{if } n > 1, \\ d & \text{if } n = 1, \end{cases}$$

has $T(n) = O(n^3)$ for any solution $T(n)$.

10. Is the big O upper bound in Problem 9 actually a big Θ bound?

*11. Give an example (different from any in the text) of a function for which $f(bx) = O(f(x))$. Give an example (different from any in the text) of a function for which $f(bx)$ is not $O(f(x))$. a/h

12. Give the best big O upper bound you can for the solution to the recurrence

$$T(n) = 2T\left(\frac{n}{3} - 3\right) + n$$

(making an informed guess is not a bad idea here). Then prove by induction that your upper bound is correct.

13. Find the best big O upper bound you can to any solution to the recurrence defined on nonnegative integers by

$$T(n) \leq 2T\left(\left\lceil \frac{n}{2} \right\rceil + 1\right) + cn .$$

(There is nothing wrong with informed guesswork.) Prove by induction that your answer is correct. a/h

4.6 RECURRENCES AND SELECTION

The Idea of Selection

One common problem that arises in algorithms is that of **selection**. In this situation, we are given n distinct data items from some set that has an underlying order. That is, given any two items a and b from that set, we can determine whether $a < b$. (Integers satisfy this property, but colors do not.) Given these n items and some value i with $1 \leq i \leq n$, we are asked to find the ith-smallest item in the set. For example, in the set

$$S = \{3, 2, 8, 6, 4, 11, 7\} , \tag{4.28}$$

the first smallest ($i = 1$) is 2, the third smallest ($i = 3$) is 4, and the seventh smallest ($i = n = 7$) is 11. An important special case is that of finding the **median**, which is the case of $i = \lceil n/2 \rceil$. Another important special case is finding percentiles; for example, the 90th percentile is the case $i = \lceil 0.9n \rceil$. As this suggests, i is frequently given as some fraction of n.

---

*This problem depends on material from a subsection marked with an asterisk.

How do you find the minimum ($i = 1$) or maximum ($i = n$) in a set? What is the running time? How do you find the second-smallest element? Does this approach extend to finding the ith smallest? What is the running time?

Give the fastest algorithm you can to find the median ($i = \lceil n/2 \rceil$).

In Exercise 4.6-1, the simple $O(n)$ time algorithm of going through the list and keeping track of the minimum value seen so far will suffice to find the minimum. Similarly, if we want to find the second smallest, we can go through the list once to find the smallest, remove it, and then go through the new list to find the smallest. This takes $O(n+n-1) = O(n)$ time. If we extend this to finding the ith smallest, the algorithm will take $O(in)$ time. Thus, for finding the median, this method takes $O(n^2)$ time. In fact, it takes $\Theta(n^2)$ time.

A better idea for finding the median is first to sort the items and then to take the item in position $n/2$. Because we can sort in $O(n \log n)$ time, this algorithm will take $O(n \log n)$ time. Thus, if $i = O(\log n)$, we might want to run the algorithm of the previous paragraph; otherwise, we would run this algorithm.[9]

All of these approaches, when applied to the median, take at least some multiple of $(n \log n)$ units of time.[10] The best sorting algorithms take $O(n \log n)$ time also, and one can prove every comparison-based sorting algorithm takes $\Omega(n \log n)$ time. This raises the natural question of whether it is possible to do selection any faster than sorting. In other words, is finding the median element or finding the ith-smallest element of a set significantly easier than ordering (sorting) the whole set?

A Recursive Selection Algorithm

Suppose that we magically knew how to find the median in $O(n)$ time. That is, we have a routine MagicMedian that returns the median when given a set A as input. We could then use this routine in a divide-and-conquer algorithm for Select, as follows.

```
Select(A,i,n)

// Selects the ith-smallest element in set A, where n = |A|
(1)   if (n == 1)
(2)         return the one item in A
(3)   else
(4)         p = MagicMedian(A)
(5)         Let H be the set of elements greater than p
(6)         Let L be the set of elements less than or equal to p
```

[9]We also note (for those who know about heaps) that the running time can be improved to $O(n + i \log n)$ by first creating a *heap,* which takes $O(n)$ time, and then performing a delete-min operation i times.

[10]An alternate notation for $f(x) = O\big(g(x)\big)$ is $g(x) = \Omega\big(f(x)\big)$. Notice the change in roles of f and g. In this notation, we say that all these algorithms take $\Omega(n \log n)$ time. (In analytic number theory, Ω is used in several different contexts with somewhat different meanings.)

```
(7)          if (i ≤ |L|)
(8)              Return Select(L,i,|L|)
(9)          else
(10)             Return Select(H,i-|L|,|H|)
```

By H, we do not mean the elements that come after p in the list; rather, we mean the elements of the list that are larger than p in the underlying ordering of our set. This algorithm is based on the following simple observation: If we could divide the set A into a "lower" half (L) and an "upper" half (H), then we know in which of these two sets the ith-smallest element in A will be. Namely, if $i \leq \lceil n/2 \rceil$, it will be in L, and otherwise, it will be in H. Thus, we can recursively look in one or the other set. We can easily partition the data into two sets by making one pass through the data, copying the numbers less than or equal to p into L, and copying the numbers larger than p into H.[11]

The only additional detail is that if we look in H, then we no longer look for the ith smallest. Instead, we look for the $i - \lceil n/2 \rceil$th smallest, because H is formed by removing the $\lceil n/2 \rceil$-smallest elements from A.

For example, if the input is the set given in Equation 4.28, and if $p = 6$, then the set L would be $\{3, 1, 6, 4\}$, and H would be $\{8, 11, 7\}$. If i were 2, we would recurse on the set L, with $i = 2$. On the other hand, if i were 6, we would recurse on the set H, with $i = 6 - 4 = 2$. Observe that the second-smallest element in H is 8, as is the sixth-smallest element in S.

We can express the running time of Select by the following recurrence:

$$T(n) \leq T\left(\frac{n}{2}\right) + cn \ .$$

From the master theorem, we know that any function that satisfies this recurrence has $T(n) = O(n)$.

So, we can conclude that if we already know how to find the median in linear time, we can design a divide-and-conquer algorithm that will solve the selection problem in linear time.[12] However, this is nothing to write home about (yet!).

Selection without Knowing the Median in Advance

Sometimes a knowledge of solving recurrences can help us design algorithms. What kinds of recurrences do we know about that have solutions $T(n)$ with $T(n) = O(n)$? In particular, consider recurrences of the form $T(n) \leq T(n/b) + cn$, and ask when they have solutions with $T(n) = O(n)$. Using the master theorem, we see that because $\log_b 1 = 0 < 1$ for any b, then for any b allowed by the master theorem, all solutions to this recurrence will have $T(n) = O(n)$. (Note that b does not have to be an integer.) If we let $b' = 1/b$, then we can say equivalently that as long as we can solve a problem of size n by solving (recursively) a problem of size $b'n$ for some $b' < 1$ and by doing

---

[11] We can do this more efficiently, and "in place," using the partition algorithm of quicksort.

[12] We say an algorithm runs in linear time if its running time on an input of size n is $O(n)$.

$O(n)$ additional work, our algorithm will run in $O(n)$ time. Interpreting this in the selection problem, it says that as long as we can choose p in $O(n)$ time to ensure that both L and H have size at most $b'n$, then we will have a linear-time algorithm. (You might ask, "What about actually dividing our set into L and H? Doesn't that take some time, too?" Yes it does, but we already know we can do the division into H and L in $O(n)$ time; so, if we can find p in $O(n)$ time as well, then we can do both these things in $O(n)$ time.)

In particular, suppose that we can choose p in $O(n)$ time to ensure that both L and H have size at most $(3/4)n$. Then the running time is described by the recurrence $T(n) \leq T(3n/4) + O(n)$, and we will be able to solve the selection problem in linear time.

To see why $(3/4)n$ is relevant, suppose that instead of the "black box" MagicMedian, we have a much weaker magic black box that only guarantees that it will return some number in the middle half of our set in $O(n)$ time. In other words, it will return a number that is guaranteed to be somewhere between the ($n/4$th-smallest number and the ($3n/4$th-smallest number. If we use the number given by this magic box to divide our set into H and L, then neither set will have size more than $3n/4$. We will call this black box a MagicMiddle box, and we use it in the following algorithm:

```
Select1(A, i, n)

// Selects the ith-smallest element in set A, where n = |A|
(1)    if (n == 1)
(2)        return the one item in A
(3)    else
(4)        p = MagicMiddle(A)
(5)        Let H be the set of elements greater than p
(6)        Let L be the set of elements less than or equal to p
(7)        if (i ≤ |L|)
(8)            Return Select1(L, i, |L|)
(9)        else
(10)           Return Select1(H, i − |L|, |H|)
```

The Select1 algorithm is similar to Select. The only difference is that p is now only guaranteed to be in the middle half. When we recurse in Select1, we decide whether to recurse on L or H based on whether i is less than or equal to $|L|$. The element p is called a **partition element** because it is used to partition our set A into the two sets L and H.

This is progress, because now we don't need to assume that we can find the median in order to have a linear-time algorithm; we only need to assume that we can find one number in the middle half of the set. This problem seems simpler than the original problem, and, conceptually, it is. Thus, our knowledge of which recurrences have solutions that are $O(n)$ led us toward a more plausible algorithm.

An Algorithm to Find an Element in the Middle Half

It takes a clever algorithm to find an item in the middle half of our set. We now describe such an algorithm in which we first choose a subset of the numbers and then *recursively* find the median of that subset. (The condition that $n < 60$ in Line 2 is a technical condition that will be justified later.)

```
MagicMiddle(A)

(1)   Let n = |A|
(2)   if (n < 60)
(3)       use sorting to return the median of A
(4)   else
(5)       Break A into k = ⌈n/5⌉ groups G₁,...,Gₖ
          with ⌊n/5⌋ of size 5 and perhaps one of smaller size
(6)       for i = 1 to k
(7)           find mᵢ, the median of Gᵢ (by sorting)
(8)       Let M = {m₁,...,mₖ}
(9)       return Select1(M,⌈k/2⌉,k)
```

We first give a visual description of why the median of medians is in the middle half of A in the special case where the size of A is a multiple of 5; then we prove in general that it is. Assume $|A|$ is a multiple of 5. Then $|A| = 5k$.

Consider arranging the elements as follows. List each set G_i of 5 vertically in sorted order, with the smallest element on top. Then line up all $n/5$ of these lists, with those with median less than the median of the medians on the left (and those with median larger than the median of the medians on the right). We get the picture in Figure 4.10. In this figure, the medians are in white and the median of medians is in blue. The figure includes all the inequalities that we know from the ordering information we have. We use arrows to indicate that the medians on the left are less than the median of medians and those on the right are greater than the median of medians.

Figure 4.10: *Dividing a set into n/5 parts of size 5, finding the median of each part, and finding the median of the medians*

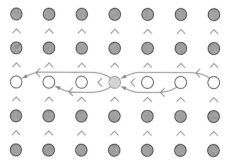

We use m^* to denote the median of medians, which is returned by MagicMiddle. We show that m^* must be in the middle half of set A when $|A|$ is large enough by considering a set S of elements guaranteed to be smaller than m^* and a set B of elements guaranteed to be bigger. Then we determine how large $|A|$ must be to ensure that $|S|$ and $|B|$ are always at least $|A|/4$.

We call the medians smaller than m^* "small medians" and those bigger "big medians." If m_i is a small median, then m_i and the two elements less than it in G_i (and thus above it in Figure 4.10) are less than m^*. The two elements above m^* in its column are less than m^*. In Figure 4.11, we draw a curve around the set S of elements.

Figure 4.11: *The enclosed elements are less than the median of the medians*

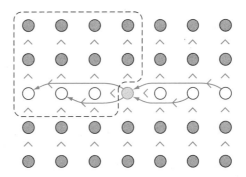

Symmetrically, every big median is larger than m^*, as are the two elements below m^* in its column. The set B of elements is enclosed in a curve in Figure 4.12.

Figure 4.12: *The enclosed elements are greater than the median of the medians*

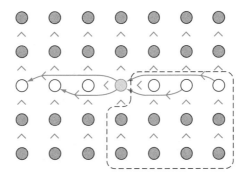

If we can choose n so that S and B each have at least one-fourth of the elements of A, we will know that m^*, which cannot be in either S or B, must be in the middle half of A. For this reason, we try to compute the sizes of S and B in terms of n.

Using k as in MagicMiddle, m^* is in column $\lceil k/2 \rceil$ of Figure 4.11. Therefore, S has three elements from each of the first $\lceil k/2 \rceil - 1$ columns and two from column $\lceil k/2 \rceil$. Using $k = n/5$, we get

$$|S| = 3 \left(\left\lceil \frac{n}{2 \cdot 5} \right\rceil - 1 \right) + 2 = 3 \left\lceil \frac{n}{10} \right\rceil - 1 .$$

Because $\lceil n/10 \rceil > n/10$, we can make $|S| \geq n/4$ by making $3n/10 - 1 \geq n/4$, which gives us $0.05n \geq 1$, or $n \geq 20$.

The number of columns to the right of column $\lceil k/2 \rceil$ is $k - \lceil k/2 \rceil$, so the size of B is

$$3 \left(k - \left\lceil \frac{k}{2} \right\rceil \right) + 2 = 3 \left(\frac{n}{5} - \left\lceil \frac{n}{10} \right\rceil \right) + 2 .$$

Because $\lceil n/10 \rceil < 1 + n/10$, we have

$$|B| > 3 \left(\frac{n}{5} - \frac{n}{10} - 1 \right) + 2 = \frac{3n}{10} - 1 .$$

Thus, we can make $|B| > n/4$ by making $0.3n - 1 \geq 0.25n$, or $0.05n \geq 1$, which gives us $n \geq 20$. Therefore, in the case where n is divisible by 5, as long as $n \geq 20$, we have that m^* is in the middle half of A. We now turn to the general case in which n need not be a multiple of 5.

| **Lemma 4.20** | The value returned by MagicMiddle(A) is in the middle half of A. |

Proof: We let m^* denote the output of MagicMiddle(A), so that m^* is the $\lceil k/2 \rceil$th element of the m_i's in sorted order. Thus, $\lceil k/2 \rceil - 1$ medians m_i are less than m^*, as are the elements in G_i less than m_i. Choose j so that $m^* \in G_j$. Then the elements of G_j less than m_j are less than m^*. However, for all but perhaps one G_i (including G_j) with $m_i \leq m^*$, there are two elements less than m_i, so that the set S' of elements less than m^* has size at least $3(\lceil k/2 \rceil - 1)$. Because k is at least $n/5$ and $\lceil n/10 \rceil \geq n/10$, we have that

$$|S'| \geq 3 \left(\left\lceil \frac{n}{10} \right\rceil - 1 \right) \geq 3 \left(\frac{n}{10} - 1 \right).$$

Thus, if we choose n so that

$$3 \left(\frac{n}{10} - 1 \right) = 0.3n - 3 \geq \frac{n}{4}, \tag{4.29}$$

we will have $S' \geq n/4$. But Equation 4.29 gives us $0.3n - 3 \geq 0.25n$, or $n \geq 60$. Now because there are $k - \lceil k/2 \rceil$ medians m_i greater than m^*, we have, as with S', that if B' is the set of elements of A larger than m^*, then B' has at least $3(k - \lceil k/2 \rceil)$ elements. Because $\lceil k/2 \rceil < k/2 + 1$, we have

$$|B| \geq 3 \left(k - \frac{k}{2} - 1 \right) = 3 \left(\frac{k}{2} - 1 \right) = 3 \left(\frac{1}{2} \left\lceil \frac{n}{5} \right\rceil - 1 \right) \geq 3 \frac{n}{10} - 3 = 0.3n - 3.$$

Thus, if we choose n so that Equation 4.29 holds—that is, so that $n \geq 60$—then we have both $|S'| > n/4$ and $|B'| > n/4$. Therefore, m^* is in the middle half of A. ■

Note that we don't actually identify all the nodes that are guaranteed to be, say, less than the median of medians; we are just guaranteed that the proper number exists.

Because we only have the guarantee that MagicMiddle gives an element in the middle half of the set if the set has at least 60 elements, we modify Select1 to start by checking whether $n < 60$ and then sorting the set to find the element in position i if $n < 60$. Because 60 is a constant, sorting and finding the desired element takes, at most, a constant amount of time.

An Analysis of the Revised Selection Algorithm

Exercise 4.6-3 Let $T(n)$ be the running time of the modified Select1 on n items. How can you express the running time of MagicMiddle in terms of $T(n)$?

Exercise 4.6-4 What is a recurrence for the running time of Select1? (*Hint:* How could Exercise 4.6-3 help you?)

Exercise 4.6-5 Can you prove by induction that each solution to the recurrence for Select1 is $O(n)$?

For Exercise 4.6-3, we have the following steps.

1. Divide the items into sets of five; this takes $O(n)$ time.

2. Find the median of each five-element set. (We can find this median by any straightforward method we choose and still only take, at most, a constant amount of time; we don't use recursion here.) There are $n/5$ sets, and we spend no more than some constant time per set, so the total time is $O(n)$.

3. Recursively call Select1 to find the median of medians; this takes $T(n/5)$ time.

4. Partition A into those elements less than or equal to the "magic middle" and those that are not, which takes $O(n)$ time.

Thus, the total running time is $T(n/5) + O(n)$, which implies that for some n_0 there is a constant $c_0 > 0$ such that the running time is no more than $c_0 n$ for all $n > n_0$. Even if $n_0 > 60$, there are only finitely many cases between 60 and n_0, which means there is a constant c such that the running time of MagicMiddle is no more than $T(n/5) + cn$ for $n \geq 60$.

We now get a recurrence for the running time of Select1. Note that for $n \geq 60$, Select1 has to call MagicMiddle and then recurse on either L or H, each of which has size at most $3n/4$. For $n < 60$, note that it takes no more than some constant amount d of time to find the median by sorting. Therefore, we get the following recurrence for the running time of Select1:

$$T(n) \leq \begin{cases} T(3n/4) + T(n/5) + c'n & \text{if } n \geq 60 \,, \\ d & \text{if } n < 60 \,. \end{cases}$$

This answers Exercise 4.6-4.

As Exercise 4.6-5 requests, we can now verify by induction that $T(n) = O(n)$. What we want to prove is that there is a constant k such that $T(n) \leq kn$. What the recurrence tells us is that there are constants c and d such that $T(n) \leq T(3n/4) + T(n/5) + cn$ if $n \geq 60$; otherwise, $T(n) \leq d$. For the base case, we have $T(n) \leq d \leq dn$ for $n < 60$, so we choose k to be at least d; then $T(n) \leq kn$ for $n < 60$. We now assume that

$n \geq 60$ and $T(m) \leq km$ for values $m < n$. We get

$$T(n) \leq T\left(\frac{3n}{4}\right) + T\left(\frac{n}{5}\right) + cn$$

$$\leq \frac{3kn}{4} + \frac{kn}{5} + cn$$

$$= \frac{19}{20kn} + cn$$

$$= kn + \left(c - \frac{k}{20}\right)n .$$

As long as $k \geq 20c$, this is at most kn; so we simply choose k this big, and by the principle of mathematical induction, we have $T(n) < kn$ for all positive integers n.

Theorem 4.21

The revised Select1 algorithm runs in time $T(n) = O(n)$.

Proof: The proof is given in the discussion of Exercises 4.6-3 through 4.6-5. ∎

Uneven Divisions

The kind of recurrence we found for the running time of Select1 is actually an instance of a more general class, which we will now explore.

Exercise 4.6-6

We already know that when $g(n) = O(n)$, every solution of $T(n) = T(n/2) + g(n)$ satisfies $T(n) = O(n)$. Use the master theorem to find a big O bound to the solution of $T(n) = T(cn) + g(n)$ for any constant $c < 1$, assuming that $g(n) = O(n)$.

Exercise 4.6-7

Use the master theorem to find big O bounds to all solutions of $T(n) = 2T(cn) + g(n)$ for any constant $c < 1/2$, assuming $g(n) = O(n)$.

Exercise 4.6-8

Suppose $g(n) = O(n)$ and you have a recurrence of the form $T(n) = T(an) + T(bn) + g(n)$ for some nonnegative constants a and b. What conditions on a and b guarantee that all solutions to this recurrence have $T(n) = O(n)$?

Using the master theorem for Exercise 4.6-6, we get $T(n) = O(n)$, because $\log_{1/c} 1 < 1$. We also get $T(n) = O(n)$ for Exercise 4.6-7, because $\log_{1/c} 2 < 1$ for $c < 1/2$. You might now guess that as long as $a + b < 1$, any solution to the recurrence $T(n) \leq T(an) + T(bn) + cn$ has $T(n) = O(n)$. We will now see why this is the case.

First, let's return to the recurrence $T(n) = T(3n/4) + T(n/5) + g(n)$, where $g(n) = O(n)$. Let's try to draw a recursion tree. This recurrence doesn't quite fit our model for recursion trees, because the two subproblems have unequal size (thus, we can't even write the problem size on the left), but we will try to draw a recursion tree in Figure 4.13 anyway and see what happens.

Figure 4.13: *Attempting a recursion tree for $T(n) = T(3/4)n + T(n/5) + g(n)$*

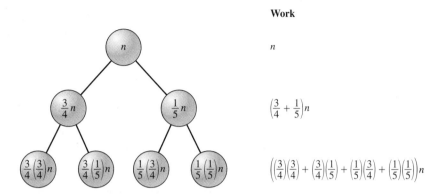

As we draw Levels 1 and 2, we see that at Level 1, we have $(3/4 + 1/5)n$ work. At Level 2, we have

$$\left(\left(\frac{3}{4}\right)^2 + 2\left(\frac{3}{4}\right)\left(\frac{1}{5}\right) + \left(\frac{1}{5}\right)^2 \right) n$$

work. Were we to work out the third level, we would see that we have

$$\left(\left(\frac{3}{4}\right)^3 + 3\left(\frac{3}{4}\right)^2\left(\frac{1}{5}\right) + 3\left(\frac{3}{4}\right)\left(\frac{1}{5}\right)^2 + \left(\frac{1}{5}\right)^3 \right) n .$$

Thus, we can see a pattern emerging. At Level 1, we have $(3/4 + 1/5)n$ work. At Level 2, we have, by the binomial theorem, $(3/4 + 1/5)^2 n$ work. At Level 3, we have, by the binomial theorem, $(3/4 + 1/5)^3 n$ work. And, similarly, at Level i, we have $((3/4) + (1/5))^i n = (19/20)^i n$ work. Thus, when we sum over all levels, we get

$$\sum_{i=0}^{O(\log n)} \left(\frac{19}{20}\right)^i n \leq \left(\frac{1}{1 - \frac{19}{20}}\right) n = 20n$$

for an upper bound on the total work. We have actually ignored one detail here. In contrast to a recursion tree in which all subproblems at a level have equal size, the "bottom" of the tree is more complicated. Different branches of the tree will reach problems of size 1 and terminate at different levels. For example, the branch that follows all 3/4s will bottom out after $\log_{4/3} n$ levels, while the one that follows all 1/5s will bottom out after $\log_5 n$ levels. However, the analysis above *overestimates the work*—that is, it assumes that nothing bottoms out until everything bottoms out, which occurs at $\log_{20/19} n$ levels. In fact, the upper bound we gave on the sum is what we would get by assuming that the recurrence never bottoms out.

We see here something general happening. It seems as if to understand a recurrence of the form $T(n) = T(an) + T(bn) + g(n)$, with $g(n) = O(n)$, we can study the simpler recurrence $T(n) = T((a + b)n) + g(n)$ instead. For a more precise formulation, see

Problem 4, which solves Exercise 4.6-8. This simplifies things enough (in particular, it lets us use the master theorem) to let us analyze a larger class of recurrences. Turning to the median algorithm, it tells us that the important thing that happened there was that the sizes of the two recursive calls, namely, $3n/4$ and $n/5$, summed to a proper fraction of n. As long as an algorithm has a recurrence of the form $T(n) = T(an)+T(bn)+g(n)$, has $a + b < 1$, and has $g(n) = O(n)$, the algorithm will work in $O(n)$ time.

Important Concepts, Formulas, and Theorems

1. *Median.* The *median* of a set (with an underlying order) of n elements is the element that would be in position $\lceil n/2 \rceil$ if the set were listed in order.

2. *Percentile.* The pth percentile of a set (with an underlying order) is the element that would be in position $\lceil (p/100)n \rceil$ if the set were listed in order.

3. *Selection.* Given an n-element set with some underlying order, the problem of *selection* of the ith-smallest element is that of finding the element that would be in the ith position if the set were listed in order. Note that i is often expressed as a fraction of n.

4. *Partition element.* A *partition element* in an algorithm is an element of a set (with an underlying order) that is used to divide the set into two parts, those that come before or are equal to the element (in the underlying order) and then the remaining elements. Notice that the order in which the set is given to the algorithm is not necessarily (in fact, not usually) the underlying order.

5. *Linear-time algorithms.* If the running time of an algorithm satisfies a recurrence of the form $T(n) \leq T(an) + cn$, with $0 \leq a < 1$, or a recurrence of the form $T(n) \leq T(an) + T(bn) + cn$, with a and b nonnegative and $a + b < 1$, then $T(n) = O(n)$.

6. *Finding a good partition element.* If a set (with an underlying order) has 60 or more elements, then the procedure of breaking the set into pieces of size 5 (plus one leftover piece, if necessary), finding the median of each piece, and then finding the median of the medians gives an element guaranteed to be in the middle half of the set.

7. *Selection algorithm.* The selection algorithm that runs in linear time sorts a set of size less than 60 to find the element in the ith position; otherwise,

   • it recursively uses the median of medians of five to find a partition element,

   • it uses that partition element to divide the set into two pieces, and

   • then it looks for the appropriate element in the appropriate piece recursively.

Problems

1. Find the best big O bound you can on $T(n)$ if it satisfies the recurrence $T(n) \leq T(n/4) + T(n/2) + n$, with $T(n) = 1$ if $n < 4$. a/h

2. In the MagicMiddle algorithm, suppose you broke the data into $n/7$ sets of size 7. What would the running time of Select1 be?

3. Let
$$T(n) = \begin{cases} T(n/3) + T(n/2) + n & \text{if } n \geq 6, \\ 1 & \text{otherwise}, \end{cases}$$

   and let
$$S(n) = \begin{cases} S(5n/6) + n & \text{if } n \geq 6, \\ 1 & \text{otherwise}. \end{cases}$$

   Draw recursion trees for T and S. What are the big O bounds on solutions to the recurrences? Use the recursion trees to argue that $T(n) \leq S(n)$ for all n. a/h

4. Suppose you are given that a and b are nonnegative real numbers, with $a + b < 1$, and that c is a nonnegative real number. Explain why it is that if $T(n) \leq T(an) + T(bn) + cn$, then $T(n) = O(n)$. Explain how this solves Exercise 4.6-8.

5. Find a big Θ bound (the best you know how to get) on solutions to the recurrence $T(n) = T(n/3) + T(n/6) + T(n/4) + n$, with $T(1) = 1$.

6. Find a big Θ bound to solutions to the recurrences $T(n/4) + T(3n/4) + dn \leq T(n) \leq T(n/4) + T(3n/4) + cn$.

7. In the MagicMiddle algorithm, suppose you broke your data into $n/3$ sets of size 3. What would the running time of Select1 be? a/h

8. Find a big O upper bound (the best you know how to get) on solutions to the recurrence $T(n) = T(n/4) + T(n/2) + n^2$, with $T(n) = 1$ if $n < 4$. a/h

9. Note that we have chosen the median of an n-element set to be the element in position $\lceil n/2 \rceil$. We have also chosen to put the median of the medians into the set L of algorithm Select1. Show that this allows you to prove that $T(n) \leq T(3n/4) + T(n/5) + cn$ for $n \geq 40$ rather than $n \geq 60$. (You will need to analyze separately the case where $\lceil n/5 \rceil$ is even and the case where it is odd.) Is 40 the least value possible?

5 PROBABILITY

5.1 INTRODUCTION TO PROBABILITY

Why Study Probability?

You have probably studied hashing as a way to store data so that it is possible to access that data quickly. But for those of you who have not, we will explain it here by telling a true story about a catalog order store where two of this book's authors used to shop.

Customers would come to the store and fill out order forms for items from the catalog (or they would call in their orders). The store employees would then order the items, which would be delivered from a warehouse some days later. When merchandise was delivered to the store, the customer who ordered it would be telephoned and would eventually come to pick it up. Meanwhile dozens (or, in busy times, hundreds) of order forms would accumulate at the order pickup desk. It would be impractical to search through all of the forms to find a customer's order.

The store came up with an ingenious solution. Behind the desk were 100 cubbyholes, numbered 00 through 99, for holding order forms. Order forms were put into the cubbyhole corresponding to the last two digits of the customer's phone number. A customer arriving at the desk was asked for the last two digits of his or her phone number. The clerk would then look through the order forms in the corresponding cubbyhole. There would never be more than a few forms, even when there were hundreds of total orders. Therefore, filing a form into a cubbyhole and finding a particular form were both fast and easy.

Hash tables in a computer use the same idea. Instead of cubbyholes, there is a table with m numbered locations. Each location, called a *bucket* or *slot,* holds a list of data items.[1] Each item has a unique identifier, called the *key*. When a data item arrives to be stored in the table, a hash function *h* that maps keys into bucket numbers gives the number of the bucket into which the data item should be inserted. (In the catalog order store example, the data items were the order forms, the keys were the phone numbers, and the hash function returned the last two digits of the phone number.) To look up the data corresponding to a particular key, you would simply compute the hash function of that key and look in the corresponding bucket.[2]

A good hash function spreads the keys evenly among the buckets. Taking the last two digits of a phone number is a good hash function. However, taking the first two digits of the phone number would be a bad choice, because most phone numbers in a local area start with one of a relatively small number of three-digit numbers. The hash function is defined for all conceivable keys, even though relatively few of them usually occur as input. If we don't know anything about the data coming in, then we can't make a good guess as to what makes a good hash function. Thus, in creating our model for hashing, we will assume that all functions from the keys we receive to the slots in the table are equally likely to result from applying the hash function.

If we have a table with 100 buckets and 50 keys to put in those buckets, it is possible that all 50 of those keys could be assigned (hashed) to the same bucket in the table. However, someone who is experienced with using hash functions will tell you that you'd never see this in a million years. But that same person might also tell you that neither would you ever see, in a million years, all the keys hash into different locations. In fact, it is far less likely that all 50 keys would hash into one place than that all 50 keys would hash into different places, but both events are quite unlikely. Being able to understand just how likely or unlikely such events are is a major reason for taking up the study of probability.

To assign probabilities to events, we need to have a clear picture of what these events are. Thus, we present a model of the kinds of situations in which it is reasonable to assign probabilities, and then we recast our questions about probabilities into questions about this model. We use the phrase **sample space** to refer to the set of possible outcomes of a process. For now, we deal with processes that have finite sample spaces, such as a game of cards, a sequence of hashes into a hash table, a sequence of tests on a number to see if it fails to be a prime, a roll of a die, a series of coin flips, a laboratory experiment, a survey, or any of many other possibilities.

As with all sets, the items in the sample space are called **elements**. For example, if a professor starts each class with a three-question true-false quiz, then the sample space of all possible patterns of correct answers is

$$\{TTT, TTF, TFT, FTT, TFF, FTF, FFT, FFF\},$$

---

[1] It is common for the data items in a bucket to be stored in a linked list, but all we need to know at this stage is that the items are in a list.

[2] The scheme we have described for hashing is called *open hashing*. Other schemes are possible. For example, the table may consist of slots that can hold a single item; if the hash function says to put a second item into a slot that is already full, then further computation finds an empty slot in the table. Analyzing such schemes is beyond the scope of this book.

and TTT is the element of the sample space corresponding to all answers being true. A set of elements in a sample space is called an **event**. The event of the first two answers being true is {TTT, TTF}.

To compute probabilities, we assign a **probability weight** $P(x)$ to each element of the sample space so that the weight represents what we believe to be the relative likelihood of that outcome. There are two rules in assigning weights. First, the weights must be nonnegative numbers, and second, the sum of the weights of all the elements in a sample space must be 1. We define the **probability** $P(E)$ of the event E to be the sum of the weights of the elements of E. Algebraically, we write

$$P(E) = \sum_{x:x\in E} P(x) . \tag{5.1}$$

We read this as "$P(E)$ equals the sum, over all x such that x is in E, of $P(x)$." In particular, we have just defined the probability of the set $\{x\}$, denoted by $P(\{x\})$, to equal the weight $P(x)$, which makes our notation consistent.

Notice that a probability function P on a sample space S satisfies the following rules:[3]

1. $P(A) \geq 0$ for any $A \subseteq S$.

2. $P(S) = 1$.

3. $P(A \cup B) = P(A) + P(B)$ for any two disjoint events A and B.

The first two rules reflect our rules for assigning weights. We say that two events A and B are **disjoint** if $A \cap B = \emptyset$. The third rule follows directly from the definition of *disjoint* and our definition of the probability of an event. A function P that satisfies these rules is called a **probability distribution** or a **probability measure**.

In the case of the professor's three-question quiz, it is natural to expect each sequence of trues and falses to be equally likely. (If a professor showed any pattern of preferences, then a student who observed this pattern could use it in educated guessing.) Thus, it is natural to assign an equal weight of $1/8$ to each of the eight elements of our quiz sample space. We defined the probability of an event E, which we denote by $P(E)$, is the sum of the weights of its elements. Thus, the probability of the event "the first answer is true" is

$$\frac{1}{8} + \frac{1}{8} + \frac{1}{8} + \frac{1}{8} = \frac{1}{2} .$$

The event "there is exactly one true" is {TFF, FTF, FFT}; so, P(there is exactly one true) is $3/8$.

---

[3] These rules are often called the *axioms of probability*. For a finite sample space, we could show that if we started with these axioms, our definition of *probability* in terms of the weights of individual elements of S is the only definition possible. That is, for any other definition, the probabilities we would compute would still be the same if we take $w(x) = P(\{x\})$.

Some Examples of Probability Computations

Exercise 5.1-1 Try flipping a coin five times. Did you get at least one head? Repeat five coin flips a few more times. What is the probability of getting at least one head in five flips of a coin? What is the probability of no heads?

Exercise 5.1-2 Find a good sample space for rolling two dice. What weights are appropriate for the members of your sample space? What is the probability of getting a total of 6 or 7 on the two dice? Assume the dice are red and green. What is the probability of getting less than 3 on the red one and more than 3 on the green one?

Exercise 5.1-3 Suppose you hash a list of n keys into a hash table with 20 locations. What is an appropriate sample space, and what is an appropriate weight function? (Assume the keys and the hash function are not in any special relationship to the number 20.) If $n = 3$, what is the probability that all three keys hash to different locations? If you hash ten keys into the table, what is the probability that at least two keys have hashed to the same location? We say two keys **collide** if they hash to the same location. How big does n have to be to ensure that the probability is at least $1/2$ that there has been at least one collision?

In Exercise 5.1-1, a good sample space is the set of all 5-tuples of H's and T's. There are 32 elements in the sample space, and no element has any reason to be more likely than any other. Thus, a natural weight to use is $1/32$ for each element of the sample space. The event of at least one head is the set of all elements except TTTTT. Because there are 31 elements in this set, its probability is $31/32$, which suggests that you should have observed at least one head fairly often!

Complementary Probabilities

The probability of no heads is the probability of the set {TTTTT}, which is $1/32$. Notice that the probabilities of the event "no heads" and the opposite event "at least one head" add to 1. This observation suggests a theorem. The **complement** of an event E in a sample space S, denoted by $S - E$, is the set of all outcomes in S except those in E. The theorem tells us how to compute the probability of the complement of an event from the probability of the event. We say that two events E and F are **complementary** if E is the complement of F in the sample space.

Theorem 5.1 If two events E and F are complementary, then

$$P(E) = 1 - P(F) \, .$$

Proof: The sum of all the probabilities of all the elements of the sample space is 1. Because we can break this sum into the sum of the probabilities of the elements of E plus the sum of the probabilities of the elements of F, we have

$$P(E) + P(F) = 1 ,$$

which gives us $P(E) = 1 - P(F)$. ∎

For Exercise 5.1-2, a good sample space would be pairs of numbers (a, b), where $(1 \leq a, b \leq 6)$. By the product principle (see Section 1.1), the size of this sample space is $6 \cdot 6 = 36$. Thus, a natural weight for each ordered pair is $1/36$. How do we compute the probability of getting a sum of 6 or 7? There are five ways to roll a 6 and six ways to roll a 7, so our event has 11 elements, each of weight $1/36$. Thus, the probability of our event is $11/36$. For the question about the red and green dice, there are two ways for the red one to turn up less than 3 and three ways for the green one to turn up more than 3. Thus, the event of getting less than 3 on the red one and greater than 3 on the green one is a set of size $2 \cdot 3 = 6$, by the product principle. Because each element of the event has weight $1/36$, the event has probability $6/36$, or $1/6$.

Probability and Hashing

In Exercise 5.1-3, an appropriate sample space is the set of n-tuples of numbers between 1 and 20. The first entry in an n-tuple is the position our first key hashes to, the second entry is the position our second key hashes to, and so on. Thus, each n-tuple represents a possible hash function, and each hash function, applied to our keys, would give us one n-tuple. The size of the sample space is 20^n (why?), so an appropriate weight for an n-tuple is $1/20^n$. To compute the probability of a collision, we first compute the probability that all keys hash to different locations; we then apply Theorem 5.1, which tells us to subtract this probability from 1 to get the probability of a collision.

To compute the probability that all keys hash to different locations, we consider the event that all keys hash to different locations. This is the set of n-tuples in which all entries are different. (In the terminology of functions, these n-tuples correspond to one-to-one hash functions). There are 20 choices for the first entry of an n-tuple in our event. Because the second entry has to be different, there are 19 choices for the second entry of this n-tuple. Similarly, there are 18 choices for the third entry (it has to be different from the first two), 17 for the fourth, and, in general, $20 - i + 1$ possibilities for the ith entry of the n-tuple. Thus, we have

$$(20)(19)(18) \cdots (20 - n + 1) = 20^{\underline{n}}$$

elements of our event.[4] Because each element of this event has weight $1/20^n$, the probability that all the keys hash to different locations is

$$\frac{(20)(19)(18)\cdots(20-n+1)}{20^n} = \frac{20^{\underline{n}}}{20^n}.$$

In particular, if n is 3, the probability is $(20 \cdot 19 \cdot 18)/20^3 = .855$.

Table 5.1 shows the values of this function for n between 0 and 20. Note how quickly the probability of getting a collision grows. As you can see with $n = 10$, the probability that there have been no collisions is about .065, so the probability of at least one collision is .935. If $n = 5$, then this probability is about .58, and if $n = 6$, then it is about .43. By Theorem 5.1, the probability of a collision is 1 minus the probability

Table 5.1: *The probabilities that all elements of a set hash to different entries of a hash table of size 20*

n	Probability of Empty Slot	Probability of No Collisions
1	1	1
2	.95	.95
3	.9	.855
4	.85	.72675
5	.8	.5814
6	.75	.43605
7	.7	.305235
8	.65	.19840275
9	.6	.11904165
10	.55	.065472908
11	.5	.032736454
12	.45	.014731404
13	.4	.005892562
14	.35	.002062397
15	.3	.000618719
16	.25	.00015468
17	.2	.0000309359
18	.15	.00000464039
19	.1	.000000464039
20	.05	.000000023202

[4]Here, we use the notation for falling factorial powers introduced in Section 1.2.

that all the keys hash to different locations. Thus, if we hash six items into our table, the probability of a collision is more than $1/2$. Our first intuition might well have been that we would need to hash ten items into our table to have probability $1/2$ of a collision. This example shows the importance of supplementing intuition with careful computation!

If we created a similar table for hashing keys into a table with 100 slots, we would see that for hashing 50 keys into 100 slots, the probability that all 50 items go to different slots is about .0000003, or three ten-millionths. Thus, if we repeated the experiment of hashing 50 items into 100 slots ten million times, we should not be surprised if on one or more of the repeats, all keys went to different slots. So, even though the probability of all keys going to different slots is small, a person who says we would never see this in a million years is wrong, even if we just do one experiment per month.

The technique of computing the probability of an event of interest by first computing the probability of its complementary event and then subtracting that from 1 is very useful. You will see many opportunities to use it, perhaps because about half the time, it is easier to compute directly the probability that an event doesn't occur than it is to compute the probability that it does. We stated Theorem 5.1 as a theorem to emphasize the importance of this technique.

The Uniform Probability Distribution

In the previous three exercises, it was appropriate to assign the same weight to all members of our sample space. We say that P is the **uniform probability measure** or **uniform probability distribution** when we assign the same probability to all members of our sample space. The computations in the exercises suggest the following useful theorem.

Theorem 5.2	Suppose P is the uniform probability measure defined on a sample space S. Then for any event E,

$$P(E) = \frac{|E|}{|S|},$$

which is the size of E divided by the size of S.

Proof: Let $S = \{x_1, x_2, \ldots, x_{|S|}\}$. Because P is the uniform probability measure, there must be some value p such that $P(x_i) = p$ for each $x_i \in S$. Combining this fact with the second and third probability rules, we obtain

$$\begin{aligned}
1 &= P(S) \\
&= P(x_1 \cup x_2 \cup \cdots \cup x_{|S|}) \\
&= P(x_1) + P(x_2) + \cdots + P(x_{|S|}) \\
&= p|S| .
\end{aligned}$$

Equivalently,

$$p = \frac{1}{|S|} . \tag{5.2}$$

E is a subset of S with $|E|$ elements and, therefore,

$$P(E) = \sum_{x_i \in E} p(x_i) = |E|p . \tag{5.3}$$

Combining Equations 5.2 and 5.3 gives

$$P(E) = |E|p = |E|(1/|S|) = |E|/|S|. \; \blacksquare$$

<table>
<tr><td>**Exercise 5.1-4**</td><td>What is the probability of an odd number of heads in three tosses of a coin? Use Theorem 5.2, which states that with the uniform probability measure, for any event E,

$$P(E) = \frac{|E|}{|S|} ,$$

which is the size of E divided by the size of S.</td></tr>
</table>

Using a sample space similar to that of the first example (with T and F replaced with H and T, respectively), we see there are three sequences with one H and there is one sequence with three H's. Thus, we have four sequences in the event "an odd number of heads come up." Because there are eight sequences in the sample space, the probability is $4/8 = 1/2$ by Theorem 5.2.

The fact that we got $1/2$ shows a symmetry inherent in this problem. In flipping coins, heads and tails are equally likely. Further, if we are flipping three coins, an odd number of heads implies an even number of tails. Therefore, the probabilities of the following events must all be the same.

- an odd number of heads

- an even number of heads

- an odd number of tails

- an even number of tails

A word of caution is appropriate here. Theorem 5.2 applies only to probabilities that come from the equiprobable weighting function. The next exercise shows that the theorem does not apply in general.

<table>
<tr><td>**Exercise 5.1-5**</td><td>A sample space consists of the numbers 0, 1, 2, and 3. We assign weight $1/8$ to 0, $3/8$ to 1, $3/8$ to 2, and $1/8$ to 3. What is the probability that an element of the sample space is positive? Show that this is not the result we would obtain if we used the formula of Theorem 5.2.</td></tr>
</table>

The event "x is positive" is the set $E = \{1, 2, 3\}$. The probability of E is

$$P(E) = P(1) + P(2) + P(3) = \frac{3}{8} + \frac{3}{8} + \frac{1}{8} = \frac{7}{8}.$$

However, $|E|/|S| = 3/4$.

Exercise 5.1-5 may seem to be "cooked up" in an unusual way just to prove a point. However, that sample space and that probability measure could easily arise in studying something as simple as coin flipping.

Exercise 5.1-6   Use the set $\{0, 1, 2, 3\}$ as a sample space for the process of flipping a coin three times and counting the number of heads. Determine the appropriate probability weights $P(0)$, $P(1)$, $P(2)$, and $P(3)$.

There is one way to get no heads, namely, tails on each flip. There are, however, three ways to get one head and three ways to get two heads. Thus, $P(1)$ and $P(2)$ should each be 3 times $P(0)$. There is one way to get three heads—heads on each flip. Thus, $P(3)$ should equal $P(0)$. We can change these statements into the following equations:

$$P(1) = 3P(0)$$
$$P(2) = 3P(0)$$
$$P(3) = P(0)$$

We also have the equation saying all the weights add to 1:

$$P(0) + P(1) + P(2) + P(3) = 1 .$$

There is one and only one solution to these equations, namely,

$$P(0) = \frac{1}{8}$$
$$P(1) = \frac{3}{8}$$
$$P(2) = \frac{3}{8}$$
$$P(3) = \frac{1}{8} .$$

Do you notice a relationship between $P(x)$ and the binomial coefficient $\binom{3}{x}$ here? Can you predict the probabilities of zero, one, two, three, and four heads in four flips of a coin?

Together, the previous two exercises demonstrate that we must be careful not to apply Theorem 5.2 unless we are using the uniform probability measure.

Important Concepts, Formulas, and Theorems

1. *Sample space.* A *sample space* is the set of possible outcomes of a process.

2. *Event.* A set of elements in a sample space is called an *event.*

3. *Disjoint.* Two events E and F are said to be *disjoint* if $E \cap F = \emptyset$.

4. *Probability.* To compute probabilities, we assign a weight to each element of the sample space so that the weight represents what we believe to be the relative likelihood of that outcome. We must follow two rules in assigning weights. First, the weights must be nonnegative numbers, and second, the sum of the weights of all the elements in a sample space must be 1. We define the *probability* $P(E)$ of the event E to be the sum of the weights of the elements of E. The function P is called a *probability measure.*

5. *The axioms of probability.* A probability measure on a finite sample space must satisfy the following three rules. (Alternately, these rules could be used to define what we mean by probability.)

   a. $P(A) \geq 0$ for any $A \subseteq S$.

   b. $P(S) = 1$.

   c. $P(A \cup B) = P(A) + P(B)$ for any two disjoint events A and B.

6. *Probability distribution.* A function that assigns a probability to each member of a sample space is called a (discrete) *probability distribution.*

7. *Complement.* The *complement* of an event E in a sample space S, denoted by $S - E$, is the set of all outcomes in S but not in E. We say that the events E and F are *complementary* events if E is the complement of F in S.

8. *The probabilities of complementary events.* If two events E and F are complementary, then

$$P(E) = 1 - P(F) .$$

9. *Collision/Collide (in hashing).* Two keys *collide* if they hash to the same location.

10. *Uniform probability distribution.* We say P is the *uniform probability measure* or *uniform probability distribution* when we assign the same probability to all members of our sample space.

11. *Computing probabilities with the uniform distribution.* Suppose P is the uniform probability measure defined on a sample space S. Then for any event E, we have $P(E) = |E|/|S|$, which is the size of E divided by the size of S. This *does not* apply to general probability distributions.

Problems

1. What is the probability of exactly three heads when you flip a coin five times? What is the probability of three or more heads when you flip a coin five times? a/h

2. When you roll two dice, what is the probability of getting a sum of 4 or less on the tops?

3. If you hash three keys into a hash table with ten slots, what is the probability that all three keys hash to different slots? How big does n have to be so that if n keys hash to a table with ten slots, the probability is at least $1/2$ that some slot has at least two keys hash to it? How many keys do you need to have probability at least $2/3$ that some slot has at least two keys hash to it? a/h

4. What is the probability of an odd sum when you roll three dice?

5. Suppose you use the numbers 2 through 12 as your sample space for rolling two dice and adding the numbers on top. What would you get for the probability of a sum of 2, 3, or 4, if you used the equiprobable measure on this sample space? Does your answer make sense? a/h

6. Two pennies, a nickel, and a dime are placed in a cup. You draw a first coin and a second coin.

    a. Assuming you are sampling without replacement (that is, you don't replace the first coin before taking the second), write the sample space of all ordered pairs of letters P, N, and D that represent the outcomes. What would you say are the appropriate weights for the elements of the sample space?

    b. What is the probability of getting 11 cents?

7. Why is the probability of five heads in ten flips of a coin equal to $63/256$? a/h

8. Using five-element sets as a sample space, determine the probability that a hand of 5 cards, chosen from an ordinary deck of 52 cards, will consist of cards of the same suit.

9. Using five-element permutations as a sample space, determine the probability that a hand of 5 cards, chosen from an ordinary deck of 52 cards, will have all the cards from the same suit. a/h

10. How many five-card hands chosen from a standard deck of playing cards consist of five cards in a row (such as the nine of diamonds, ten of clubs, jack of clubs, queen of hearts, and king of spades)? Such a hand is called a straight. What is the probability that a five-card hand is a straight? Explore whether you get the same answer by using five-element sets as your model of hands or five-element permutations as your model of hands.

11. A student taking a ten-question, true-false diagnostic test knows none of the answers and must guess at each one. Compute the probability that the student gets a score of 80 or higher. What is the probability that the grade is 70 or lower? a/h

12. A die is made of a cube with a square painted on one side, a circle on two sides, and a triangle on three sides. If the die is rolled twice, what is the probability that the two shapes you see on top are the same?

13. Are the following two events equally likely? Event 1 consists of drawing an ace and a king when you draw two cards from among the 13 spades in a deck of cards. Event 2 consists of drawing an ace and a king when you draw two cards from the whole deck. a/h

14. There is a retired professor who used to love to go into a probability class of 30 or more students and announce, "I will give even money odds that there are two people in this classroom with the same birthday." With 30 students in the room, what is the probability that all have different birthdays? What is the minimum number of students that must be in the room so that the professor has probability at least $1/2$ of winning the bet? What is the probability that he wins his bet if there are 50 students in the room? Does this probability make sense to you? (There is no wrong answer to this last question!) Explain why or why not. (A programmable calculator, spreadsheet, computer program, or computer algebra system will be helpful in this problem.)

15. Which is more likely, or are both equally likely?

    a. Drawing an ace and a king when you draw 2 cards from among the 13 spades, or drawing an ace and a king when you draw 2 cards from an ordinary deck of 52 playing cards? a/h

    b. Drawing an ace and a king of the same suit when you draw 2 cards from a deck, or drawing an ace and a king when you draw 2 cards from among the 13 spades? a/h

5.2 UNIONS AND INTERSECTIONS

The Probability of a Union of Events

Exercise 5.2-1 If you roll two dice, what is the probability of either an even sum or a sum of 8 or more (or both)?

Exercise 5.2-2 In Exercise 5.2-1, let E be the event "even sum" and let F be the event "8 or more." We found the probability of the union of the events E and F. Why isn't it the case that $P(E \cup F) = P(E) + P(F)$? What weights appear twice in the sum $P(E) + P(F)$? Find a formula for $P(E \cup F)$ in terms of the probabilities of E, F, and $E \cap F$. Apply this formula to Exercise 5.2-1. What is the value of expressing one probability in terms of three?

Exercise 5.2-3 What is $P(E \cup F \cup G)$ in terms of probabilities of the events E, F, and G and their intersections?

In the sum $P(E) + P(F)$, the weights of elements of $E \cap F$ each appear twice, while the weights of all other elements of $E \cup F$ each appear once. We can see this by looking at a diagram called a Venn diagram (see Figure 5.1). In a **Venn diagram**, the rectangle represents the sample space, and the circles represent the events.

Figure 5.1: *A Venn diagram for two events*

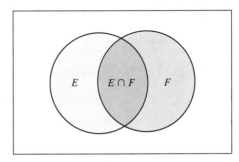

If we were to shade both E and F, we would wind up shading the region $E \cap F$ twice. In Figure 5.2, we represent this situation by putting numbers in the regions, indicating how many times they are shaded. This illustrates why the sum $P(E) + P(F)$ includes the probability weight of each element of $E \cap F$ twice. Thus, to get a sum that includes the probability weight of each element of $E \cup F$ exactly once, we have to subtract the weight of $E \cap F$ from the sum $P(E) + P(F)$. This is why

$$P(E \cup F) = P(E) + P(F) - P(E \cap F). \tag{5.4}$$

Figure 5.2: *If we shade each of E and F once, then we shade $E \cap F$ twice*

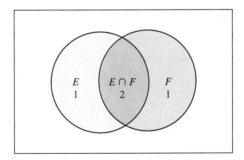

We can now apply this equation to Exercise 5.2-1 by noting that the probability of an even sum is $1/2$, while the probability of a sum of 8 or more is

$$\frac{1}{36} + \frac{2}{36} + \frac{3}{36} + \frac{4}{36} + \frac{5}{36} = \frac{15}{36}.$$

From a similar sum, the probability of an *even* sum of 8 or more is $9/36$, so the probability of a sum that is even or is 8 or more is

$$\frac{1}{2} + \frac{15}{36} - \frac{9}{36} = \frac{2}{3}.$$

In this case, our computation merely illustrates the formula; with less work, we could add the probability of an even sum to the probability of a sum of 9 or 11. In many cases, however, probabilities of individual events and their intersections are more straightforward to compute than probabilities of unions (we will see such examples later in this section), and in such cases, our formula is quite useful.

Now let's consider the case for three events. We draw a Venn diagram and fill in the numbers for shading E, F, and G. To avoid crowding the figure, we use EF to label the region corresponding to $E \cap F$ and similarly label other regions. This gives Figure 5.3.

Figure 5.3: *The number of times the intersections are shaded when we shade E, F, and G*

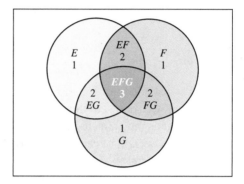

Thus, we have to figure out a way to subtract from $P(E) + P(F) + P(G)$ the weights of elements in the regions $E \cap F$, $F \cap G$, and $E \cap G$ but not $E \cap F \cap G$ (labeled EF, FG, and EG, respectively) once, and then the weight of elements in the region labeled EFG twice. Subtracting the weights of elements of each $E \cap F$, $F \cap G$, and $E \cap G$ does more than we wanted to do, because this subtracts the weights of elements in EF, FG, and EG once but the weights of elements in EFG three times, leaving us with Figure 5.4. We see that all we have left to do is to add weights of elements in $E \cap F \cap G$ back into our sum. Thus, we have

$$P(E \cup F \cup G) = P(E) + P(F) + P(G) - P(E \cap F) - P(E \cap G) - P(F \cap G)$$
$$+ P(E \cap F \cap G).$$

Figure 5.4: *The result of removing the weights of each intersection of two sets*

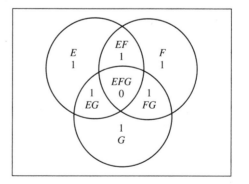

Principle of Inclusion and Exclusion for Probability

From the previous two exercises, it is natural to guess the formula

$$P\left(\bigcup_{i=1}^{n} E_i\right) = \sum_{i=1}^{n} P(E_i) - \sum_{i=1}^{n-1} \sum_{j=i+1}^{n} P(E_i \cap E_j)$$

$$+ \sum_{i=1}^{n-2} \sum_{j=i+1}^{n-1} \sum_{k=j+1}^{n} P(E_i \cap E_j \cap E_k) - \cdots. \qquad (5.5)$$

All the sum signs in this notation suggest that we need some new notation to describe sums. We are now going to make what we hope is a small leap of abstraction in our notation and introduce notation capable of describing compactly the sum in Equation 5.5. This notation is an extension of the one we introduced in Equation 5.1. We use

$$\sum_{\substack{i_1, i_2, \ldots, i_k: \\ 1 \le i_1 < i_2 < \cdots < i_k \le n}} P(E_{i_1} \cap E_{i_2} \cap \cdots E_{i_k})$$

to stand for the sum, over all increasing sequences $i_1, i_2, \ldots, i_k$ of integers between 1 and n, of the probabilities of the sets $E_{i_1} \cap E_{i_2} \cdots \cap E_{i_k}$. More generally,

$$\sum_{\substack{i_1, i_2, \ldots, i_k: \\ 1 \le i_1 < i_2 < \cdots < i_k \le n}} f(i_1, i_2, \ldots, i_k)$$

is the sum of $f(i_1, i_2, \ldots, i_k)$ over all increasing sequences of k numbers between 1 and n.

Exercise 5.2-4

To practice with notation, what is

$$\sum_{\substack{i_1, i_2, i_3: \\ 1 \le i_1 < i_2 < i_3 \le 4}} i_1 + i_2 + i_3 \ ?$$

The sum in Exercise 5.2-4 is $(1+2+3) + (1+2+4) + (1+3+4) + (2+3+4) = 3(1+2+3+4) = 30$.

With this understanding of the notation in hand, we can now write a formula that captures the idea in Equation 5.5 more concisely. Notice that Equation 5.5 includes probabilities of single sets with a plus sign, probabilities of intersections of two sets with a minus sign, and, in general, probabilities of intersections of any even number of sets with a minus sign and probabilities of intersections of any odd number of sets (including the odd number 1) with a plus sign. Thus, if we are intersecting k sets, the proper coefficient for the probability of the intersection of these sets is $(-1)^{k+1}$. (It would be equally good to use $(-1)^{k-1}$ and correct, but unconventional, to use $(-1)^{k+3}$.) This lets us translate the formula of Equation 5.5 to the equation in the theorem called the **principle of inclusion and exclusion for probability**, which follows. We give two completely different proofs of the theorem—one of which is a nice counting argument but is a bit on the abstract side, and one of which is straightforward induction but is complicated by the fact that it takes a lot of notation to say what is happening.

| Theorem 5.3 | **(Principle of Inclusion and Exclusion for Probability)** The probability of the union $E_1 \cup E_2 \cup \cdots \cup E_n$ of events in a sample space S is given by |

$$P\left(\bigcup_{i=1}^{n} E_i\right) = \sum_{k=1}^{n}(-1)^{k+1} \sum_{\substack{i_1, i_2, \ldots, i_k: \\ 1 \le i_1 < i_2 < \cdots < i_k \le n}} P\left(E_{i_1} \cap E_{i_2} \cap \cdots \cap E_{i_k}\right). \qquad (5.6)$$

Proof 1: Consider an element x of $\bigcup_{i=1}^{n} E_i$. Let $E_{i_1}, E_{i_2}, \ldots, E_{i_h}$ be the set of all events E_i of which x is a member. Let $H = \{i_1, i_2, \ldots, i_h\}$. Then x is in another event $E_{j_1} \cap E_{j_2} \cap \cdots \cap E_{j_m}$ if and only if $\{j_1, j_2, \ldots, j_m\} \subseteq H$. Why is this? If there is a j_r that is not in H, then $x \notin E_{j_r}$ and thus $x \notin E_{j_1} \cap E_{j_2} \cap \cdots \cap E_{j_m}$. Notice that every x in $\bigcup_{i=1}^{n} E_i$ is in at least one E_i, so it is in at least one of the sets $E_{i_1} \cap E_{i_2} \cap \cdots \cap E_{i_k}$, namely, E_i.

Recall that we define $P\left(E_{j_1} \cap E_{j_2} \cap \cdots \cap E_{j_k}\right)$ to be the sum of the probability weights $P(x)$ for $x \in E_{j_1} \cap E_{j_2} \cap \cdots \cap E_{j_k}$. Suppose we substitute this sum of probability weights for $P\left(E_{j_1} \cap E_{j_2} \cap \cdots \cap E_{j_k}\right)$ on the right side of Equation 5.6. Then the right side becomes a sum of terms, each of which is plus or minus a probability weight. The sum of all the terms involving $P(x)$ on the right side of Equation 5.6 includes a term involving $P(x)$ for each nonempty subset $\{j_1, j_2, \ldots, j_m\}$ of H and no other terms involving $P(x)$. The coefficient of the probability weight $P(x)$ in the term for the subset $\{j_1, j_2, \ldots, j_m\}$ is $(-1)^{m+1}$. Because there are $\binom{h}{m}$ subsets of H of size m, the sum of the terms involving $P(x)$ will be

$$\sum_{m=1}^{h}(-1)^{m+1} \binom{h}{m} P(x) = \left(-\sum_{m=0}^{h}(-1)^{m} \binom{h}{m} P(x)\right) + P(x)$$

$$= 0 \cdot P(x) + P(x) = P(x).$$

We got the term $0 \cdot P(x)$ by using the fact that $h \ge 1$, so that by the binomial theorem, $\sum_{j=0}^{h} \binom{h}{j} (-1)^j = (1-1)^h = 0$. This proves that for each x, the sum of all the terms involving $P(x)$ after we substitute the sum of probability weights into Equation 5.6 is exactly $P(x)$. We noted above that every x in $\bigcup_{i=1}^{n} E_i$ appears in at least one of the sets $E_{i_1} \cap E_{i_2} \cap \cdots \cap E_{i_k}$. Thus, the right side of Equation 5.6 is the sum of every $P(x)$ such that x is in $\bigcup_{i=1}^{n} E_i$, which, by definition, is the left side of Equation 5.6. ∎

Proof 2: The proof is simply an application of mathematical induction using Equation 5.4. When $n = 1$, the formula is true because it says $P(E_1) = P(E_1)$. Now suppose inductively that for any family of $n - 1$ sets $F_1, F_2, \ldots, F_{n-1}$, we have

$$P\left(\bigcup_{i=1}^{n-1} F_i\right) = \sum_{k=1}^{n-1} (-1)^{k+1} \sum_{\substack{i_1, i_2, \ldots, i_k: \\ 1 \leq i_1 < i_2 < \cdots < i_k \leq n-1}} P\left(F_{i_1} \cap F_{i_2} \cap \cdots \cap F_{i_k}\right). \quad \textbf{(5.7)}$$

If in Equation 5.4 we let $E = E_1 \cup \cdots \cup E_{n-1}$ and $F = E_n$, then we may apply Equation 5.4 to compute $P\left(\bigcup_{i=1}^{n} E_i\right)$ as follows:

$$P\left(\bigcup_{i=1}^{n} E_i\right) = P\left(\bigcup_{i=1}^{n-1} E_i\right) + P(E_n) - P\left(\left(\bigcup_{i=1}^{n-1} E_i\right) \cap E_n\right). \quad \textbf{(5.8)}$$

By the distributive law,

$$\left(\bigcup_{i=1}^{n-1} E_i\right) \cap E_n = \bigcup_{i=1}^{n-1} (E_i \cap E_n).$$

Substituting this into Equation 5.8 gives us

$$P\left(\bigcup_{i=1}^{n} E_i\right) = P\left(\bigcup_{i=1}^{n-1} E_i\right) + P(E_n) - P\left(\bigcup_{i=1}^{n-1}(E_i \cap E_n)\right).$$

Now we use the inductive hypothesis (Equation 5.7) in two places to get

$$P\left(\bigcup_{i=1}^{n} E_i\right) = \left(\sum_{k=1}^{n-1} (-1)^{k+1} \sum_{\substack{i_1, i_2, \ldots, i_k: \\ 1 \leq i_1 < i_2 < \cdots < i_k \leq n-1}} P\left(E_{i_1} \cap E_{i_2} \cap \cdots \cap E_{i_k}\right)\right) + P(E_n)$$

$$- \sum_{k=1}^{n-1} (-1)^{k+1} \sum_{\substack{i_1, i_2, \ldots, i_k: \\ 1 \leq i_1 < i_2 < \cdots < i_k \leq n-1}} P\left(E_{i_1} \cap E_{i_2} \cap \cdots \cap E_{i_k} \cap E_n\right).$$

The first summation on the right side sums $(-1)^{k+1} P\left(E_{i_1} \cap E_{i_2} \cap \cdots \cap E_{i_k}\right)$ over all lists $i_1, i_2, \ldots, i_k$ that *do not* contain n, while the $P(E_n)$ and the second summation work together to sum $(-1)^{k+1} P\left(E_{i_1} \cap E_{i_2} \cap \cdots \cap E_{i_k}\right)$ over all lists $i_1, i_2, \ldots, i_k$ that *do* contain n. Therefore,

$$P\left(\bigcup_{i=1}^{n} E_i\right) = \sum_{k=1}^{n} (-1)^{k+1} \sum_{\substack{i_1, i_2, \ldots, i_k: \\ 1 \leq i_1 < i_2 < \cdots < i_k \leq n}} P\left(E_{i_1} \cap E_{i_2} \cap \cdots \cap E_{i_k}\right).$$

Thus, by the principle of mathematical induction, this formula holds for all integers $n > 0$. ∎

At a fancy restaurant, n students check their backpacks. They are the only ones in the restaurant to check backpacks. A child visits the checkroom and plays with the check tickets for the backpacks so they are all mixed up. If there are five students named Judy, Sam, Pat, Jill, and Jo, in how many ways may the backpacks be returned so that Judy gets the correct backpack (and maybe some other students do, too)? What is the probability that this happens? What is the probability that Sam gets the correct backpack (and maybe some other students do, too)? What is the probability that Judy and Sam both get the correct backpacks (and maybe some other students do, too)? For any particular two-element set of students, what is the probability that these two students get the correct backpacks (and maybe some other students do, too)? What is the probability that at least one student gets his or her own backpack? What is the probability that no students get their own backpacks? What do you expect the answer will be for the last two questions for n students? Because this classic problem is often stated using hats rather than backpacks (quaint, isn't it?), it is called the **hatcheck problem**. It is also known as the **derangement problem**—a *derangement* of a set is a one-to-one function from a set onto itself (i.e., a bijection) that sends each element to something not equal to it.

For Exercise 5.2-5, let E_i be the event that person i on our list gets the right backpack. Thus, E_1 is the event that Judy gets the correct backpack, and E_2 is the event that Sam gets the correct backpack. The event $E_1 \cap E_2$ is the event that Judy *and* Sam get the correct backpacks (and maybe some other people do, too). In Exercise 5.2-5, there are 4! ways to return the backpacks so that Judy gets her own (as with Sam or any other single student). Thus, $P(E_1) = P(E_i) = 4!/5!$. For any particular two-element subset, such as Judy and Sam, there are 3! ways that these two people may get the correct backpacks. Therefore, $P(E_i \cap E_j) = 3!/5!$ for each i and j. For a particular group of k students, the probability that each one of these k students gets his or her own backpack is $(5-k)!/5!$. Here is another way to say the same things: If E_i is the event that student i gets his or her own backpack, then the probability of an intersection of k of these events is $(5-k)!/5!$. The probability that at least one person gets his or her own backpack is the probability of $E_1 \cup E_2 \cup E_3 \cup E_4 \cup E_5$. Then, by the principle of inclusion and exclusion, the probability that at least one person gets his or her own backpack is

$$P(E_1 \cup E_2 \cup E_3 \cup E_4 \cup E_5)$$
$$= \sum_{k=1}^{5} (-1)^{k+1} \sum_{\substack{i_1, i_2, \ldots, i_k: \\ 1 \le i_1 < i_2 < \cdots < i_k \le 5}} P\left(E_{i_1} \cap E_{i_2} \cap \cdots \cap E_{i_k}\right). \qquad (5.9)$$

As we argued above, for a set of k people, the probability that all k people get their backpacks is $\frac{(5-k)!}{5!}$. In symbols, $P\left(E_{i_1} \cap E_{i_2} \cap \cdots \cap E_{i_k}\right) = \frac{(5-k)!}{5!}$. Recall that there are $\binom{5}{k}$ sets of k people chosen from our five students. That is, there are $\binom{5}{k}$ lists $i_1, i_2, \ldots, i_k$ with $1 < i_1 < i_2 < \cdots < i_k \le 5$. Thus, we can rewrite the right side of

Equation 5.9 as

$$\sum_{k=1}^{5}(-1)^{k+1}\binom{5}{k}\frac{(5-k)!}{5!}.$$

This gives us

$$
\begin{aligned}
P(E_1 \cup E_2 \cup E_3 \cup E_4 \cup E_5) &= \sum_{k=1}^{5}(-1)^{k-1}\binom{5}{k}\frac{(5-k)!}{5!} \\
&= \sum_{k=1}^{5}(-1)^{k-1}\frac{5!}{k!(5-k)!}\frac{(5-k)!}{5!} \\
&= \sum_{k=1}^{5}(-1)^{k-1}\frac{1}{k!} \\
&= 1-\frac{1}{2}+\frac{1}{3!}-\frac{1}{4!}+\frac{1}{5!}.
\end{aligned}
$$

The probability that nobody gets his or her own backpack is 1 minus the probability that someone does, or

$$\frac{1}{2}-\frac{1}{3!}+\frac{1}{4!}-\frac{1}{5!}.$$

To do the general case of n students, we simply substitute n for 5 and get that the probability of at least one person getting his or her own backpack is

$$\sum_{i=1}^{n}(-1)^{i-1}\frac{1}{i!} = 1-\frac{1}{2}+\frac{1}{3!}-\cdots+\frac{(-1)^{n-1}}{n!},$$

and the probability that nobody gets his or her own backpack is 1 minus the probability above, or

$$\sum_{i=2}^{n}(-1)^{i}\frac{1}{i!} = \frac{1}{2}-\frac{1}{3!}+\cdots+\frac{(-1)^{n}}{n!}. \tag{5.10}$$

If you learned about power series in calculus, you may recall the power series representation of e^x, namely,

$$e^x = 1+x+\frac{x^2}{2!}+\frac{x^3}{3!}+\cdots = \sum_{i=0}^{\infty}\frac{x^i}{i!}.$$

Thus, the expression in Equation 5.10 is the approximation to e^{-1}, which we get by substituting -1 for x in the power series and stopping the series at $i = n$. Note that the result depends very lightly on n; as long as we have at least four or five people, then no matter how many people we have, the probability that no one gets his or her own backpack (or hat) remains at roughly e^{-1}. Our intuition might have suggested that as the number of students increases, the probability that *someone* gets his or her own backpack approaches 1 rather than $1 - e^{-1}$. Thus, this is another example of why it is important to use computations, instead of intuition, with the rules of probability!

The Principle of Inclusion and Exclusion for Counting

Exercise 5.2-6 How many functions from an n-element set N to an m-element set $M = \{y_1, y_2, \ldots, y_m\}$ map nothing to y_1? Another way to say this is if I have n distinct candy bars and m children (Sam, Mary, Pat, etc.), in how ways may I pass out the candy bars so that Sam doesn't get any candy (and maybe some other children don't either)?

Exercise 5.2-7 How many functions map nothing to a k-element subset K of M? Another way to say this is if I have n distinct candy bars and m children (Sam, Mary, Pat, etc.), in how ways may I pass out the candy bars so that some particular k-element subset of the children don't get any (and maybe some other children don't either)?

Exercise 5.2-8 How many functions from an n-element set N to an m-element set M map nothing to at least one element of M? Another way to say this is if I have n distinct candy bars and m children (Sam, Mary, Pat, etc.), in how ways may I pass out the candy bars so that some child doesn't get any (and maybe some other children don't either)?

Exercise 5.2-9 On the basis of Exercises 5.2-6–5.2-8, how many functions are there from an n-element set onto an m-element set?

The number of functions from an n-element set to an m-element set $M = \{y_1, y_2, \ldots, y_m\}$ that map nothing to y_1 is simply $(m - 1)^n$, because we have $m - 1$ choices of where to map each of our n elements. Similarly, the number of functions that map nothing to a particular set K of k elements will be $(m - k)^n$. This warms us up for Exercise 5.2-8.

In Exercise 5.2-8, we need an analog of the principle of inclusion and exclusion for the size of a union of m sets. Because we can make the same argument about the size of the union of two or three sets that we made about probabilities of unions of two or three sets, we have a very natural analog. Because events are sets, we might be able to get an analog simply by changing the probabilities of the events E_i to the sizes of the sets E_i (here, set E_i is the set of functions that map nothing to element i of the set M—that is, the event that a function maps nothing to i). The analog is the **principle of inclusion and exclusion for counting**:

$$\left| \bigcup_{i=1}^{m} E_i \right| = \sum_{k=1}^{m} (-1)^{k+1} \sum_{\substack{i_1, i_2, \ldots, i_k: \\ 1 \le i_1 < i_2 < \cdots < i_k \le m}} \left| E_{i_1} \cap E_{i_2} \cap \cdots \cap E_{i_k} \right|.$$

In fact, this formula is proved by induction or by a counting argument in virtually the same way. Applying this formula to the number of functions from N to M that map nothing to at least one element of K gives us

$$\left| \bigcup_{i=1}^{m} E_i \right| = \sum_{k=1}^{m} (-1)^{k+1} \sum_{\substack{i_1, i_2, \ldots, i_k: \\ 1 \le i_1 < i_2 < \cdots < i_k \le m}} \left| E_{i_1} \cap E_{i_2} \cap \cdots \cap E_{i_k} \right|$$

$$= \sum_{k=1}^{m} (-1)^{k+1} \binom{m}{k} (m-k)^n, \qquad (5.11)$$

where $\left| E_{i_1} \cap E_{i_2} \cap \cdots \cap E_{i_k} \right|$ is the number of functions that map nothing to the k-element set $\{i_1, i_2, \ldots, i_k\}$. By our solution to Exercise 5.2-7, the number of functions that map nothing to the k-element set $\{i_1, i_2, \ldots, i_k\}$ is $(m-k)^n$. The number in Equation 5.11 is the number of functions from N that map nothing to at least one element of M. The total number of functions from N to M is m^n. Thus, the number of onto functions is

$$m^n - \sum_{k=1}^{m} (-1)^{k+1} \binom{m}{k} (m-k)^n = \sum_{k=0}^{m} (-1)^k \binom{m}{k} (m-k)^n,$$

where the equality results because $\binom{m}{0}$ is 1, $(m-0)^n$ is m^n, and $-(-1)^{k+1} = (-1)^k$.

Theorem 5.4

The number of functions from an n-element set onto an m element set is

$$\sum_{k=0}^{m} (-1)^k \binom{m}{k} (m-k)^n.$$

Proof: Given above. ∎

Important Concepts, Formulas, and Theorems

1. *Venn diagram.* To draw a Venn diagram for two or three sets, we draw a rectangle that represents the sample space and two or three mutually overlapping circles to represent the events.

2. *Probability of a union of two events.* $P(E \cup F) = P(E) + P(F) - P(E \cap F)$.

3. *Probability of a union of three events.* $P(E \cup F \cup G) = P(E) + P(F) + P(G) - P(E \cap F) - P(E \cap G) - P(F \cap G) + P(E \cap F \cap G)$.

4. *A summation notation.* The sum of $f(i_1, i_2, \ldots, i_k)$ over all increasing sequences of k numbers between 1 and n is denoted by

$$\sum_{\substack{i_1, i_2, \ldots, i_k: \\ 1 \le i_1 < i_2 < \cdots < i_k \le n}} f(i_1, i_2, \ldots, i_k).$$

5. *Principle of inclusion and exclusion for probability.* The probability of the union $E_1 \cup E_2 \cup \cdots \cup E_n$ of events in a sample space S is given by

$$P\left(\bigcup_{i=1}^{n} E_i\right) = \sum_{k=1}^{n} (-1)^{k+1} \sum_{\substack{i_1, i_2, \ldots, i_k: \\ 1 \le i_1 < i_2 < \cdots < i_k \le n}} P\left(E_{i_1} \cap E_{i_2} \cap \cdots \cap E_{i_k}\right).$$

6. *Hatcheck problem.* The *hatcheck problem,* or *derangement problem,* asks for the probability that a bijection of an n-element set maps no element to itself. The answer is

$$\sum_{i=2}^{n} (-1)^i \frac{1}{i!} = \frac{1}{2} - \frac{1}{3!} + \cdots + \frac{(-1)^n}{n!},$$

which is the result of truncating the power series expansion of e^{-1} at the $(-1)^n/n!$ term. Thus, the result is very close to $1/e$, even for relatively small values of n.

7. *Principle of inclusion and exclusion for counting.*

$$\left| \bigcup_{i=1}^{n} E_i \right| = \sum_{k=1}^{n} (-1)^{k+1} \sum_{\substack{i_1, i_2, \ldots, i_k: \\ 1 \le i_1 < i_2 < \cdots < i_k \le n}} \left| E_{i_1} \cap E_{i_2} \cap \cdots \cap E_{i_k} \right|.$$

Problems

1. Compute the probability that in three flips of a coin, the coin comes up heads on the first flip or on the last flip. a/h

2. The eight kings and queens are removed from a deck of cards, and then two of these cards are selected. What is the probability that the king or queen of spades is among the cards selected?

3. Two dice are rolled. What is the probability that you get a die with six dots on top? a/h

4. A bowl contains two red, two white, and two blue balls. If you remove two balls, what is the probability that at least one is red or white? Compute the probability that at least one is red.

5. Remove one card from an ordinary deck of cards. What is the probability that it is an ace, a diamond, or black? a/h

6. Give a formula for the probability of $P(E \cup F \cup G \cup H)$ in terms of the probabilities of E, F, G, and H and their intersections.

7. What is

$$\sum_{\substack{i_1, i_2, i_3: \\ 1 \le i_1 < i_2 < i_3 \le 4}} i_1 i_2 i_3 \ ?$$ a/h

8. What is
$$\sum_{\substack{i_1, i_2, i_3: \\ 1 \le i_1 < i_2 < i_3 \le 5}} i_1 + i_2 + i_3 \ ?$$

9. The boss asks the secretary to stuff n letters into envelopes, forgetting to mention that he has been adding notes to the letters and, in the process, has rearranged the letters but not the envelopes. In how many ways can the letters be stuffed into the envelopes so that nobody gets the letter intended for him or her? What is the probability that nobody gets the letter intended for him or her? a/h

10. If you are hashing n keys into a hash table with k locations, what is the probability that every location gets at least one key?

11. From Theorem 5.2, find a formula for $S(n, m)$, which is defined in Problem 12 of Section 1.4. These numbers are called **Stirling numbers (of the second kind)**. a/h

12. If you roll eight dice, what is the probability that each of the numbers 1 through 6 appear on top at least once? What about with nine dice?

13. Explain why the number of ways of distributing k identical apples to n children is $\binom{n+k-1}{k}$. In how many ways may you distribute the apples to the children so that Sam gets more than m apples? In how many ways may you distribute the apples to the children so that no child gets more than m apples? a/h

14. A group of n married couples sits around a circular table for a discussion of marital problems. The counselor assigns each person to a seat at random. What is the probability that no husband and wife are side-by-side? a/h

15. Suppose you have a collection of m objects and a set P of p "properties." (We won't define the term "property," but note that a property is something the objects may or may not have.) For each subset S of the set P of all properties, define $N_a(S)$ to be the number of objects in the collection that have *at least* the properties in S (a is for "at least"). Thus, for example, $N_a(\emptyset) = m$. In a typical application, formulas for $N_a(S)$ for other sets $S \subseteq P$ are not difficult to figure out. Define $N_e(S)$ to be the number of objects in our collection that have *exactly* the properties in S (e is for "exactly"). Show that

$$N_e(\emptyset) = \sum_{K:K \subseteq P} (-1)^{|K|} N_a(K) \ .$$

Explain how this formula could be used to compute the number of onto functions in a more direct way than we did when using unions of sets. How would this formula apply to Problem 9?

*16. In Problem 14, two people of the same sex could sit side-by-side. If in addition to the condition that no husband and wife are side-by-side we require that no two people of the same sex are side-by-side, we obtain a famous problem known as the *mènage problem*. Solve this problem. a/h

17. In how many ways may you place n distinct books on j shelves so that Shelf 1 gets at least m books? (See Problem 7 in Section 1.4.) In how many ways may you place n distinct books on j shelves so that no shelf gets more than m books? a/h

18. In Problem 15, what is the probability that an object has none of the properties, assuming all objects are equally likely? How would this apply to Problem 10?

5.3 CONDITIONAL PROBABILITY AND INDEPENDENCE

Conditional Probability

Exercise 5.3-1

Two cubical dice each have a triangle painted on one side, a circle painted on two sides, and a square painted on three sides. The probability of seeing at least one circle on top is the probability of a circle on the top of the first die or of a circle on top of the second die. Applying the principle of inclusion and exclusion, we can compute that the probability of seeing a circle on at least one top when we roll the dice is $1/3 + 1/3 - 1/9 = 5/9$. We are experimenting to see if reality agrees with our computation. We throw the dice onto the floor, and they bounce a few times before landing in the next room. Our friend in the next room tells us both top sides are the same. What is the probability that our friend sees a circle on at least one top?

Intuitively, it may seem as if the chance of getting circles ought to be four times the chance of getting triangles, and the chance of getting squares ought to be nine times the chance of getting triangles. We could turn this into the algebraic statements that $P(\text{circles}) = 4P(\text{triangles})$ and $P(\text{squares}) = 9P(\text{triangles})$. These two equations and the one that says the probabilities sum to 1 are enough to conclude that the probability that our friend saw two circles is $2/7$. But does this analysis make sense? To convince ourselves, let's start with a sample space for the original experiment and see what natural assumptions about probability we can make to determine the new probabilities. In the process, we will be able to replace intuitive calculations with a formula we can use in similar situations. This is a good thing, because we have already seen situations where our intuitive idea of probability did not always agree with what the rules of probability give us.

Let us take as our sample space for this experiment the ordered pairs shown in Table 5.2, along with their probabilities.

---

*Although this problem can be solved by extending the technique of Problem 14, it does require more insight than the other problems in this section.

Table 5.2: *Rolling two unusual dice*

TT	TC	TS	CT	CC	CS	ST	SC	SS
$\frac{1}{36}$	$\frac{1}{18}$	$\frac{1}{12}$	$\frac{1}{18}$	$\frac{1}{9}$	$\frac{1}{6}$	$\frac{1}{12}$	$\frac{1}{6}$	$\frac{1}{4}$

We know that the event {TT, CC, SS} happened. Thus, we would say that although this event used to have probability

$$\frac{1}{36} + \frac{1}{9} + \frac{1}{4} = \frac{14}{36} = \frac{7}{18},$$ (5.12)

it now has probability 1. Given this, what probability would we now assign to the event of seeing a circle? Notice that the event of seeing a circle has become the event CC. Should we expect CC to become more or less likely in comparison with TT or SS just because we now know that one of these three outcomes has occurred? Nothing has happened to make us expect that, so whatever new probabilities we assign to these two events, they should have the same ratios as the old probabilities.

Multiplying all three old probabilities by 18/7 to get our new probabilities will preserve the ratios and make the three new probabilities add to 1. (Is there any other way to get the three new probabilities to add to 1 and make the new ratios the same as the old ones?) This gives us that the probability of two circles is $(1/9)(18/7) = 2/7$. Notice that nothing we have learned about probability so far told us what to do; we just made a decision based on common sense. When faced with similar situations in the future, it would make sense to use common sense in the same way. However, do we really need to go through the process of constructing a new sample space and reasoning about its probabilities again? Fortunately, our entire reasoning process can be captured in a formula. We wanted the probability of an event E given that the event F happened. We figured out what the event $E \cap F$ was and then multiplied its probability by $1/P(F)$. We summarize this process in a definition.

The **conditional probability** of E given F, denoted by $P(E|F)$ and read as "the probability of E given F," is

$$P(E|F) = \frac{P(E \cap F)}{P(F)}.$$ (5.13)

Whenever we want the probability of E, knowing that F has happened, we compute $P(E|F)$. (If $P(F) = 0$, then we cannot divide by $P(F)$; but F gives us no new information about our situation. For example, if our friend in the next room says, "A pentagon is on top," we have no information except that the student isn't looking at the dice we rolled. Thus, because we have no reason to change our sample space or the probability weights of its elements, we define $P(E|F) = P(E)$ when $P(F) = 0$.)

Notice, we did not prove that the probability of E given F is what we said it is. We simply defined it in this way, because in the process of making the derivation, we made an additional assumption that the relative probabilities of the outcomes in the event F don't change when F happens. This assumption led us to Equation 5.13. Then we chose that equation as our definition of the new concept of the conditional probability of E given F.[5]

In the preceding example, we can let E be the event that there is more than one circle and F be the event that both dice are the same. Then $E \cap F$ is the event that both dice are circles, and $P(E \cap F)$ is, from Table 5.2, 1/9. $P(F)$ is, from Equation 5.12, 7/18. Dividing, we get the probability $P(E|F)$, which is $(1/9)/(7/18) = 2/7$.

Exercise 5.3-2

When we roll two ordinary dice, what is the probability that the sum of the tops comes out even, given that the sum is greater than or equal to 10? Use the definition of conditional probability in solving the problem.

Exercise 5.3-3

We say E is **independent** of F if $P(E|F) = P(E)$. Show that when we roll two dice, one red and one green, the event "the total number of dots on top is odd" is independent of the event "the red die has an odd number of dots on top."

Exercise 5.3-4

Sometimes information about conditional probabilities is given to us indirectly in the statement of a problem, and we have to derive information about other probabilities or conditional probabilities. Here is such an example: If a student knows 80% of the material in a course, what do you expect her grade to be on a (well-balanced) 100-question short-answer test about the course? What is the probability that she answers a question correctly on a 100-question true-false test if she guesses at each question for which she does not know the answer? (We assume she knows what she knows—that is, if she thinks she knows the answer, then she really does.) What do you expect her grade to be on a 100-question true-false test?

For Exercise 5.3-2, let E be the event that the sum is even and F be the event that the sum is greater than or equal to 10. Using a sample space of ordered pairs, each of weight 1/36, $P(F) = 1/6$ and $P(E \cap F) = 1/9$, because the latter is the probability that the roll is either 10 or 12. Dividing $P(E \cap F)$ by $P(F)$, we get 2/3.

In Exercise 5.3-3, the event that the total number of dots is odd has probability 1/2. Similarly, given that the red die has an odd number of dots, the probability of an odd sum is 1/2, because this event corresponds exactly to getting an even roll on the green die. That is,

$$P(\text{even number of dots on top} \mid \text{red die is odd}) = \frac{3}{6} = \frac{1}{2}.$$

Thus, by the definition of independence, the event of an odd number of dots on the red die and the event that the total number of dots is odd are independent.

---

[5] For those who like to think in terms of axioms of probability, note that if F is an event in a sample space S and $P(F) \neq 0$, then the function of E given by $P'(E \cap F)/P(F)$ satisfies the axioms of probability on S. Thus, the function is a probability measure on S. We then define it to be the conditional probability of E given F.

In Exercise 5.3-4, if a student knows 80% of the material in a course, we would hope that her grade on a well-designed test of the course would be around 80%. But what if the test is a true-false test? Let R be the event that she gets the right answer, K be the event that she knows that right answer, and $\overline{K}$ be the event that she guesses. Then, $R = (R \cap K) \cup (R \cap \overline{K})$. Because R is a union of two disjoint events, its probability would be the sum of the probabilities of these two events. How do we get the probabilities of these two events? The statement of the problem implicitly gives us the conditional probability $P(R|K)$—namely, 1—that she gets the right answer given that she knows the answer. It also gives us the probability $P(R|\overline{K})$—namely, 1/2—that she gets the right answer if she doesn't know the answer. The problem also tells us explicitly that $P(K) = .8$ and $P(\overline{K}) = .2$. How can we make use of this information? Notice that we are given two of the three terms of Equation 5.13, where E is R and F is either K or $\overline{K}$. Thus, we can use the equation

$$P(E \cap F) = P(E|F)P(F)$$

to compute $P(R \cap K)$ and $P(R \cap \overline{K})$. In symbols,

$$
\begin{aligned}
P(R) &= P(R \cap K) + P(R \cap \overline{K}) \\
&= P(R|K)P(K) + P(R|\overline{K})P(\overline{K}) \\
&= 1 \cdot .8 + .5 \cdot .2 = .9 \,.
\end{aligned}
$$

We have shown that the probability of her getting the right answer is .9. Thus, we would expect her to get a grade of 90%.

Independence

We said in Exercise 5.3-3 that E is independent of F if $P(E|F) = P(E)$. The **product principle for independent probabilities** gives another test for independence:

Theorem 5.5	**(Product Principle for Independent Probabilities)** Suppose E and F are events in a sample space. Then E is independent of F if and only if $P(E \cap F) = P(E)P(F)$.

Proof: First, consider the case when F is nonempty. Then, from our definition in Exercise 5.3-3 (recall the convention of using "if" in a definition even though we mean "if and only if"),

$$E \text{ is independent of } F \quad \Leftrightarrow \quad P(E|F) = P(E) \,. \qquad (5.14)$$

Starting with the right side of Implication 5.14 and using the definition of $P(E|F)$ in Equation 5.13, we get

$$
\begin{aligned}
P(E|F) &= P(E) \\
\Leftrightarrow \quad \frac{P(E \cap F)}{P(F)} &= P(E) \\
\Leftrightarrow \quad P(E \cap F) &= P(E)P(F)
\end{aligned}
$$

Because every step in this proof is an "if and only if" statement, we have completed the proof for the case when F is nonempty.

If F is empty, then E is independent of F and both $P(E)P(F)$ and $P(E \cap F)$ are zero. Thus, in this case as well, E is independent of F if and only if $P(E \cap F) = P(E)P(F)$. ∎

Corollary 5.6 E is independent of F if and only if F is independent of E.

When we flip a coin twice, we think of the second outcome as being independent of the first. It would be a sorry state of affairs if our definition of independence did not capture this intuitive idea! Let's compute the relevant probabilities to see if it does. For flipping a coin twice, our sample space is {HH, HT, TH, TT}, and we weight each outcome $1/4$. To say the second outcome is independent of the first, we must mean that getting an H second is independent of whether we get an H or a T first; the same is true for getting a T second. Because each element of our sample space has weight $1/4$, $P(\text{H first}) = 1/4 + 1/4 = 1/2$ and $P(\text{H second}) = 1/2$, while $P(\text{H first and H second}) = 1/4$. Note that

$$P(\text{H first}) P(\text{H second}) = \frac{1}{2} \cdot \frac{1}{2} = \frac{1}{4} = P(\text{H first and H second}) .$$

By Theorem 5.5, this means that the event "H second" is independent of the event "H first." We can make a similar computation for each possible combination of outcomes for the first and second flip, and so we see that our definition of independence captures our intuitive idea of independence in this case. Clearly the same sort of computation applies to rolling dice.

Exercise 5.3-5 What sample space and probabilities have we been using when discussing hashing? Using these, show that the event "key i hashes to position r" and the event "key j hashes to position q" are independent when $i \neq j$. Are they independent if $i = j$?

In Exercise 5.3-5, if we have a list of n keys to hash into a table of size k, our sample space consists of all n-tuples of numbers between 1 and k. The event that key i hashes to some number r consists of all n-tuples with r in the ith position, so its probability is $k^{n-1}/k^n = 1/k$. The probability that key j hashes to some number q is also $1/k$. If $i \neq j$, then the event that key i hashes to r and key j hashes to q has probability $k^{n-2}/k^n = 1/k^2$, which is the product of the probabilities that key i hashes to r and key j hashes to q. Therefore, these two events are independent. If $i = j$, the probability of key i hashing to r and key j hashing to q is 0, unless $r = q$, in which case it is 1. Thus, if $i = j$, these events are not independent.

Independent Trials Processes

Coin flipping and hashing are examples of processes called "independent trials processes." Suppose we have a process that occurs in stages. (For example, we might flip a coin n times.) Let us use x_i to denote the outcome at stage i. (For flipping a coin n times, $x_i = $ H means that the outcome of the ith flip is a head.) We let S_i stand for the set of possible outcomes of stage i. (Thus, if we flip a coin n times, $S_i = \{H, T\}$ for each i, $1 \le i \le n$.) A process that occurs in stages is called an **independent trials process** if

$$P(x_i = a_i | x_1 = a_1, \ldots, x_{i-1} = a_{i-1}) = P(x_i = a_i) \qquad (5.15)$$

for each sequence $a_1, a_2, \ldots, a_n$, with $a_i \in S_i$. Letting E_i be the event that $x_i = a_i$, we can rewrite Equation 5.15 as

$$P(E_i | E_1 \cap E_2 \cap \cdots \cap E_{i-1}) = P(E_i) . \qquad (5.16)$$

In words, an independent trials process has the property that the outcome of stage i is independent of the outcomes of stages 1 through $i - 1$. By the product principle for independent probabilities (Theorem 5.5), Equation 5.16 implies that

$$P(E_1 \cap E_2 \cap \cdots E_{i-1} \cap E_i) = P(E_1 \cap E_2 \cap \cdots E_{i-1})P(E_i) . \qquad (5.17)$$

Theorem 5.7	In an independent trials process, the probability of a sequence $a_1, a_2, \ldots, a_n$ of outcomes is $P(\{a_1\})P(\{a_2\}) \cdots P(\{a_n\})$.

Proof: To prove this theorem, we apply mathematical induction and Equation 5.17. ∎

How do independent trials relate to coin flipping? When flipping coins, our sample space consists of sequences of n H's and T's, and the event that we have an H (or a T) on the ith flip is independent of the event that we have an H (or a T) on each of the first $i - 1$ flips. In particular, the probability of an H on the ith flip is $2^{n-1}/2^n = .5$, and the probability of an H on the ith flip, given a particular sequence on the first $i - 1$ flips, is $2^{n-i-1}/2^{n-i} = .5$.

How do independent trials relate to hashing a list of keys? As in Exercise 5.3-5, if we have a list of n keys to hash into a table of size k, our sample space consists of all n-tuples of numbers between 1 and k. The probability

$$P(\text{key } i \text{ hashes to } r \text{ and keys 1 through } i - 1 \text{ hash to } q_1, q_2, \ldots, q_{i-1})$$

is

$$\frac{k^{n-i}}{k^n} = k^{-i} = \frac{1}{k^i} .$$

The probability

$$P(\text{keys 1 through } i - 1 \text{ hash to } q_1, q_2, \ldots, q_{i-1})$$

is

$$\frac{k^{n-i+1}}{k^n} = k^{1-i} \; .$$

By the definition of conditional probability, we get

$$P(\text{key } i \text{ hashes to } r \,|\, \text{keys 1 through } i - 1 \text{ hash to } q_1, q_2, \ldots, q_{i-1}) = \frac{k^{n-i}/k^n}{k^{n-i+1}/k^n}$$

$$= \frac{1}{k} \; .$$

Consequently, the event that key i hashes to some number r is independent of the event that the first $i - 1$ keys hash to some numbers $q_1, q_2, \ldots, q_{i-1}$. Thus, our model of hashing is an independent trials process.

Exercise 5.3-6	Suppose we draw a card from a standard deck of 52 cards, replace it, draw another card, and continue for a total of ten draws. Is this an independent trials process?

Exercise 5.3-7	Suppose we draw a card from a standard deck of 52 cards, discard it (i.e., we do not replace it), draw another card, and continue for a total of ten draws. Is this an independent trials process?

In Exercise 5.3-6, we have an independent trials process because the probability that we draw a given card at one stage does not depend on what cards we have drawn in earlier stages. However, in Exercise 5.3-7, we don't have an independent trials process. In the first draw, we have 52 cards to draw from, while in the second draw, we have 51. In particular, we do not have the same cards to draw from on the second draw as on the first; so, the probabilities for each possible outcome on the second draw depend on whether that outcome was the result of the first draw.

Tree Diagrams

When we have a sample space that consists of sequences of outcomes, it is often helpful to visualize the outcomes with a tree diagram. As an example of what we mean, let's look at creating a tree diagram of the following experiment. We have one nickel, two dimes, and two quarters in a cup. We draw a first and second coin. Figure 5.5 shows our diagram for this process. Notice that in probability theory, it is standard to have trees open to the right, rather than opening up or down.

Figure 5.5: *A tree diagram illustrating a two-stage process*

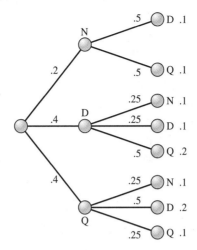

Each level of the tree corresponds to one stage of the process of generating a sequence in our sample space. We label each vertex with one of the possible outcomes at the stage it represents. We label each edge with a conditional probability—the probability of getting the outcome at the edge's right end given the sequence of outcomes that have occurred so far. Because no outcomes have occurred at Stage 0, we label the edges from the root to the first-stage vertices with the probabilities of the outcomes at the first stage. Each path from the root to the far right of the tree represents a possible sequence of outcomes for our process. We label each leaf node with the probability of the sequence that corresponds to the path from the root to that node. By the definition of conditional probabilities, the probability of a path is the product of the probabilities along its edges. We can draw a tree diagram, also known as a probability tree, for any (finite) sequence of successive trials in this way.

Sometimes a tree diagram provides a very effective way of answering questions about a process. For example, what is the probability of having a nickel in our coin experiment? We see in Figure 5.5 that there are four paths containing an N, and the sum of their weights is .4. So, the probability that one of our two coins is a nickel is .4.

Exercise 5.3-8   How can we recognize from a tree diagram whether it is the tree diagram of an independent trials process?

Exercise 5.3-9   Exercise 5.3-4 asked (among other things), if a student knows 80% of the material in a course, what is the probability that she answers a question correctly on a 100-question true-false test (assuming that she guesses on any question for which she does not know the answer)? (We assume she knows what she knows—that is, if she thinks she knows the answer, then she really does.) Show how we can use a tree diagram to answer this question.

Exercise 5.3-10	A test for a disease that affects 0.1% of the population is 99% effective on people with the disease (that is, the test says they have the disease with probability .99). The test gives a false reading (saying that a person who does not have the disease is affected with it) for 2% of the population without the disease. We can think of choosing someone and testing them for the disease as a two-stage process. In Stage 1, we either choose someone with the disease or we don't. In Stage 2, the test is either positive or it isn't. Give a tree diagram for this process. What is the probability that someone selected at random and given a test for the disease tests positive? What is the probability that someone who tests positive in fact has the disease?

A tree for an independent trials process has the property that at each level, for each node at that level, the labeled tree consisting of that node and all its children is identical to each labeled tree consisting of another node at that level and all its children. If we have such a tree, then it automatically satisfies the definition of an independent trials process.

In Exercise 5.3-9, if a student knows 80% of the material in a course, we expect that she has probability .8 of knowing the answer to any given question of a well-designed true-false test. We regard her work on a question as a two-stage process. In Stage 1, she determines whether she knows the answer, and in Stage 2, either she answers correctly, with probability 1, or she guesses, in which case she answers correctly with probability $1/2$ or incorrectly with probability $1/2$. As we see in Figure 5.6, there are two root-leaf paths corresponding to her getting a correct answer. One of these paths has probability .8 and the other has probability .1. Thus, she actually has probability .9 of getting a right answer if she guesses the answer to each question for which she does not know the answer.

Figure 5.6: *The probability of getting a right answer is .9*

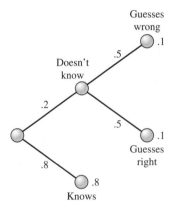

Figure 5.7 shows the tree diagram for thinking of Exercise 5.3-10 as a two-stage process. In the first stage, a person either has or doesn't have the disease. In the second stage, we administer the test, and its result is either positive or negative. We use D to stand for having the disease and ND to stand for not having the disease. We use pos to stand for a positive test and neg to stand for a negative test. We assume that a test is either positive or negative. The question asks for the conditional probability that

Figure 5.7: *A tree diagram illustrating Exercise 5.3-10*

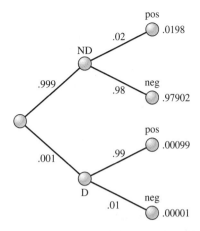

someone has the disease, given that he or she tests positive:

$$P(\text{D}|\text{pos}) = \frac{P(\text{D} \cap \text{pos})}{P(\text{pos})} \, .$$

From the tree, we read that $P(\text{D} \cap \text{pos}) = .00099$, because this event consists of just one root-leaf path. The event pos consists of two root-leaf paths, whose probabilities total $.0198 + .00099 = .02097$. Thus, $P(\text{D}|\text{pos}) = P(\text{D} \cap \text{pos})/P(\text{pos}) = .00099/.02097 = .0472$. Given a disease this rare and a test with this error rate, a positive result only gives a roughly 5% chance of having the disease. Here is another instance where a probability analysis shows something we might not have expected initially. This explains why doctors often don't want to administer a test to someone unless that person is already showing some symptoms of the disease being tested for.

We can also solve Exercise 5.3-10 purely algebraically. We are given that

$$P(\text{D}) = .001 \, , \tag{5.18}$$

$$P(\text{pos}|\text{D}) = .99 \, , \text{ and} \tag{5.19}$$

$$P(\text{pos}|\text{ND}) = .02 \, . \tag{5.20}$$

We wish to compute $P(\text{D}|\text{pos})$. We use Equation 5.13 to write

$$P(\text{D}|\text{pos}) = \frac{P(\text{D} \cap \text{pos})}{P(\text{pos})} \, . \tag{5.21}$$

How do we compute the numerator? Using $P(\text{D} \cap \text{pos}) = P(\text{pos} \cap \text{D})$ and Equation 5.13 again, we can write

$$P(\text{pos}|\text{D}) = \frac{P(\text{pos} \cap \text{D})}{P(\text{D})} \, .$$

Plugging Equations 5.18 and 5.19 into this equation, we get

$$.99 = \frac{P(\text{pos} \cap \text{D})}{.001} \, ,$$

or $P(\text{pos} \cap \text{D}) = (.001)(.99) = .00099$.

To compute the denominator of Equation 5.21, we observe that because each person either has the disease or doesn't, we can write

$$P(\text{pos}) = P(\text{pos} \cap D) + P(\text{pos} \cap ND) . \tag{5.22}$$

We have already computed $P(\text{pos} \cap D)$, and we can compute the probability $P(\text{pos} \cap ND)$, the probability of a positive result and no disease, in a similar manner. Writing

$$P(\text{pos}|ND) = \frac{P(\text{pos} \cap ND)}{P(ND)} ,$$

observing that $P(ND) = 1 - P(D)$, and plugging in the values from Equations 5.18 and 5.20, we get that $P(\text{pos} \cap ND) = (.02)(1 - .001) = .01998$. We now have the two components of the right side of Equation 5.22, and thus, $P(\text{pos}) = .00099 + .01998 = .02097$. Finally, we have all the pieces in Equation 5.21 and conclude that

$$P(D|\text{pos}) = \frac{P(D \cap \text{pos})}{P(\text{pos})} = \frac{.00099}{.02097} = .0472 .$$

Clearly, using the tree diagram mirrors these computations, but it both simplifies the thought process and reduces the amount we have to write.

Primality Testing

Exercise 5.3-10 illustrates the problems we might face in determining whether a number is likely to be prime. We have so far discussed the idea that a nonprime will fail 3/4 of the primality tests we use on it. Thus, if we use five such independent tests, the chance that a nonprime will fail to be certified nonprime is only $1/4^5$, or about $1/1000$. In Section 5.4, we will see what assumptions are required to get $1/4^5$. We have noted that the expected number of primes in an interval of length $\ln n$ centered around the number n is 1. So, if we are going to choose a number n randomly, the probability that n is prime is about $1/\ln n$. Because $\ln n$ grows quite slowly with n, this probability is not too small, even for reasonably large values of n. But if n has on the order of 150 digits, then $\ln n$ is about 350. (For RSA, we suggested that choosing primes of about 150 digits is sufficient.) Thus, if we are going to choose a number with 150 digits, the probability that it is prime is about $1/350$. The tree diagram for testing a prime is similar to Figure 5.6. Being prime corresponds to the lowest branch of the tree and has probability $1/350$. Not being prime corresponds to the upper branch of the tree and has probability $349/350$ (in place of the .2 in Figure 5.6). Testing nonprime corresponds to guessing right and has probability approximately $999/1000$; being composite but not testing so corresponds to being wrong and has probability approximately $1/1000$. This gives us that the probability of a random number not testing as nonprime is $1/350 + 349/(350 \cdot 1000)$. Thus, by the methods of Exercise 5.3-10, the probability that a number that doesn't test nonprime *is* prime is

$$\frac{1/350}{1/350 + 349/(350 \cdot 1000)} \approx \frac{1/350}{1/350 + 1/1000} ,$$

which is about .74. So, the fact that a number fails to test nonprime in five tests is pretty poor evidence that it is prime.

Suppose we use $5k$ tests for some integer $k > 1$. Then the probability that a nonprime fails to test nonprime in this many tests is about $1/1000^k$. The probability that a number that fails $5k$ tests is prime is then approximately

$$\frac{1/350}{1/350 + 1/1000^k} = \frac{1}{1 + 350/1000^k} .$$

This formula is slightly clumsy to work with. However, when $x < 1$, we have

$$\frac{1}{1 - x^2} > 1 ,$$

so that

$$\frac{1}{1 + x} > 1 - x .$$

This gives us

$$\frac{1}{1 + 350/1000^3} > 1 - \frac{350}{1000^k} > 1 - \frac{1}{1000^{(k-1)}} ,$$

which shows that we can guarantee that the probability a randomly chosen number is prime, given that it fails $5(k + 1)$ tests, is at least $1 - 1000^{-k}$. The same bound holds if we replace 350 with 1000. So, reversing the process that led us from 150 digits to an interval of 350 numbers tells us we can apply this guarantee to numbers with $\log_{10} e^{1000} \approx 435$ digits. To guarantee that a randomly chosen number of up to 435 digits is prime with probability less than 1000^{-k} of being wrong, we need only run $5(k+1)$ nonprimality tests.

Important Concepts, Formulas, and Theorems

1. *Conditional probability.* The *conditional probability* of E given F, denoted by $P(E|F)$ and read as "the probability of E given F," is defined by

$$P(E|F) = \frac{P(E \cap F)}{P(F)}$$

when $P(F) \neq 0$.

2. *Independent.* We say E is *independent* of F if $P(E|F) = P(E)$.

3. *Product principle for independent probabilities.* The product principle for independent probabilities (Theorem 5.5) gives another test for independence. Suppose E and F are events in a sample space. Then E is independent of F if and only if $P(E \cap F) = P(E)P(F)$.

4. *Symmetry of independence.* The event E is independent of the event F if and only if F is independent of E.

5. *Independent trials process.* A process that occurs in stages is called an *independent trials process* if, for each sequence $a_1, a_2, \ldots, a_n$ with $a_i \in S_i$,

$$P(x_i = a_i | x_1 = a_1, \ldots, x_{i-1} = a_{i-1}) = P(x_i = a_i).$$

6. *Probabilities of outcomes in independent trials.* In an independent trials process, the probability of a sequence $a_1, a_2, \ldots, a_n$ of outcomes is $P(\{a_1\})P(\{a_2\}) \cdots P(\{a_n\})$.

7. *Coin flipping.* Repeatedly flipping a coin is an independent trials process.

8. *Hashing.* Hashing a list of n keys into k slots is an independent trials process with n stages.

9. *Tree diagram.* In a tree diagram for a multistage process, each level of the tree corresponds to one stage of the process. Each vertex is labeled with one of the possible outcomes at the stage it represents. Each edge is labeled with a conditional probability—the probability of getting the outcome at its right end given the sequence of outcomes that have occurred so far. Each path from the root to a leaf represents a sequence of outcomes and is labeled with the product of the probabilities along that path. This is the probability of that sequence of outcomes.

Problems

1. In three flips of a coin, what is the probability that two flips in a row are heads, given that there is an even number of heads? a/h

2. In three flips of a coin, is the event that two flips in a row are heads independent of the event that there is an even number of heads?

3. In three flips of a coin, is the event of getting at most one tail independent of the event that not all flips are identical? a/h

4. What is the sample space that you use for rolling two dice, a first one and then a second one? Using this sample space, explain why the event "i dots are on top of the first die" and the event "j dots are on top of the second die" are independent if you roll two dice.

5. If you flip a coin twice, is the event of having an odd number of heads independent of the event that the first flip comes up heads? Is it independent of the event that the second flip comes up heads? Would you say that the three events are mutually independent? (The term "mutually independent" hasn't been defined, so the question is one of opinion. However, you should back up your opinion with a reason that makes sense.) a/h

6. Assume that on a true-false test, students will answer correctly any question on a subject that they know. Assume students guess at answers they do not know. For students who know 60% of the material in a course, what is the probability that they will answer a question correctly? What is the probability that they will know the answer to a question they answer correctly?

7. A nickel, two dimes, and two quarters are in a cup. You draw three coins, one at a time, without replacement. Draw the tree diagram that represents the process. Use the tree to determine the probability of getting a nickel on the last draw. Use the tree to determine the probability that the first coin is a quarter, given that the last coin is a quarter. `a/h`

8. Write a formula for the probability that a bridge hand (which is 13 cards chosen from an ordinary deck) has four aces, given that it has (at least) one ace. Write a formula for the probability that a bridge hand has four aces, given that it has the ace of spades. Which of these probabilities is larger?

9. A nickel, two dimes, and three quarters are in a cup. You draw three coins, one at a time, without replacement. What is the probability that the first coin is a nickel? What is the probability that the second coin is a nickel? What is the probability that the third coin is a nickel? `a/h`

10. If a student knows 75% of the material in a course, and if a 100-question multiple-choice test with five choices per question covers the material in a balanced way, what is the student's probability of getting a right answer to a question, given that the student guesses at the answer to each question whose answer he does not know?

11. Suppose E and F are events with $E \cap F = \emptyset$. Describe when E and F are independent, and explain why. `a/h`

12. In a family consisting of a mother, father, and two children of different ages, what is the probability that the family has two girls, given that one of the children is a girl? What is the probability that the children are both boys, given that the older child is a boy?

13. You are a contestant on the TV game show *Let's Make a Deal*. In this game show, there are three curtains. Behind one of the curtains is a new car, and behind the other two are cans of Spam. You get to pick one of the curtains. After you pick one of the curtains, the emcee, Monty Hall, who we assume knows where the car is, reveals what is behind one of the curtains that you did not pick, showing you some cans of Spam. He then asks you if you would like to switch your choice of curtain. Should you switch? Why or why not? Please answer this question carefully. You have all the tools needed to answer it, but several math Ph.D.s are on record (in *Parade* magazine) giving the wrong answer. `a/h`

5.4 RANDOM VARIABLES

What Are Random Variables?

A **random variable** for an experiment with a sample space S is a *function* that *assigns a number* to each element of S. Typically, instead of using f to stand for such a function, we use X. (At first, a random variable was conceived of as a variable related to an experiment, explaining the use of X, but it is very helpful in understanding the mathematics to realize that X is actually a function on the sample space.)

For example, if we consider the process of flipping a coin n times, we have the set of all sequences of n H's and T's as our sample space. The "number of heads" random variable takes a sequence and tells us how many heads are in that sequence. For example, if we let X be the number of heads in five flips of a coin, then $X(\text{HTHHT}) = 3$ while $X(\text{THTHT}) = 2$. It may be jarring to see X used to stand for a function, but it is the standard notation for a random variable.

For a sequence of hashes of n keys into a table with k locations, we might have a random variable X_i that is the number of keys hashed to location i of the table or a random variable X that counts the number of collisions (hashes to a location that already has at least one key). For an n-question test on which each answer is either right or wrong (for example, a short-answer, true-false, or multiple-choice test), we could have a random variable that gives the number of right answers in a particular sequence of answers to the test. For a meal at a restaurant, we might have a random variable that gives the price of any particular sequence of choices of menu items.

| Exercise 5.4-1 | Give several random variables that might be of interest to a doctor whose sample space is her patients. |

| Exercise 5.4-2 | If you flip a coin six times, how many heads do you expect? |

A doctor might be interested in patients' ages, weights, temperatures, blood pressures, cholesterol levels, and so on.

For Exercise 5.4-2, in six flips of a coin, it is natural to expect three heads. We might argue that if we average the number of heads over all possible outcomes, the average should be half the number of flips. Because the probability of any given sequence equals that of any other, it is reasonable to say that this average is what we expect. Thus, we expect the number of heads to be half the number of flips. We will explore this concept more formally later.

Binomial Probabilities

When we study an independent trials process with two outcomes at each stage, it is traditional to refer to those outcomes as successes and failures. When we are flipping a coin, we are often interested in the number of heads. When we are analyzing student performance on a test, we are interested in the number of correct answers. When we are analyzing the outcomes in drug trials, we are interested in the number of trials where the drug was successful in treating the disease. This suggests a natural random variable associated with an independent trials process that has two outcomes at each stage—namely, the number of successes in n trials. We analyze, in general, the probability of exactly k successes in n independent trials with probability p of success (and thus probability $1 - p$ of failure) on each trial. It is standard to call such an independent trials process a **Bernoulli trials process**.

Exercise 5.4-3 Suppose we have five Bernoulli trials, with probability p of success on each trial. What is the probability of success on the first three trials and failure on the last two? Failure on the first two trials and success on the last three? Success on Trials 1, 3, and 5, and failure on the other two? Success on any particular three trials and failure on the other two?

Because the probability of a sequence of outcomes is the product of the probabilities of the individual outcomes, the probability of any sequence of three successes and two failures is $p^3(1-p)^2$. More generally, in n Bernoulli trials, the probability of a given sequence of k successes and $n-k$ failures is $p^k(1-p)^{n-k}$. However, this is not the probability of having k successes, because many different sequences could have k successes.

How many sequences of n successes and failures have exactly k successes? The number of ways to choose the k places out of n where the successes occur is $\binom{n}{k}$. Therefore, the number of sequences with k successes is $\binom{n}{k}$. This paragraph and the paragraph that precedes it give us the following theorem.

Theorem 5.8 The probability of having exactly k successes in a sequence of n independent trials with two outcomes and probability p of success on each trial is given by

$$P(\text{exactly } k \text{ successes}) = \binom{n}{k} p^k(1-p)^{n-k}.$$

Proof: The proof follows from the two paragraphs preceding the theorem. ■

Because of the connection between these probabilities and the binomial coefficients, the probabilities of Theorem 5.8 are called **binomial probabilities**, or the **binomial probability distribution**.

Exercise 5.4-4 A student takes a ten-question objective test.[6] Suppose that a student who knows 80% of the course material has probability .8 of success on any question, independent of how he did on any other problem. What is the probability that he earns a grade of 80 or better (out of 100)?

---

[6]By an objective test, we mean one in which the answer is either right or wrong and guessing is not possible (i.e., the test is not true-false or multiple choice).

Recall the primality testing algorithm from Section 2.4. In it, we said we could choose a random number less than or equal to n in order to perform a test on n, such that if n was not prime (in other words, n was composite), the number would certify this fact with probability 3/4. Suppose we perform 20 of these tests. It is reasonable to assume that each test is independent of the others. What is the probability that a composite number is certified to be composite?

Because a grade of 80 or better on a ten-question test corresponds to eight, nine, or ten successes in ten trials, in Exercise 5.4-4, we have

$$P(80 \text{ or better}) = \binom{10}{8} (.8)^8(.2)^2 + \binom{10}{9} (.8)^9(.2)^1 + \binom{10}{10} (.8)^{10}(.2)^0 .$$

Some work with a calculator gives us that this sum is approximately .678.

In Exercise 5.4-5, we first compute the probability that a composite number is not certified to be composite. If we think of success as being times when the number is certified composite and failure when it isn't, then we see that the only way to fail to certify a number is to have 20 failures. Using our formula, we see that the probability of a composite number not being certified composite is $\binom{20}{20} (.25)^{20} = \frac{1}{1099511627776}$. Thus, the chance of this happening is less than one in a trillion, and the chance of certifying the composite as composite is $1 - \frac{1}{1099511627776}$. Therefore, the probability that a composite number will be certified composite is $\frac{1099511627775}{1099511627776}$, which is more than .999999999999, so, a composite number is almost sure to be certified composite.

A Taste of Generating Functions

We note a nice connection between the probability of having exactly k successes and the binomial theorem. Consider, as an example, the polynomial $(H + T)^3$. Using the binomial theorem, we get that this is

$$(H + T)^3 = \binom{3}{0} H^3 + \binom{3}{1} H^2T + \binom{3}{2} HT^2 + \binom{3}{3} T^3 .$$

We can interpret this equation as telling us that if we flip a coin three times, with outcomes heads or tails each time, then there are

- $\binom{3}{0} = 1$ way of getting 3 heads,

- $\binom{3}{2} = 3$ ways of getting two heads and one tail,

- $\binom{3}{1} = 3$ ways of getting one head and two tails, and

- $\binom{3}{3} = 1$ way of getting 3 tails.

Similarly, if we replace H and T with px and $(1 - p)y$, respectively, we would get

$$(px + (1 - p)y)^3 = \binom{3}{0} p^3 x^3 + \binom{3}{1} p^2 (1 - p)x^2 y + \binom{3}{2} p(1 - p)^2 xy^2$$
$$+ \binom{3}{3} (1 - p)^3 y^3 .$$

Generalizing this to n repeated trials, where the probability of success in each trial is p, we see that by taking $(px + (1 - p)y)^n$, we get

$$(px + (1 - p)y)^n = \sum_{k=0}^{k} \binom{n}{k} p^k (1 - p)^{n-k} x^k y^{n-k} .$$

Taking the coefficient of $x^k y^{n-k}$ from this sum, we get exactly the formula of Theorem 5.8.

This connection is a simple case of a very powerful tool known as **generating functions**. We say that the polynomial $(px + (1 - p)y)^n$ *generates* the binomial probabilities. In fact, we don't even need the y, because

$$(px + 1 - p)^n = \sum_{i=0}^{n} \binom{n}{i} p^i (1 - p)^{n-i} x^i .$$

In general, the generating function for the sequence $a_0, a_1, a_2, \ldots, a_n$ is $\sum_{i=1}^{n} a_i x^i$, and the generating function for an infinite sequence $a_0, a_1, a_2, \ldots, a_n, \ldots$ is the infinite series $\sum_{i=1}^{\infty} a_i x^i$.

Expected Value

In Exercise 5.4-2, we asked what value you would expect a random variable (in this case, the number of heads in six flips of a coin) to have. Although we haven't yet defined what we mean by the value we expect, it seems to make sense to ask about it. If we say we expect one head if we flip a coin twice, we can explain our reasoning by taking an average. There are four outcomes—one with no heads, two with one head, and one with two heads—giving us an average of

$$\frac{0 + 1 + 1 + 2}{4} = 1 .$$

Notice that using averages compels us to have some expected values that are impossible to achieve. For example, in three flips of a coin, the eight possibilities for the number of heads are 0, 1, 1, 1, 2, 2, 2, 3, giving us for our average

$$\frac{0 + 1 + 1 + 1 + 2 + 2 + 2 + 3}{8} = 1.5 .$$

An interpretation in games and gambling makes it clear that it makes sense to expect a random variable to have a value that is not one of the possible outcomes. Suppose that I proposed the following game: You pay me some money, and then you flip three coins. I will pay you $1.00 for every head that comes up. Would you play this game if you had to pay me $2.00? What if you had to pay me $1.00? For this game to be fair, how much do you think it should cost?

Because you expect to get 1.5 heads, you expect to make $1.50. Therefore, it is reasonable to play this game as long as the cost is at most $1.50.

Certainly, averaging our variable over all elements of our sample space by adding one result for each element of the sample space, as we did in our solution to Exercise 5.4-6, is impractical, even when we are talking about something as simple as ten flips of a coin. However, for ten flips of a coin, we can ask how many times each possible number of heads arises and then multiply the number of heads by the number of times it arises to get that the average number of heads is

$$\frac{0\binom{10}{0} + 1\binom{10}{1} + 2\binom{10}{2} + \cdots + 9\binom{10}{9} + 10\binom{10}{10}}{1024} = \frac{\sum_{i=0}^{10} i\binom{10}{i}}{1024}. \qquad (5.23)$$

Have we seen a formula for $\sum_{i=0}^{n} i\binom{n}{i}$? Perhaps we have, but in any case, the binomial theorem and a bit of calculus or a proof by induction (see Problem 14) show that

$$\sum_{i=0}^{n} i\binom{n}{i} = 2^{n-1}n,$$

giving us $(512 \cdot 10)/1024 = 5$ for the fraction in Equation 5.23. If you are asking, "Does it have to be that hard?" then good for you. Once we know a bit about the theory of expected values of random variables, computations like this will be replaced by far simpler ones.

In addition to the nasty computations to which our simple question led us, the average value of a random variable on a sample space need not have anything to do with the result we expect. For instance, if we replace heads and tails with right and wrong, we get the sample space of possible results that a student will get when taking a ten-question test with probability .9 of getting the right answer on any one question. Thus, if we compute the average number of right answers in all the possible patterns of test results, we get an average of five right answers. This is not the number of right answers we expect, because averaging has nothing to do with the underlying process that gave us our probability. If we analyze the ten coin flips a bit more carefully, we can resolve this disconnection. We can rewrite Equation 5.23 as

$$0\frac{\binom{10}{0}}{1024} + 1\frac{\binom{10}{1}}{1024} + 2\frac{\binom{10}{2}}{1024} + \cdots + 9\frac{\binom{10}{9}}{1024} + 10\frac{\binom{10}{10}}{1024} = \sum_{i=0}^{10} i\frac{\binom{10}{i}}{1024}. \qquad (5.24)$$

In Equation 5.24, we see that we can compute the average number of heads by multiplying each value of our "number of heads" random variable by the probability that our random variable equals that value and then adding the results. This gives us a weighted average of the values of our random variable, with each value weighted by its probability. Because the idea of weighting a random variable by its probability comes up so much in probability theory, there is a special notation that has developed for using this weight in equations. We use $P(X = x_i)$ to stand for the probability that the random variable X equals the value x_i. We call the function that assigns $P(x = x_i)$ to the number x_i for each i the **distribution function** of the random variable X. Thus, for example, the binomial probability distribution is the distribution function for the "number of successes" random variable in Bernoulli trials.

We define the **expected value**, or **expectation**, of a random variable X whose values are the set $\{x_1, x_2, \ldots, x_k\}$ to be

$$E(X) = \sum_{i=1}^{k} x_i P(X = x_i) .$$

For someone taking a ten-question test with probability .9 of getting the correct answer on each question, the expected number of right answers is

$$\sum_{i=0}^{10} i \binom{10}{i} (.9)^i (.1)^{10-i} .$$

In Problem 17, we show a technique (which could be considered an application of generating functions) that allows us to compute this sum directly by using the binomial theorem and calculus. We now proceed to develop a less direct, but easier, way to compute this and many other expected values.

Exercise 5.4-7   Show that if a random variable X is defined on a sample space S (you may assume that X has values $x_1, x_2, \ldots, x_k$, as above), then the expected value of X is given by

$$E(X) = \sum_{s:s \in S} X(s) P(s) .$$

(In words, we take each member of the sample space, compute its probability, multiply the probability by the value of the random variable, and add the results.)

Exercise 5.4-7 asks for a proof of the following fundamental lemma.

Lemma 5.9   If a random variable X is defined on a (finite) sample space S, then its expected value is given by

$$E(X) = \sum_{s:s \in S} X(s) P(s) . \tag{5.25}$$

Proof: Assume that the values of the random variable are $x_1, x_2, \ldots, x_k$. Let F_i stand for the event that the value of X is x_i, so that $P(F_i) = P(X = x_i)$. Then, in the sum on the right side of Equation 5.25, we can take the items in the sample space, group them together into the events F_i, and rework the sum into the definition of expectation, as follows:

$$\sum_{s:s \in S} X(s)P(s) = \sum_{i=1}^{k} \sum_{s:s \in F_i} X(s)P(s)$$

$$= \sum_{i=1}^{k} \sum_{s:s \in F_i} x_i P(s)$$

$$= \sum_{i=1}^{k} x_i \sum_{s:s \in F_i} P(s)$$

$$= \sum_{i=1}^{k} x_i P(F_i)$$

$$= \sum_{i=1}^{k} x_i P(X = x_i) = E(X) . \blacksquare$$

The proof of the lemma need not be so formal and symbolic as what we just wrote; in English, it simply says that when we compute the sum in Lemma 5.9, we can group together all elements of the sample space that have X-value x_i and add their probabilities. This gives us $x_i P(x_i)$, which leads us to the definition of the expected value of X.

Expected Values of Sums and Numerical Multiples

Another important point about expected value follows naturally from what we think about when we use the word "expect" in English. If a paper grader expects to earn $10 grading papers today and expects to earn $20 grading papers tomorrow, then she expects to earn $30 grading papers in these two days. We could use X_1 to stand for the amount of money she makes grading papers today and X_2 to stand for the amount of money she makes grading papers tomorrow, so we are saying

$$E(X_1 + X_2) = E(X_1) + E(X_2) .$$

This formula holds for any sum of a pair of random variables and, more generally, for any sum of random variables on the same sample space.

Theorem 5.10 Suppose X and Y are random variables on the (finite) sample space S. Then

$$E(X + Y) = E(X) + E(Y) .$$

Proof: From Lemma 5.9, we may write

$$E(X + Y) = \sum_{s:s \in S} \big(X(s) + Y(s) \big) P(s)$$

$$= \sum_{s:s \in S} X(s) P(s) + \sum_{s:s \in S} Y(s) P(s)$$

$$= E(X) + E(Y) . \blacksquare$$

If we double the credit we give for each question on a test, we would expect students' scores to double. Thus, our next theorem should be no surprise. In it, we use the notation cX for the random variable we get from X by multiplying all its values by the number c.

<table>
<tr><td>Theorem 5.11</td><td>Suppose X is a random variable on a sample space S. Then for any number c, we have

$$E(cX) = cE(X) .$$
</td></tr>
</table>

Proof: The proof of this theorem is left as Problem 15. ■

Theorems 5.10 and 5.11 are very useful in proving facts about random variables. Taken together, they are typically called **linearity of expectation**. (The idea that the expectation of a sum is the same as the sum of expectations is called the **additivity of expectation**.) The idea of linearity will often allow us to work with expectations much more easily than if we had to work with the underlying probabilities.

For example, on one flip of a coin, our expected number of heads is .5. Suppose we flip a coin n times and let X_i be the number of heads we see on flip i, so that X_i is either 0 or 1. (For example, in five flips of a coin, $X_2(\text{HTHHT}) = 0$ and $X_3(\text{HTHHT}) = 1$.) Then X, the total number of heads in n flips, is given by

$$X = X_1 + X_2 + \cdots + X_n , \tag{5.26}$$

which is the sum of the number of heads on the first flip, the number of heads on the second flip, and so on through the number of heads on the last flip. But the expected value of each X_i is .5. We can take the expectation of both sides of Equation 5.26 and apply Theorem 5.10 repeatedly (or use induction) to get that

$$E(X) = E(X_1 + X_2 + \cdots + X_n)$$

$$= E(X_1) + E(X_2) + \cdots + E(X_n)$$

$$= .5 + .5 + \cdots + .5$$

$$= .5n .$$

Thus, in n flips of a coin, the expected number of heads is $.5n$. Compare the ease of this method with the effort needed earlier to deal with the expected number of heads in ten flips! Dealing with probability $.9$ or with probability p, in general, poses no problem.

Exercise 5.4-8

Use the additivity of expectation to determine the expected number of correct answers a student will get on an n-question fill-in-the-blanks test if he knows 90% of the material in the course and the questions on the test are an accurate and uniform sampling of the material in the course. (Assume the student does not guess.)

In Exercise 5.4-8, because the questions accurately sample the material in the course, the most natural probability for us to assign to the event of the student getting a correct answer on a given question is $.9$. If we let X_i be the number of correct answers on Question i (that is, either 1 or 0, depending on whether the student gets the correct answer), then the expected number of right answers is the expected value of the sum of the variables X_i. From Theorem 5.10, we see that in n trials with probability $.9$ of success, we expect to have $.9n$ successes. This gives us that the expected number of right answers on a ten-question test with probability $.9$ of getting each question right is 9, as we expected. This is a special case of our next theorem, which is proved by the same kind of computation.

Theorem 5.12

In a Bernoulli trials process with n trials in which each experiment has two outcomes and probability p of success, the expected number of successes is np.

Proof: Let X_i be the number of successes in the ith trial of n independent trials. The expected number of successes on the ith trial (i.e., the expected value of X_i) is, by definition,

$$p \cdot 1 + (1 - p) \cdot 0 = p .$$

The number of successes X in all n trials is the sum of the random variables X_i. Then, by Theorem 5.10, the expected number of successes in n independent trials is the sum of the expected values of the n random variables X_i, which is np. ∎

Indicator Random Variables

Notice that in the proof of Theorem 5.12, we made use of a random variable that is 1 if the ith trial is a success; otherwise, it is 0. To make it more natural to think about such a random variable, we described X_i as the number of successes on trial i, a number that happens to be 0 or 1. We used the same kind of computation device in computing the number of heads in a sequence of coin flips or the number of correct answers in a quiz. A random variable that is 1 if a certain event happens and 0 otherwise is called an **indicator random variable**. These variables have the very nice property that

$$E(X_i) = P(X_i = 1) = P(\text{the event occurs}) . \tag{5.27}$$

As in the examples we have already seen, we use sums of indicator random variables to count the number of times an event happens. The expected value of the sum is the expected number of times the event happens. In a multistage process, we might be interested in different events at different stages. We can still count them by summing appropriate indicator random variables and computing their expected values as expected values of sums. Because of the linearity of expectation, there is no need for the events to be independent. In Exercise 5.2-5, we considered the hatcheck problem, where n students check their backpacks (or hats) and each is then given a backpack at random. (In other words, the backpacks are returned according to a random permutation.[7]) We considered the probability that nobody had his or her own backpack returned. We now consider the expected value of the random variable X (the number of people who get their own backpack returned).

We let X_i be the indicator variable for the event E_i that person i gets the correct backpack returned (that is, $X_i = 1$ if person i gets the correct backpack; otherwise, $X_i = 0$.) If

$$X = X_1 + X_2 + \cdots + X_n ,$$

then X is the total number of students who get their own backpacks. Note that the events E_i are not independent. For example, if $n = 2$, either both students or neither of the students get their own backpacks returned. Nonetheless, by linearity of expectation, we have

$$E(X) = E(X_1) + E(X_2) + \cdots + E(X_n) .$$

What is $E(X_i)$ for a given i? By Equation 5.27, it is $P(\text{person } i \text{ gets the correct backpack})$. Because there are $n!$ permutations of n people and $(n-1)!$ permutations in which person i's backpack is returned, $E(X_i) = 1/n$. Thus, $E(X) = n(1/n) = 1$ for any number of people.

Indicator random variables are very useful in analyzing algorithms. Here is an example.

Exercise 5.4-9

Consider the following procedure for computing the minimum of an array of items.

```
FindMin(A, n)

// Finds the smallest element in Array A, where n = |A|.
(1)   min = A[1]
(2)   for i = 2 to n
(3)        if (A[i] < min)
(4)            min = A[i]
(5)   Return min
```

If Array A contains a random permutation of the integers 1 to n, what is the expected number of times that min is assigned a value?

---

[7] To say we have a random permutation means we have chosen the permutation from the sample space of all permutations of a set, and we were equally likely to have chosen any permutation.

We solve this problem by letting X be the number of times that min is assigned a value and X_i be the indicator random variable for the event that $A[i]$ is assigned to min. Then $X = X_1 + X_2 + \cdots + X_n$, and $E(X_i)$ is the probability that $A[i]$ is the smallest element in the set $\{A[1], A[2], \ldots, A[i]\}$. Because $(i-1)!$ of the $i!$ permutations of these elements have $A[i]$ as the smallest element, $E(X_i) = 1/i$. Thus,

$$E(X) = \sum_{i=1}^{n} \frac{1}{i} .$$

In Section 5.5, we will see that this sum is $\Theta(\log n)$.

The Number of Trials until the First Success

Exercise 5.4-10

How many times should we expect to have to flip a coin until we first see a head? Why? How many times should we expect to have to roll two dice until we see a sum of 7? Why?

Our intuition suggests that we should have to flip a coin twice to see a head. However, we could conceivably flip a coin forever without seeing a head, so should we really expect to see a head in two flips? The probability of getting a 7 on two dice is $1/6$. Does that mean we should expect to have to roll the dice six times before we see a 7?

To analyze this kind of question, we have to realize that we are stepping out of the realm of independent trials processes on finite sample spaces. Instead, we consider the process of repeating independent trials with probability p of success until we have a success and then stopping. Now, for our multistage process, the possible outcomes are the infinite set

$$\{S, FS, FFS, \ldots, F^i S, \ldots\} ,$$

in which we have used the notation $F^i S$ to stand for the sequence of i failures followed by a success. Because we have an infinite sequence of outcomes, it makes sense to think about whether we can assign an infinite sequence of probability weights to its members so that the resulting sequence of probabilities adds to 1. If so, then all our definitions make sense, and, in fact, the proofs of all our theorems remain valid.[8] There is only one way to assign weights that is consistent with our knowledge of (finite) independent trials processes; namely,

$$P(S) = p, \quad P(FS) = (1-p)p, \quad \ldots, \quad P(F^i S) = (1-p)^i p, \quad \ldots .$$

---

[8]For those familiar with the concept of convergence for infinite sums (i.e., infinite series), it is worth noting that the fact that probability weights cannot be negative and must add to 1 is what makes all the sums we need to deal with, for all the theorems we have proved so far, converge. This doesn't mean that all sums that we might want to deal with will converge; some random variables defined on the sample space we have described will have infinite expected value. However, those we need to deal with for the expected number of trials until success do converge.

Thus, we have to hope that these weights add to 1. In fact, their sum is

$$\sum_{i=0}^{\infty}(1-p)^i p = p\sum_{i=0}^{\infty}(1-p)^i = p\frac{1}{1-(1-p)} = \frac{p}{p} = 1\,.$$

With this, we have a legitimate assignment of probabilities. The set of sequences

$$\{F, FS, FFS, FFFS, \ldots, F^i S, \ldots\}$$

is a sample space with these probability weights. The probability distribution $P(F^i S) = (1-p)^i p$ is called a **geometric distribution** because of the geometric series we used in proving that the probabilities sum to 1.

Theorem 5.13	Suppose we have a sequence of trials in which each trial has two outcomes, success and failure, and in which the probability of success at each step is p and $p > 0$. Then the expected number of trials until the first success is $1/p$.

Proof: We consider the random variable X, which is i if the first success is on Trial i. (In other words, $X(F^{i-1}S) = i$.) The probability that the first success is on Trial i is $(1-p)^{i-1}p$, because for this to happen, there must be $i-1$ failures followed by 1 success. The expected number of trials is the expected value of X, which is, by the definition of expected value and the previous two sentences,

$$E(\text{number of trials}) = \sum_{i=0}^{\infty} p(1-p)^{i-1}i$$

$$= p\sum_{i=0}^{\infty}(1-p)^{i-1}i$$

$$= \frac{p}{1-p}\sum_{i=0}^{\infty}(1-p)^i i$$

$$= \frac{p}{1-p}\frac{1-p}{p^2}$$

$$= \frac{1}{p}\,.$$

To go from the third to the fourth line in the previous sequence of equations, we used the fact that

$$\sum_{j=0}^{\infty} jx^j = \frac{x}{(1-x)^2}\,, \tag{5.28}$$

which is true for x with absolute value less than 1. We proved a finite version of this equation as Theorem 4.6; the infinite version is even easier to prove. ∎

Applying Theorem 5.13, we see that the expected number of times we need to flip a coin until we get a head is two, and the expected number of times we need to roll two dice until we get a 7 is six.

Important Concepts, Formulas, and Theorems

1. *Random variable.* A *random variable* for an experiment with a sample space S is a function that assigns a number to each element of S.

2. *Bernoulli trials process.* An independent trials process with two outcomes, success and failure, at each stage and probability p of success and $1 - p$ of failure at each stage is called a *Bernoulli trials process.*

3. *Probability of a sequence of Bernoulli trials.* In n Bernoulli trials with probability p of success, the probability of a given sequence of k successes and $n - k$ failures is $p^k(1 - p)^{n-k}$.

4. *The probability of k successes in n Bernoulli trials.* The probability of having exactly k successes in a sequence of n independent trials with two outcomes and probability p of success on each trial is given by

$$P(\text{exactly } k \text{ successes}) = \binom{n}{k} p^k(1 - p)^{n-k} .$$

5. *Binomial probability distribution.* The probabilities of k successes in n Bernoulli trials, $\binom{n}{k} p^k(1 - p)^{n-k}$, are called *binomial probabilities,* or the *binomial probability distribution.*

6. *Generating function.* The generating function for the sequence $a_0, a_1, a_2, \ldots, a_n$ is

$$\sum_{i=1}^{n} a_i x^i ,$$

and the generating function for an infinite sequence $a_0, a_1, a_2, \ldots, a_n, \ldots$ is the infinite series

$$\sum_{i=1}^{\infty} a_i x^i .$$

The polynomial $(px + 1 - p)^n$ is the generating function for the binomial probabilities for n Bernoulli trials with probability p of success.

7. *Distribution function.* The function that assigns $P(X = x_i)$ to the event $X = x_i$ is called the *distribution function* of the random variable X.

8. *Expected value.* The *expected value,* or *expectation,* of a random variable X, whose values are the set $\{x_1, x_2, \ldots, x_k\}$, is defined by

$$E(X) = \sum_{i=1}^{k} x_i P(X = x_i) .$$

9. *Another formula for expected values.* If a random variable X is defined on a (finite) sample space S, then its expected value is given by

$$E(X) = \sum_{s:s \in S} X(s)P(s) .$$

10. *Expected value of a sum.* Suppose X and Y are random variables on the (finite) sample space S. Then

$$E(X + Y) = E(X) + E(Y) .$$

This is called the *additivity of expectation.*

11. *Expected value of a numerical multiple.* Suppose X is a random variable on a sample space S. Then $E(cX) = cE(X)$ for any number c. This result and the additivity of expectation are called the *linearity of expectation.*

12. *Expected number of successes in Bernoulli trials.* In a Bernoulli trials process, the expected number of successes is np.

13. *Indicator random variables.* A random variable that is 1 if a certain event happens and 0 otherwise is called an *indicator random variable.*

14. *Expected number of trials until success.* Suppose we have a sequence of trials in which each trial has two outcomes (success and failure) and in which the probability of success at each step is p. Then the expected number of trials until the first success is $1/p$.

15. *Geometric distribution.* The probability distribution given by $P(F^i S) = (1 - p)^i p$ is called a *geometric distribution.*

Problems

1. Give several random variables that might be of interest to someone rolling five dice (as one does, for example, in the game Yahtzee).

2. In an independent trials process consisting of six trials with probability p of success, what is the probability that the first three trials are successes and the last three are failures? The probability that the last three trials are successes and the first three are failures? The probability that Trials 1, 3, and 5 are successes and Trials 2, 4, and 6 are failures? What is the probability of three successes and three failures? a/h

3. What is the probability of exactly eight heads in ten flips of a coin? Of eight or more heads? a/h

4. Assuming that the process of answering the questions on a five-question quiz is an independent trials process and that a student has a probability .8 of answering any given question correctly, what is the probability of one particular sequence of four correct answers and one incorrect answer? What is the probability that a student answers exactly four questions correctly?

5. Suppose I offer to play the following game with you if you will pay me some money. You roll a die, and I give you a dollar for each dot that is on top. What is the maximum amount of money a rational person might be willing to pay me to play this game? a/h

6. What is the expected sum of the tops of n dice when you roll them?

7. How many sixes do you expect to see on top if you roll 24 dice? a/h

8. If you randomly choose 26 cards from a deck of 52 ordinary playing cards, one at a time, is the event of having a king on the ith draw independent of the event of having a king on the jth draw? How many kings do you expect to see?

9. How many times do you expect to have to roll a die until you see a six on the top face? a/h

10. What is the expected value of the constant random variable X that has $X(s) = c$ for every member s of the sample space? (We frequently use c to stand for this random variable. Thus, this question is asking for $E(c)$.)

11. A student is taking a true-false test and guessing when he doesn't know the answer. We are going to compute a score by subtracting a percentage of the number of incorrect answers from the number of correct answers. That is, for some number y, the student's corrected score will be

    (number of correct answers) $- y$ (number of incorrect answers) .

    When we convert this "corrected score" to a percentage score, we want its expected value to be the percentage of the material being tested that the student knows. How can we do this? a/h

12. Solve Problem 11 for the case of a student taking a multiple-choice test with five choices for each answer and randomly guessing when she doesn't know the answer.

13. Suppose you have ten independent trials with three outcomes called "good," "bad," and "indifferent," with probabilities p, q, and r, respectively. What is the probability of three goods, two bads, and five indifferents? In n independent trials with three outcomes A, B, and C, with probabilities p, q, and r, what is the probability of i A's, j B's, and k C's? (In this problem, assume $p + q + r = 1$ and $i + j + k = n$.) a/h

14. In as many ways as you can, prove that

$$\sum_{i=0}^{n} i \binom{n}{i} = n2^{n-1} .$$ a/h

15. Prove Theorem 5.11.

16. Two nickels, two dimes, and two quarters are in a cup. You draw three coins, one after the other, without replacement. What is the expected amount of money you draw on the first draw? On the second draw? What is the expected

value of the total amount of money you draw? Does this expected value change if you draw the three coins all at once? a/h

17. Evaluate the sum

$$\sum_{i=0}^{10} i \binom{10}{i} (.9)^i (.1)^{10-i} ,$$

which arose in computing the expected number of right answers a person would have on a ten-question test with probability .9 of answering each question correctly. First, use the binomial theorem and calculus to show that

$$10(.1 + x)^9 = \sum_{i=0}^{10} i \binom{10}{i} (.1)^{10-i} x^{i-1} .$$

Substituting $x = .9$ almost gives the sum you want on the right side of the equation, except that in every term of the sum, the power on .9 is one too small. Use some simple algebra to fix this and then explain why the expected number of right answers is 9.

18. Give an example of two random variables X and Y such that $E(XY) \neq E(X)E(Y)$. Here XY is the random variable with $(XY)(s) = X(s)Y(s)$. a/h

19. Let X and Y be independent in the sense that the event "$X = x$" and the event "$Y = y$" are independent for each pair of values x of X and y of Y. Prove that $E(XY) = E(X)E(Y)$. See Problem 18 for a definition of XY.

20. Use calculus and the sum of a geometric series to show that if $-1 < x < 1$, then

$$\sum_{j=0}^{\infty} jx^j = \frac{x}{(1 - x)^2} ,$$

as in Equation 5.28. a/h

21. Give an example of a random variable on the sample space $\{S, FS, FFS, \ldots, F^i S, \ldots\}$ with an infinite expected value, using a geometric distribution for probabilities of $F^i S$. a/h

5.5 PROBABILITY CALCULATIONS IN HASHING

In this section, we will use our knowledge of probability and expected value to analyze several interesting quantities that arise when using hashing. Recall that in (open) hashing, each item hashes to a particular location in an array and locations can hold more than one item. We analyze the following quantities:

1. expected number of items per location

2. expected time for a search

3. expected number of collisions

4. expected number of empty locations

5. expected time until all locations have at least one item

6. expected maximum number of items per location

Expected Number of Items per Location

Exercise 5.5-1

We are going to compute the expected number of items that hash to any particular location in a hash table. Our model of hashing n items into a table of size k allows us to think of the process as n independent trials, each with k possible outcomes (the k locations in the table). On each trial, we hash another key into the table. If we hash n items into a table with k locations, what is the probability that any one item hashes into Location 1? Let X_i be the indicator random variable that is 1 if, in the ith trial, the item hashes to Location 1; otherwise, let it be 0. What is the expected value of X_i? Let X be the random variable $X_1 + X_2 + \cdots + X_n$. What is the expected value of X? What is the expected number of items that hash to Location 1? Is the fact that we are talking about Location 1 special in any way? That is, does the same expected value apply to every location?

Exercise 5.5-2

Again, we are hashing n items into k locations. Our model of hashing is the same as that of Exercise 5.5-1. What is the probability that a location is empty? What is the expected number of empty locations? Suppose we now hash n items into the same number n of locations. What limit does the expected fraction of empty places approach as n gets large?

In Exercise 5.5-1, the probability that any one item hashes into Location 1 is $1/k$, because all k locations are equally likely. It follows that the expected value of X_i is $1/k$. The expected value of X is then n/k, or the sum of n terms each equal to $1/k$. Of course, the same expected value applies to any location. Thus, we have proved the following theorem.

Theorem 5.14

In hashing n items into a hash table of size k, the expected number of items that hash to any one location is n/k.

Expected Number of Empty Locations

In Exercise 5.5-2, the probability that Location i will be empty after we hash one item into the table will be $1 - 1/k$. (Why?) In fact, we can think of our process as an independent trials process with two outcomes: the key hashes to Location i or it doesn't. From this point of view, it is clear that the probability of nothing hashing to Location i in n trials is $(1 - 1/k)^n$. Now consider the original sample space again, and let $X_i = 1$ if Location i is empty for a given sequence of hashes; otherwise, let it be 0.

Then the number of empty slots for a given sequence of hashes is $X_1 + X_2 + \cdots + X_k$, evaluated at that sequence. Therefore, the expected number of empty slots is, by Theorem 5.10, $k(1 - 1/k)^n$. Thus, we have proved another nice theorem about hashing.

Theorem 5.15

In hashing n items into a hash table with k locations, the expected number of empty locations is $k(1 - 1/k)^n$.

Proof: The proof for this theorem is given above. ∎

If we have the same number of slots as places, the expected number of empty slots is $n(1 - 1/n)^n$, so the expected fraction of empty slots is $(1 - 1/n)^n$. What does this fraction approach as n grows? You may recall that $\lim_{n \to \infty}(1 + 1/n)^n = e$, the base for the natural logarithm. In Problem 13, we show you how to use this to derive that $\lim_{n \to \infty}(1 - 1/n)^n = e^{-1}$. Thus, for a reasonably large hash table, if we hash in as many items as we have slots, we expect $1/e$ of those slots to remain empty. In other words, we expect n/e empty slots. On the other hand, we expect $n/n = 1$ items per location, which suggests that we should expect each slot to have an item, and therefore, we expect to have no empty locations. Is something wrong? No; we simply have to accept that our expectations about expectation don't always hold true. What went wrong in this apparent contradiction is that our definition of expected value doesn't imply that if we have an expectation of one key per location then every location must have a key. It only implies that empty locations have to be balanced by locations with more than one key. When we want to make a statement about expected values, we must use either our definitions or theorems to back it up. This is another example of why we have to use careful analysis to support our intuition about probability.

Expected Number of Collisions

We say that we have a **collision** when we hash an item to a location that already contains an item. How can we compute the expected number of collisions? The number of collisions will be the number n of keys hashed minus the number of occupied locations, because each occupied location will contain one key that will not have collided in the process of being hashed. Thus, by Theorems 5.10 and 5.11,

$$E(\text{collisions}) = n - E(\text{occupied locations})$$
$$= n - k + E(\text{empty locations}),\tag{5.29}$$

where the last equality follows because the expected number of occupied locations is k minus the expected number of unoccupied locations. This gives us yet another theorem.

In hashing n items into a hash table with k locations, the expected number of collisions is $n - k + k(1 - 1/k)^n$.

Proof: We have already shown in Theorem 5.15 that the expected number of empty locations is $k(1 - 1/k)^n$. Substituting this into Equation 5.29 gives us our formula. ■

Exercise 5.5-3

In real applications, it is often the case that the hash table size is not fixed in advance, because we don't know in advance how many items we will insert. The most common heuristic for dealing with this is to start k, the hash table size, at some reasonably small value; then when n, the number of items, gets to be greater than $2k$, we double the size of the hash table. In this exercise, we propose a different idea. Suppose we waited until every single slot in the hash table had at least one item in it, and then we increased the table size. What is the expected number of items that will be in the table when we increase the size? In other words, how many items should we expect to insert into a hash table to ensure that every slot has at least one item? (*Hint:* Let X_i be the number of items added between the first time that there are $i - 1$ occupied slots and the first time that there are i occupied slots.)

For Exercise 5.5-3, the key is to let X_i be the number of items added between the time that there are $i - 1$ full slots for the first time and i full slots for the first time. Let's think about this random variable: $E(X_1) = 1$, because after one insertion, there is one full slot. In fact, X_1 itself is equal to 1.

To compute the expected value of X_2, we note that X_2 can take on any value greater than zero. In fact, what we have here (until we actually hash an item to a new slot) is an independent trials process with two outcomes, with success meaning our item hashes to an unused slot. Thus, X_2 counts the number of trials until the first success. The probability of success is $(k - 1)/k$. In asking for the expected value of X_2, we are asking for the expected number of steps until the first success. Thus, we can apply Theorem 5.13 to get that $E(X_2) = k/(k - 1)$.

Continuing, X_3 similarly counts the number of steps in an independent trials process (with two outcomes) that stops at the first success and has probability of success $(k - 2)/k$. Therefore, the expected number of steps until the first success is $k/(k - 2)$.

In general, we have that X_i counts the number of trials until success in an independent trials process with probability of success $(k - i + 1)/k$, and thus, the expected number of steps until the first success is $k/(k - i + 1)$, which is the expected value of X_i.

The total time until all slots are full is simply $X = X_1 + \cdots + X_k$. Taking expectations and using Theorem 5.13, we get

$$E(X) = \sum_{j=1}^{k} E(X_j)$$

$$= \sum_{j=1}^{k} \frac{k}{k - j + 1}$$

$$= k \sum_{j=1}^{k} \frac{1}{k - j + 1}$$

$$= k \sum_{k-j+1=1}^{k} \frac{1}{k - j + 1}$$

$$= k \sum_{i=1}^{k} \frac{1}{i} \, ,$$

where the last line follows just by switching the variable of the summation—that is, letting $k - j + 1 = i$ and summing over i.[9] The quantity $\sum_{i=1}^{k}(1/i)$ is known as a **harmonic number** and is sometimes denoted by H_k. It is well-known (and you can see why in Problem 18) that $\sum_{i=1}^{k}(1/i) = \Theta(\log k)$. More precisely,

$$\frac{1}{4} + \ln k \le H_k \le 1 + \ln k \, , \tag{5.30}$$

and in fact,

$$\frac{1}{2} + \ln k \le H_k \le 1 + \ln k$$

when k is large enough. As n gets large, $H_n - \ln n$ approaches a limit called **Euler's constant**, which is about .58. Equation 5.30 gives us that $E(X) = \Theta(k \log k)$.

Theorem 5.17	The expected number of items needed to fill all slots of a hash table of size k is between $k \ln k + k/4$ and $k \ln k + k$.

Proof: The proof of this theorem is given above. ■

So, to fill every slot in a hash table of size k, we need to hash roughly $k \ln k$ items. This problem is sometimes called the **coupon-collector's problem**. To understand the reason for this name, imagine that a brand of breakfast cereals has a promotion with five different coupons that can be redeemed by mail for five different toys, and there is one coupon in each box of cereal. The question about the number of hashes until

---

[9]Note that $k - j + 1$ runs from k to 1 as j runs from 1 to k, so we are describing exactly the same sum.

a hash table is full corresponds to asking for the expected number of boxes someone has to buy to get at least one of each coupon.

The remainder of this section, which can be skipped without loss of continuity, is devoted to proving that if we hash n items into a hash table with n slots, then the expected number of items in the slot with the most items is $O(\log n / \log \log n)$. It should be no surprise that a result of this form requires a somewhat complex proof.

Expected Maximum Number of Elements in a Location of a Hash Table*

In a hash table, the amount of time required to find an item is related to the number of items in the location where you are looking. Thus, an interesting quantity is the expected maximum length of the list of items in a location in a hash table. This quantity is more complicated than many of the others we have been computing; hence, we will only try to upper bound it rather than compute it exactly. In doing so, we will introduce a few upper bounds and techniques that appear frequently and that are useful in many areas of mathematics and computer science. We will prove that if we hash n items into a hash table of size n, the expected length of the longest list is $O(\log n / \log \log n)$. We could also prove, although we won't do it here, that there is a high probability of there being some list with $\Omega(\log n / \log \log n)$ items in it, so our bound is the best possible, up to constant factors.

Before we start, we give some useful upper bounds. The first allows us to bound terms that look like $(1 + 1/x)^x$, for any positive x, by e.

| Lemma 5.18 | For all $x > 0$, we have $(1 + 1/x)^x \leq e$. |

Proof: This follows because $\lim_{x \to \infty}(1 + 1/x)^x = e$ and $(1 + 1/x)^x$ has positive first derivative. ∎

Second, we will use the following approximation called **Stirling's formula**,

$$n! = \left(\frac{n}{e}\right)^n \sqrt{2\pi n}\left(1 + \Theta\left(\frac{1}{n}\right)\right),$$

which tells us, roughly, that $(n/e)^n$ is a good approximation for $n!$. Moreover, the constant in the $\Theta(1/n)$ term is $1/12$; so when n is moderately large, this term will be very small relative to $n!$ For our purposes, we will just say that

$$n! \approx \left(\frac{n}{e}\right)^n \sqrt{2\pi n}.$$

---

*This subsection can be skipped without loss of continuity.

(We use this equality only in our proof of Lemma 5.19. You will see that in the proof of Lemma 5.19, we make the statement that $\sqrt{2\pi} > 1$. In fact, $\sqrt{2\pi} > 2$, which is more than enough to make up for any lack of accuracy in our approximation.) Using Stirling's formula, we can get a bound on $\binom{n}{t}$.

| Lemma 5.19 | For $n > t > 0$, we have |

$$\binom{n}{t} \leq \frac{n^n}{t^t (n-t)^{n-t}} .$$

Proof:

$$\binom{n}{t} = \frac{n!}{t!(n-t)!}$$

$$= \frac{(n/e)^n \sqrt{2\pi n}}{(t/e)^t \sqrt{2\pi t} \left((n-t)/e\right)^{n-t} \sqrt{2\pi (n-t)}}$$

$$= \frac{n^n \sqrt{n}}{t^t (n-t)^{n-t} \sqrt{2\pi} \sqrt{t(n-t)}} . \qquad (5.31)$$

Now, if $1 < t < n-1$, then we have $t(n-t) \geq n$, so that $\sqrt{t(n-t)} \geq \sqrt{n}$. Further, $\sqrt{2\pi} > 1$. We can use these facts to upper bound Expression 5.31 by

$$\frac{n^n}{t^t (n-t)^{n-t}} .$$

When $t = 1$ or $t = n-1$, the inequality in the statement of the lemma is $n \leq n^n/(n-1)^{n-1}$, which is true because $n - 1 < n$. ∎

We are now ready to attack the problem at hand: the expected value of the maximum list size. Let's start with a related quantity that we already know how to compute exactly. Let H_{it} be the event that exactly t keys hash to Location i. $P(H_{it})$ is just the probability of t successes in an independent trials process with success probability $1/n$; so,

$$P(H_{it}) = \binom{n}{t} \left(\frac{1}{n}\right)^t \left(1 - \frac{1}{n}\right)^{n-t} . \qquad (5.32)$$

We relate this known quantity to the probability of the event M_t that the maximum list size is t.

Lemma 5.20	Let M_t be the event that t is the maximum list size in hashing n items into a hash table of size n. Let H_{1t} be the event that t keys hash to Location 1. Then

$$P(M_t) \leq n P(H_{1t}) .$$

Proof: We begin by letting M_{it} be the event that the maximum list size is t and this list appears in Location i. Observe that

$$P(M_{it}) \leq P(H_{it}) ,$$

because M_{it} is a subset of H_{it}. We know that, by definition,

$$M_t = M_{1t} \cup \cdots \cup M_{nt} ,$$

and so

$$P(M_t) = P(M_{1t} \cup \cdots \cup M_{nt}) .$$

Therefore, because the sum of the probabilities of the individual events must be at least as large as the probability of the union,

$$P(M_t) \leq P(M_{1t}) + P(M_{2t}) + \cdots + P(M_{nt}) . \qquad (5.33)$$

(Recall that we introduced the principle of inclusion and exclusion because the right side usually overestimates the probability of the union. However, Inequality 5.33, which is sometimes called **Boole's inequality**, holds for any union, not just this one.)

In this case, $P(M_{it}) = P(M_{jt})$ for any i and j, because there is no reason for Location i to be more likely than Location j to be the maximum. We can therefore write that

$$P(M_t) \leq n P(M_{1t}) \leq n P(H_{1t}) . \blacksquare$$

We can now use Equation 5.32 for $P(H_{1t})$ and then apply Lemma 5.19 to get that

$$\begin{aligned}
P(H_{1t}) &= \binom{n}{t} \left(\frac{1}{n}\right)^t \left(1 - \frac{1}{n}\right)^{n-t} \\
&\leq \frac{n^n}{t^t (n-t)^{n-t}} \left(\frac{1}{n}\right)^t \left(1 - \frac{1}{n}\right)^{n-t} .
\end{aligned}$$

Using algebra, $(1 - 1/n)^{n-t} \leq 1$, and Lemma 5.18, we continue and get that

$$P(H_{1t}) \leq \frac{n^n}{t^t(n-t)^{n-t}n^t}$$

$$= \frac{n^{n-t}}{t^t(n-t)^{n-t}}$$

$$= \left(\frac{n}{n-t}\right)^{n-t}\frac{1}{t^t}$$

$$= \left(1 + \frac{t}{n-t}\right)^{n-t}\frac{1}{t^t}$$

$$= \left(\left(1 + \frac{t}{n-t}\right)^{(n-t)/t}\right)^t\frac{1}{t^t}$$

$$\leq \frac{e^t}{t^t}.$$

We have shown the following:

<table>
<tr><td>Lemma 5.21</td><td>The probability that the maximum list length, $P(M_t)$, is t is at most ne^t/t^t.</td></tr>
</table>

Proof: Our sequence of equations and inequalities above showed that $P(H_{1t}) \leq e^t/t^t$. Multiplying by n and applying Lemma 5.20 gives us our result. ∎

Now that we have a bound on $P(M_t)$, we can compute a bound on the expected length of the longest list, namely,

$$\sum_{t=0}^{n} P(M_t)t.$$

However, if we think carefully about the bound in Lemma 5.21, we see that we have a problem. For example, when $t = 1$, the lemma tells us that $P(M_1) \leq ne$. This bound is vacuous, because we know that any probability is at most 1. We could make a stronger statement that $P(M_t) \leq \max\{ne^t/t^t, 1\}$, but even this wouldn't be sufficient, as it would tell us things like $P(M_1) + P(M_2) \leq 2$, which is also vacuous. All is not lost, however. Our lemma causes this problem only when t is small. We split the sum defining the expected value into two parts and bound the expectation for each part separately. The intuition is that when we restrict t to be small, $\sum P(M_t)t$ is small because t is small (and $\sum P(M_t) \leq 1$ over all t). When t gets larger, Lemma 5.21 tells us that $P(M_t)$ is very small; thus the sum doesn't get big in that case, either. We choose a way to split the sum so that this second part of the sum is bounded by a constant. In particular, we split the sum by

$$\sum_{t=0}^{n} P(M_t)t \leq \sum_{t=0}^{\lfloor 5 \log n/ \log \log n \rfloor} P(M_t)t + \sum_{t=\lceil 5 \log n/ \log \log n \rceil}^{n} P(M_t)t . \quad (5.34)$$

For the sum over the smaller values of t, we observe that in each term, $t \leq \lfloor 5 \log n/ \log \log n \rfloor$, so that

$$\sum_{t=0}^{\lfloor 5 \log n/ \log \log n \rfloor} P(M_t)t \leq \sum_{t=0}^{\lfloor 5 \log n/ \log \log n \rfloor} \frac{P(M_t) 5 \log n}{\log \log n}$$

$$= \frac{5 \log n}{\log \log n} \sum_{t=0}^{\lfloor 5 \log n/ \log \log n \rfloor} P(M_t)$$

$$\leq \frac{5 \log n}{\log \log n} . \quad (5.35)$$

(Note that we are not using Lemma 5.21 here; only the fact that the probabilities of disjoint events cannot add to more than 1.) For the rightmost sum in Equation 5.34, we want to first compute an upper bound on $P(M_t)$ for $t = 5 \log n/ \log \log n$. Using Lemma 5.21 and a rather complicated calculation outlined in Problem 17, we get that, in this case, $P(M_t) \leq 1/n^2$. Because the bound on $P(M_t)$ from Lemma 5.21 decreases as t grows and $t \leq n$, we can bound the right sum by

$$\sum_{t=\lceil 5 \log n/ \log \log n \rceil}^{n} P(M_t)t \leq \sum_{t=\lceil 5 \log n/ \log \log n \rceil}^{n} \frac{1}{n^2} n \leq \sum_{t=\lceil 5 \log n/ \log \log n \rceil}^{n} \frac{1}{n} \leq 1 . \quad (5.36)$$

Combining Equations 5.35 and 5.36 with Equation 5.34, we get the desired result.

Theorem 5.22	If we hash n items into a hash table of size n, then the expected maximum list length is $O(\log n/ \log \log n)$.

The choice to break the sum into two pieces here—and especially the breakpoint we chose—may have seemed like magic. What is so special about $\lfloor 5 \log n/ \log \log n \rfloor$? Consider the bound on $P(M_t)$. If we ask for the value of t for which the bound equals a certain value, say $1/n^2$, we get the equation $ne^t/t^t = n^{-2}$. If we try to solve the equation $ne^t/t^t = n^{-2}$ for t, then we quickly see that we get an equation of a form that we do not know how to solve. (Try typing this into a computer algebra system, such as Mathematica or Maple, to see how they try to solve this equation. At best, you will get a formula containing something called a Lambert function.) The equation we need to solve is somewhat similar to the simpler equation $t^t = n$. Although this equation does not have a closed-form solution in commonly used functions, we can show that the t that satisfies this equation is roughly $c \log n/ \log \log n$ for some constant c. This is why it makes sense to try some multiple of $\log n/ \log \log n$ as the magic value. For values much less than $\log n/ \log \log n$, the bound provided on $P(M_t)$ is fairly large. Once we get past $\log n/ \log \log n$, however, the bound on $P(M_t)$ starts to get significantly smaller. The factor of 5 was chosen by experimentation to make the second sum come out to be less than 1. We could have chosen any number between 4 and 5 to get the second sum to come out less than 1, or we could have chosen 4, and the second sum would have grown no faster than the first.

Important Concepts, Formulas, and Theorems

1. *Expected number of keys per location in a hash table.* In hashing n items into a hash table of size k, the expected number of items that hash to any one location is n/k.

2. *Expected number of empty locations in a hash table.* In hashing n items into a hash table with k locations, the expected number of empty locations is $k(1 - 1/k)^n$.

3. *Collision in hashing.* We have a *collision* when we hash an item to a location that already contains an item.

4. *The expected number of collisions in hashing.* In hashing n items into a hash table with k locations, the expected number of collisions is $n - k + k(1 - 1/k)^n$.

5. *Harmonic number.* The quantity $\sum_{i=1}^{k}(1/i)$ is known as a *harmonic number* and is sometimes denoted by H_k. It is a fact that $\sum_{i=1}^{k}(1/i) = \Theta(\log k)$, and, more precisely,

$$\frac{1}{2} + \ln k \leq H_k \leq 1 + \ln k$$

for large values of k.

6. *Euler's constant.* As n gets large, $H_n - \ln n$ approaches a limit called *Euler's constant,* which is about .58.

7. *Expected number of hashes until all locations of a hash table are occupied.* The expected number of items needed to fill all locations of a hash table of size k is between $k \ln k + k/4$ and $k \ln k + k$. (For large k, $k/4$ may be replaced with $k/2$.)

8. *Expected maximum number of keys per location.* If we hash n items into a hash table of size n, the expected maximum list length is $O(\log n / \log \log n)$.

*9. *Stirling's formula for $n!$.* $n!$ is approximately $(n/e)^n \sqrt{2\pi n}$.

Problems

1. A candy machine in a school has d different kinds of candy. Assume (for simplicity) that all these kinds of candy are equally popular and there is a large supply of each. Suppose that c children come to the machine, and each child purchases one package of candy. One of the kinds of candy is a Snackers bar.

   a. What is the probability that any given child purchases a Snackers bar?

   b. Let Y_i be the number of Snackers bars that Child i purchases—Y_i is either 0 or 1. What is the expected value of Y_i?

---

*Stirling's formula appears in a subsection marked with an asterisk.

c. Let Y be the random variable $Y_1 + Y_2 + \cdots + Y_c$. What is the expected value of Y?

d. What is the expected number of Snackers bars that are purchased?

e. Does the same result apply to any of the varieties of candy? a/h

2. As in Problem 1, c children are choosing from among ample supplies of d different kinds of candy, with one package for each child and all choices equally likely.

a. What is the probability that a given variety of candy is chosen by no child?

b. What is the expected number of kinds of candy chosen by no child?

c. Suppose that $c = d$. What happens to the expected number of kinds of candy chosen by no child?

3. In Problem 1, how many children do you expect to have to observe buying candy until someone has bought a Snackers bar? a/h

4. In Problem 1, how many children do you expect to have to observe buying candy until each type of candy has been selected at least once?

5. In Problem 1, if there are 20 kinds of candy, how many children have to buy candy for the probability to be at least $1/2$ that (at least) two children buy the same kind of candy? a/h

6. In Problem 1, what is the expected number of duplications among all the candy the children have selected?

7. Compute the values on the left and right side of the inequality in Lemma 5.19 for $n = 2, t = 0, 1, 2$, and for $n = 3, t = 0, 1, 2, 3$. a/h

8. Suppose you hash n items into k locations.

a. What is the probability that all n items hash to different locations?

b. What is the probability that the ith item is the first collision?

c. What is the expected number of items you hash until the first collision?

d. Use a computer program or spreadsheet to compute the expected number of items hashed into a hash table until the first collision, with $k = 20$ and with $k = 100$.

9. We have seen a number of occasions when our intuition about expected values or probability in general fails us. When we studied Equation 5.29, we said that the expected number of occupied locations is k minus the expected number of empty locations. Although this seems obvious, there is a short proof. Give the proof. a/h

10. Write a computer program that prints out a table of values of the expected number of collisions with n keys hashed into a table with k locations for interesting values of n and k. Does this value vary much as n and k change?

11. Suppose you hash n items into a hash table of size k. It is natural to ask how long it takes to find an item in the hash table. You can divide this into two cases, one in which the item is not in the hash table (an unsuccessful search) and one in which the item is in the hash table (a successful search). First consider the unsuccessful search. Assume the keys hashing to the same location are stored in a list, with the most recent arrival at the beginning of the list.

    a. Using the expected list length, write a bound for the expected time for an unsuccessful search. Next, consider the successful search. Recall that when you insert items into a hash table, you typically insert them at the beginning of a list; thus, the time for a successful search for Item i should depend on how many entries were inserted after Item i.

    b. Carefully compute the expected running time for a successful search. Assume that the item you are searching for is randomly chosen from among the items already in the table. (*Hint:* The unsuccessful search should take roughly twice as long as the successful one. Be sure to explain why this is the case.) a/h

*12. Suppose you hash $n \log n$ items into n buckets. What is the expected maximum number of items in a bucket?

13. The fact that $\lim_{n \to \infty} (1 + 1/n)^n = e$ (where n varies over integers) is a consequence of the fact that $\lim_{h \to 0} (1 + h)^{1/h} = e$ (where h varies over real numbers). Thus, if h varies over negative real numbers but approaches 0, the limit still exists and equals e. What does this tell you about $\lim_{n \to -\infty} (1 + 1/n)^n$? Using this and rewriting $(1 - 1/n)^n$ as $(1 + 1/-n)^n$, show that

$$\lim_{n \to \infty} \left(1 - \frac{1}{n}\right)^n = \frac{1}{e}.$$ a/h

14. What is the expected number of empty slots when you hash $2k$ items into a hash table with k slots? What is the expected fraction of empty slots close to when k is reasonably large?

†15. Using whatever methods you like (hand calculations or computer), give upper and/or lower bounds in terms of n on the value of the x that satisfies $x^x = n$. a/h

16. A professor decides that the method proposed for computing the maximum list size is much too complicated. He proposes the following solution: If we let X_i be the size of list i, then what we want to compute is $E\left(\max_i(X_i)\right)$. This means

$$E\left(\max_i(X_i)\right) = \max_i\left(E(X_i)\right) = \max_i(1) = 1 .$$

What is the flaw in his solution?

---

*This problem depends on material marked with an asterisk in the text.

†This problem relates to a subsection marked with an asterisk and requires more insight into logarithms and exponential functions than other problems in this section.

*17. In our analysis of Equation 5.34, we said that for $t = (5 \ln n / \ln \ln n)$, Lemma 5.21 gives us that $P(M_t) \leq 1/n^2$. The lemma also gives us that $P(M_t) \leq n \, (e/t)^t$. To get the bound of $1/n^2$, it suffices to show that

$$n \left(\frac{e}{t} \right)^t \leq \frac{1}{n^2} . \tag{5.37}$$

We now outline how to show this.

a. Show that Inequality 5.37 is equivalent to $t(1 - \ln t) \leq -3 \ln n$.

b. Is there a t of the form $c \ln n / \ln \ln n$ that satisfies (5.37)? Show that if there is such a t, then

$$-c \ln n + \frac{c \ln n}{\ln \ln n} (1 - \ln c + \ln \ln \ln n) \leq -3 \ln n .$$

c. You know that $\ln \ln \ln n \leq \ln \ln n$, but by how much? To find out, determine where the function $\ln \ln \ln x / \ln \ln x$ has its maximum value and what that maximum value is. (You know it has a maximum, because the function is 0 when $x = e^e$, approaches 0 as x becomes large, but is positive for $x > e^e$.)

d. Show that $\ln \ln \ln n \leq 0.4 \ln \ln n$.

e. Show that with $c = 5$, you have

$$-c \ln n + \frac{c \ln n}{\ln \ln n} (1 - \ln c + \ln \ln \ln n) \leq -3 \ln n .$$

This completes the proof of the bound $P(M_t) \leq 1/n^2$.

18. Prove as tight upper and lower bounds as you can for $\sum_{i=1}^{k} (1/i)$. For this purpose, it is useful to remember the definition of the natural logarithm as an integral involving $1/x$ and to draw rectangles and other geometric figures above and below the curve. a/h

19. Notice that $\ln n! = \sum_{i=1}^{n} \ln i$. Sketch a careful graph of $y = \ln x$, and, by drawing in geometric figures above and below the graph, show that

$$\sum_{i=1}^{n} \ln i - \frac{1}{2} \ln n \leq \int_{1}^{n} \ln x \, dx \leq \sum_{i=1}^{n} \ln i .$$

Based on your drawing, which inequality do you think is tighter? Use integration by parts to evaluate the integral. What bounds on $n!$ can you get from these inequalities? Which one do you think is tighter? How does it compare with Stirling's formula? What big O bound can you get on $n!$?

---

*This problem depends on material marked with an asterisk.

5.6 CONDITIONAL EXPECTATIONS, RECURRENCES, AND ALGORITHMS

Probability is a very important tool in algorithm design. We have already seen two important examples in which it is used: primality testing and hashing. In this section, we study several more examples of probabilistic analysis in algorithms. We focus on computing the running time of various algorithms. When the running time of an algorithm is different for different inputs of the same size, we can think of the running time of the algorithm as a random variable on the sample space of inputs, and thus, we can analyze the expected running time of the algorithm. This gives us an understanding different from studying just the worst-case running time for an input of a given size. We then consider **randomized algorithms**, which are algorithms that depend on choosing something randomly, to see how we can use recurrences to give bounds on the algorithms' expected running times.

For randomized algorithms, it will be useful to have access to a function that generates random numbers. We will assume that we have a function `randint(i,j)`, which generates a random integer uniformly between i and j (inclusive). This means the random integer is equally likely to be any number between i and j. We also have a function `rand01()`, which generates a random real number between 0 and 1 uniformly.[10] Functions such as `randint` and `rand01` are called **random number generators**. A great deal of number theory goes into the construction of good random number generators.

When Running Times Depend on More than Size of Inputs

| Exercise 5.6-1 | Let A be an array of length $n-1$ (whose elements are chosen from some ordered set), sorted into increasing order. Let b be another element of the ordered set that we want to insert into A to get a sorted array of length n. Assuming that the elements of A and the element b are chosen randomly,[11] what is the expected number of elements of A that have to be shifted one place to the right to let us insert b? |

| Exercise 5.6-2 | One of the standard methods of sorting that you have probably studied is insertion sort. We describe this technique briefly here: Let $A[1:n]$ denote the elements in Positions 1 to n of Array A. A recursive description of insertion sort is that to sort $A[1:n]$, we first sort $A[1:n-1]$, and then we insert $A[n]$ by shifting each element greater than $A[n]$ one place to the right and then inserting the original value of $A[n]$ into the place we have opened up. If $n=1$, we do nothing. |

---

[10]To say we have a random number chosen uniformly between 0 and 1 means that given any two pairs of real numbers (r_1, r_2) and (s_1, s_2) with $r_2 - r_1 = s_2 - s_1$ and r_1, r_2, s_1, and s_2 all between 0 and 1, our random number is just as likely to be between r_1 and r_2 as it is to be between s_1 and s_2.

[11]When we say the elements are chosen randomly from some finite set, we mean that all elements of the set are equally likely to be chosen. If the set from which we are choosing is infinite, we mean that for any two intervals of the same length in the ordered set, the elements are equally likely to be in either interval.

The purpose of this exercise is to analyze the expected time needed to carry out insertion sort. We consider two random variables—S_j for sorting and I_j for inserting.

- Let $S_j(A[1{:}j])$ be the time needed to sort the portion of A from Position 1 to Position j.

- Let $I_j(A[1{:}j], b)$ be the time needed to insert the element b into a sorted list originally in the first j positions of A to give a sorted list in the first $j + 1$ positions of A.

Note that S_j and I_j depend on the actual array A and not only on the value of j. Find a way to use S_{n-1} and I_{n-1} to describe the time needed to use insertion sort to sort $A[1{:}n]$ in terms of the time needed to sort $A[1{:}n-1]$. Remember that it is necessary to copy the element in Position n of A into a variable B before moving elements of A to the right to make a place for it—this moving process will write over $A[n]$. This copying will take some time c_1. We let $T(n)$ be the expected value of S_n—that is, the expected running time of insertion sort on a list of n items. Write a recurrence for $T(n)$ in terms of $T(n-1)$ by taking expected values in the equation that corresponds to your previous description of the time needed to use insertion sort on a particular array. Solve your recurrence relation in big Θ terms.

If X is the random variable with $X(A, b)$ equal to the number of items we need to move one place to the right in order to insert b into the resulting empty slot in A, then X takes on the values $0, 1, \ldots, n - 1$ with equal probability $1/n$. Thus, we have

$$E(X) = \sum_{i=0}^{n-1} i\,\frac{1}{n} = \frac{1}{n} \sum_{i=0}^{n-1} i = \frac{1}{n}\frac{(n-1)n}{2} = \frac{n-1}{2} \,.$$

We use $S_j(A[1{:}j])$ to stand for the time required to sort the portion of Array A from Positions 1 to j by insertion sort. We use $I_j(A[1{:}j], b)$ to stand for the time needed to insert b into a sorted list in the first j positions of Array A, moving all items larger than j to the right one place and putting b into the empty slot that results. In terms of S_j and I_j, we can write that for insertion sort,

$$S_n(A[1{:}n]) = S_{n-1}(A[1{:}n - 1]) + I_{n-1}(A[1{:}n - 1], A[n]) + c_1 \,.$$

We have included the constant term c_1 for the time it takes to copy the value of $A[n]$ into some variable B, because we will overwrite $A[n]$ in the process of moving items one place to the right. Using the additivity of expected values, we get

$$E(S_n) = E(S_{n-1}) + E(I_{n-1}) + E(c_1) \,.$$

Using $T(n)$ for the expected time to sort $A[1{:}n]$ by insertion sort and the result of Exercise 5.6-1, we get

$$T(n) = T(n - 1) + c_2 \frac{n - 1}{2} + c_1 \,.$$

We wrote $c_2(n-1)/2$ for $E(I_{n-1})$ because the time needed to prepare the place where we will do the insertion is proportional to the number of items we have to move. By our solution to Exercise 5.6-1, the expected number of items we need to move is $(n-1)/2$. We can say that $T(1) = 1$ (or some third constant) because with a list of size 1, we have to realize that it has size 1 and then do nothing. It might be more realistic to write

$$T(n) \leq T(n-1) + cn$$

and

$$T(n) \geq T(n-1) + c'n \, ,$$

because the time needed to do the insertion may not be exactly proportional to the number of items we need to move, but it might depend on implementation details. By iterating the recurrence or drawing a recursion tree, we see that $T(n) = \Theta(n^2)$. (We could also give an inductive proof.) Because the best-case time of insertion sort is $\Theta(n)$ and the worst-case time is $\Theta(n^2)$, it is interesting to know that the expected case is much closer to the worst case than to the best case.

Conditional Expected Values

Our next example introduces an idea that we often use in analyzing the expected running times of algorithms, especially randomized algorithms.

Exercise 5.6-3	I have two nickels and two quarters in my left pocket and four dimes in my right pocket. Suppose I flip a penny and take two coins from my left pocket if the penny comes up heads and two coins from my right pocket if it comes up tails. Assuming I am equally likely to choose any coin in my pocket at any time, what is the expected amount of money that I draw from my pocket?

We could do this problem by drawing a tree diagram or by observing that the outcomes can be modeled by 3-tuples in which the first entry is heads or tails and the second and third entries represent coins. Thus, our sample space is HNQ, HQN, HQQ, HNN, and TDD. The probabilities of these outcomes are 1/6, 1/6, 1/12, 1/12, and 1/2, respectively. Thus, our expected value is

$$30 \left(\frac{1}{6}\right) + 30 \left(\frac{1}{6}\right) + 50 \left(\frac{1}{12}\right) + 10 \left(\frac{1}{12}\right) + 20 \left(\frac{1}{2}\right) = 25 \, .$$

Here is a method that seems even simpler: If the coin comes up heads, there is an expected value of 15 cents on each draw. So, with probability 1/2, our expected value is 30 cents. If the coin comes up tails, we have an expected value of 10 cents on each draw. So, with probability 1/2, our expected value is 20 cents. Thus, it is natural to expect that our expected value is $(1/2)30 + (1/2)20 = 25$ cents. In fact, if we group the four outcomes that have an H first, we see that their contribution to the expected value is 15 cents, which is $(1/2)30$. If we look at the single element that has a T first, then its contribution to the sum is 10 cents, which is $(1/2)20$.

The intuition for this second view of the problem is as follows. We took the probability of heads times the expected value of our draws, given that the penny came up heads, plus the probability of tails times the expected value of our draws, given that the penny came up tails. In particular, we were using a new (and as yet undefined) idea of **conditional expected value**. To get the conditional expected value, given that our penny came up heads, we could have created a new sample space with four outcomes, NQ, QN, NN, QQ, with probabilities 1/3, 1/3, 1/6, and 1/6. In this sample space, the expected amount of money from two draws would be 30 cents (15 cents for the first draw plus 15 cents for the second). So, we would say the conditional expected value of our draws, given that the penny came up heads, was 30 cents. With a one-element sample space {DD}, we would say that the conditional expected value of our draws, given that the penny came up tails, is 20 cents.

How do we define conditional expected value? Rather than create a new sample space, as we did above, we use the idea of a new sample space (as we did in discovering a good definition for conditional probability) to lead us to a good definition for conditional expected value. In particular, to get the conditional expected value of X, given that an event F has happened, we use our conditional probability weights for the elements of F—namely, $P(s)/P(F)$ is the weight for the element s of F—and pretend that F is our sample space. Thus, we define the **conditional expected value** of X, given F, by

$$E(X|F) = \sum_{s:s \in F} X(s) \frac{P(s)}{P(F)} . \qquad (5.38)$$

Remember that we defined the expected value of a random variable X with values $x_1, x_2, \ldots, x_k$ by

$$E(X) = \sum_{i=1}^{k} x_i P(X = x_i) ,$$

where $X = x_i$ stands for the event that X has the value x_i. Using our standard notation for conditional probabilities, $P((X = x_i)|F)$ stands for the conditional probability of the event $X = x_i$, given that the event F occurs. This lets us rewrite Equation 5.38 as

$$E(X|F) = \sum_{i=1}^{k} x_i P((X = x_i)|F) .$$

| Theorem 5.23 | Let X be a random variable defined on a sample space S and let $F_1, F_2, \ldots, F_n$ be disjoint events whose union is S (i.e., a partition of S). Then |

$$E(X) = \sum_{i=1}^{n} E(X|F_i)P(F_i) .$$

Proof: The proof is simply an exercise in applying definitions. ■

Exercise 5.6-4

Consider an algorithm that, given a list of n numbers, prints them all out. It then picks a random integer between 1 and 3. If the number is 1 or 2, it stops. If the number is 3, it starts again from the beginning. What is the expected running time of this algorithm?

Exercise 5.6-5

Consider the following variant on the algorithm in Exercise 5.6-4.

```
funnyprint(n)

// Assumes n is a positive integer
(1)   if (n == 1)
(2)        return
(3)   for i = 1 to n
(4)        print i
(5)   x = randint(1,2)
(6)   if (x == 2)
(7)        funnyprint(n/2)
(8)   else
(9)        return
```

What is the expected running time of this algorithm?

For Exercise 5.6-4, with probability 2/3, we will print out the numbers and quit. With probability 1/3, we will run the algorithm again. Using Theorem 5.23, we see that if $T(n)$ is the expected running time on a list of length n, then there is a constant c such that

$$T(n) = \frac{2}{3}cn + \frac{1}{3}(cn + T(n)) ,$$

which gives us $(2/3)T(n) = cn$. This simplifies to $T(n) = (3/2)cn$, so $T(n) = \Theta(n)$.

Another view is that we have an independent trials process with probability 2/3 of success. In this process, we stop at the first success. We refer to a stage of the independent trials process as a **round**. For each round of the independent trials process, we spend $\Theta(n)$ time. Letting T be the running time (note that T is a random variable on the sample space $\{1, 2, 3\}$ with probabilities $1/3$ for each member) and R be the number of rounds, we have that

$$T = R \cdot \Theta(n) ,$$

and so

$$E(T) = E(R)\Theta(n) .$$

In a sense, we are applying Theorem 5.11, because in this context, $\Theta(n)$ behaves as if it were a constant,[12] because n does not depend on R. By Theorem 5.13, we have that $E(R) = 3/2$, and so $E(T) = \Theta(n)$.

In Exercise 5.6-5, because we have a recursive algorithm, it is appropriate to write a recurrence to describe the algorithm's running time. We can let $T(n)$ stand for the *expected* running time of the algorithm on an input of size n. Notice how we are changing back and forth between letting T stand for the running time of an algorithm and letting it stand for the expected running time of an algorithm. Usually, we use T to stand for the quantity of most interest to us, either running time, if that makes sense, or expected running time (or maybe worst-case running time) if the actual running time might vary over different inputs of size n. The nice thing about this is that once we write down a recurrence for the expected running time of an algorithm, the methods for solving it will be those we have already learned for solving recurrences. For the problem at hand, we immediately get that with probability $1/2$, we will spend n units of time (perhaps we should say $\Theta(n)$ time) and then stop, and with probability $1/2$, we will spend n units of time and then recurse on a problem of size $n/2$. Thus, using Theorem 5.23, we get that

$$T(n) = n + \frac{1}{2}T\left(\frac{n}{2}\right) .$$

Including a base case of $T(1) = 1$, we get that

$$T(n) = \begin{cases} (1/2)T(n/2) + n & \text{if } n > 1 , \\ 1 & \text{if } n = 1 . \end{cases}$$

A simple proof by induction shows that $T(n) = \Theta(n)$. Note that the master theorem (as we originally stated it) doesn't apply here, because $a < 1$. However, we could also observe that the solution to this recurrence is no more than the solution to the recurrence $T(n) = T(n/2) + n$ and then apply the master theorem.

Selection Revisited

We now return to the selection algorithm from Section 4.6. The purpose of the algorithm is to select the ith-smallest element in a set with some underlying order. Recall that in this algorithm, we first picked an element p in the middle half of the set—an element whose value was simultaneously larger than at least a quarter of the items and smaller than at least a quarter of the items. We used p to partition the items into two sets and then recursed on one of the two sets. If you recall, we worked very hard to find an item in the middle half so that our partitioning would work well. It is natural to try instead to pick a partition element at random, because with

---

[12]What we mean here is that $T \geq Rc_1 n$ for some constant c_1 and $T \leq Rc_2 n$ for some other constant c_2. Then we apply Theorem 5.11 to both of these inequalities, because if $X > Y$, then $E(X) > E(Y)$ as well.

probability 1/2, this element will be in the middle half. We can extend this idea to the following algorithm:

```
RandomSelect(A,i,n)
// Selects the ith-smallest element in set A, where n = |A|
(1)   if (n == 1)
(2)        return the one item in A
(3)   else
(4)        p = RandomElement(A)
(5)        Let H be the set of elements greater than p
(6)        Let L be the set of elements less than or equal to p
(7)        If (H is empty)
(8)             put p in H
(9)        if (i ≤ |L|)
(10)            Return RandomSelect(L,i,|L|)
(11)       else
(12)            Return RandomSelect(H,i-|L|,|H|).
```

Here `RandomElement(A)` returns one element from A uniformly at random. We use this element as our partition element; that is, we use it to divide A into sets L and H, with every element less than the partition element in L and every element greater than it in H. We add the special case when H is empty to ensure that both recursive problems have size strictly less than n. Although this simplifies a detailed analysis, it is not strictly necessary. At the end of this section, we will show how to get a recurrence that describes fairly precisely the time needed to carry out this algorithm. However, by being a bit less precise, we can still get the same big O upper bound with less work.

When we choose our partition element, we expect that half of the time it will be between $(1/4)n$ and $(3/4)n$. Then, when we partition our set into H and L, each of these sets will have no more than $(3/4)n$ elements. The rest of the time, each of H and L will have no more than n elements. In any case, the time to partition our set into H and L is $O(n)$. Thus, we may write

$$T(n) \leq \begin{cases} (1/2)T(3n/4) + (1/2)T(n) + bn & \text{if } n > 1, \\ d & \text{if } n = 1. \end{cases}$$

We can rewrite the recursive part of the recurrence as

$$\frac{1}{2}T(n) \leq \frac{1}{2}T\left(\frac{3}{4}n\right) + bn,$$

or

$$T(n) \leq T\left(\frac{3}{4}n\right) + 2bn = T\left(\frac{3}{4}n\right) + b'n.$$

Notice that it is possible (but unlikely) that each time our algorithm chooses a pivot element, it chooses the worst one possible, in which case the selection process could take n rounds and, thus, take time $\Theta(n^2)$. Why, then, is the algorithm of interest? It involves far less computation than finding the median of medians, and its expected running time is still $\Theta(n)$. Thus, it is reasonable to suspect that, on the average, it would be significantly faster than the deterministic process. In fact, with good implementations of both algorithms, this will be the case.

| Exercise 5.6-6 | Why does every solution to the recurrence |

$$T(n) \leq T\left(\frac{3}{4}n\right) + b'n$$

have $T(n) = O(n)$?

By the master theorem, we know that any solution to this recurrence is $O(n)$, giving a proof of our next theorem.

| Theorem 5.24 | Algorithm RandomSelect has expected running time $O(n)$. |

QuickSort

There are many algorithms that will efficiently sort a list of n numbers. The two most common sorting algorithms that are guaranteed to run in $O(n \log n)$ time are MergeSort and HeapSort. However, there is another algorithm, QuickSort, which, though having a worst-case running time of $O(n^2)$, has an expected running time of $O(n \log n)$. Moreover, when implemented well, this algorithm tends to have a faster running time than MergeSort or HeapSort. Because many computer operating systems and programs come with QuickSort built in, it has become the sorting algorithm of choice in many applications. We will now see why it has expected running time $O(n \log n)$. We will concern ourselves only with a high-level description, rather than the low-level implementation issues that make this algorithm the fastest one.

QuickSort actually works similarly to the RecursiveSelect algorithm of the previous subsection. We pick a random element and then use it to partition the set of items into two sets, L and H. In this case, we don't recurse on one or the other; instead, we recurse on both, sorting each one. After both L and H have been sorted, we concatenate them to get a sorted list. (In fact, QuickSort is usually done "in place" by pointer manipulation, and so the concatenation just happens.) Here is a pseudocode description of QuickSort:

```
QuickSort(A, n)

(1)   if (n == 1)
(2)        return the one item in A
(3)   else
(4)        p = RandomElement(A)
(5)        Let H be the set of elements greater than p; Let h = |H|
(6)        Let L be the set of elements less than or equal to p; Let ℓ = |L|
(7)        If (H is empty)
(8)             put p in H
```

```
(9)          A₁ = QuickSort(H,h)
(10)         A₂ = QuickSort(L,ℓ)
(11)         return the concatenation of A₁ and A₂
```

Based on the preceding analysis of RandomSelect, we think about modifying the algorithm a bit to make the analysis easier. First, consider what would happen if the random element was the median each time. We would be solving two subproblems of size $n/2$, and would have the recurrence

$$T(n) = \begin{cases} 2T(n/2) + O(n) & \text{if } n > 1, \\ O(1) & \text{if } n = 1, \end{cases}$$

and we know by the master theorem that all solutions to this recurrence have $T(n) = O(n \log n)$. In fact, we don't need such an even division to guarantee such performance.

Exercise 5.6-7

Suppose we had a recurrence of the form

$$T(n) \leq \begin{cases} T(a_n n) + T((1 - a_n)n) + cn & \text{if } n > 1, \\ d & \text{if } n = 1, \end{cases}$$

where a_n is between $1/4$ and $3/4$. Show that all solutions of a recurrence of this form have $T(n) = O(n \log n)$. What do we really need to assume about a_n to prove this upper bound?

In Exercise 5.6-7, we can prove that $T(n) = O(n \log n)$ by induction or via a recursion tree, noting that there are $O(\log n)$ levels and each level has at most $O(n)$ work. (The details of the recursion tree are complicated somewhat by the fact that a_n varies with n, while the details of an inductive proof simply use the fact that a_n and $1 - a_n$ are both no more than $3/4$.) As long as we know that there is some positive number $a < 1$ such that $a_n < a$ and $1 - a_n < a$ for every n, then we know

- we have at most $\log_{(1/a)} n$ levels in a recursion tree, and

- we have at most cn units of work per level for some constant c.

Thus, we have the same $T(n) = O(n \log n)$.

What does this tell us? As long as our problem splits into two pieces, each having size at least, say, a quarter of the items, QuickSort will run in $O(n \log n)$ time. Given this, we modify our algorithm to enforce this condition. That is, if at first we choose a pivot element p that is not in the middle half, we will just pick another one. This leads to the following algorithm:

```
Slower QuickSort(A,n)

(1)  if (n == 1)
(2)       return the one item in A
(3)  else
(4)       Repeat
(5)            p = RandomElement(A)
```

```
(6)              Let H be the set of elements greater than p; Let h = |H|
(7)              Let L be the set of elements less than or equal to p; Let ℓ = |L|
(8)        Until (|H| ≥ n/4) and (|L| ≥ n/4)
(9)        A₁ = Slower QuickSort(H,h)
(10)       A₂ = Slower QuickSort(L,ℓ)
(11)       return the concatenation of A₁ and A₂
```

Now let's analyze this algorithm. Let r be the number of times[13] that we execute the loop to pick p, and let $a_n \cdot n$ be the position of the pivot element.[14] If $T(n)$ is the expected running time for a list of length n, then for some constant b

$$T(n) \le E(r)bn + T(a_n n) + T((1 - a_n)n) \ ,$$

because each iteration of the loop takes $O(n)$ time. Note that we take the expectation of r, because $T(n)$ stands for the expected running time on a problem of size n. Fortunately, $E(r)$ is simple to compute; it is the expected time until the first success in an independent trials process with success probability at least $1/2$, which is 2. So we get that the running time of Slower QuickSort satisfies the recurrence

$$T(n) \le \begin{cases} T(a_n n) + T((1 - a_n))n + b'n & \text{if } n > 1 \ , \\ d & \text{if } n = 1 \ , \end{cases}$$

where a_n is between $1/4$ and $3/4$. Thus, by Exercise 5.6-7, the running time of this algorithm is $O(n \log n)$.

As another variant on the same theme, observe that looping until we have $|H| \ge n/4$ and $|L| \ge n/4$ is effectively the same as choosing p, finding H and L, and then calling Slower QuickSort(A, n) once again if either H or L has size less than $n/4$. Then, because with probability $1/2$, the element p is between $n/4$ and $3n/4$, we can write

$$T(n) \le \frac{1}{2}T(n) + \frac{1}{2}(T(a_n n) + T((1 - a_n)n) + bn) \ ,$$

which simplifies to

$$T(n) \le T(a_n n) + T((1 - a_n)n) + 2bn \ ,$$

or

$$T(n) \le T(a_n n) + T((1 - a_n)n) + b'n \ .$$

Again by Exercise 5.6-7, the running time of this algorithm is $O(n \log n)$.

---

[13] We think of r as standing for the number of *rounds,* where a round is a loop through the algorithm.

[14] Each choice of a pivot element chooses some fraction of n. We use a_n to denote this fraction. The reason we choose to set up the problem in this way is that we know that half of the time, a_n will be between $1/4$ and $3/4$.

Further, it is straightforward to see that the expected running time of Slower QuickSort is no less than half of that of QuickSort (and, incidentally, no more than twice that of QuickSort) and so we have proved our next theorem.

Theorem 5.25   QuickSort has expected running time $O(n \log n)$.

A More Careful Analysis of RandomSelect*

Recall that our analysis of RandomSelect was based on using $T(n)$ as an upper bound for $T(|H|)$ or $T(|L|)$ if either the set H or the set L had more than $3n/4$ elements. Here we show how to avoid this assumption. The kinds of computations we do here are the kind we would need to do if we wanted to try to get bounds on the constants implicit in our big O bounds.

Exercise 5.6-8   Explain why, if we pick the kth element as the random element in RandomSelect ($k \neq n$), our recursive problem is of size no more than $\max\{k, n - k\}$.

If we pick the kth element, then we recurse either on the set L, which has size k, or on the set H, which has size $n - k$. Both of these sizes are at most $\max\{k, n - k\}$. (If we pick the nth element, then $k = n$. Thus, because of Line 8 of RandomSelect, L actually has size $k - 1$ and H has size $n - k + 1$. But because $\max\{n, n - n\} = n$, both sizes are at most this maximum.)

Now, let X be the random variable equal to the rank of the chosen random element (e.g., if the random element is the third smallest, then $X = 3$). Using Theorem 5.23 and the solution to Exercise 5.6-8, we can write

$$
T(n) \leq
\begin{cases}
\left(\sum_{k=1}^{n-1} P(X = k) \Big(T\left(\max\{k, n - k\} \right) + bn \Big) \right) \\
\quad + P(X = n) \Big(T\left(\max\{1, n - 1\} \right) + bn \Big) & \text{if } n > 1, \\
d & \text{if } n = 1.
\end{cases}
$$

Because X is chosen uniformly between 1 and n, we have that $P(X = k) = 1/n$ for all k. Ignoring the base case for a minute, we get that

$$
T(n) \leq \sum_{k=1}^{n-1} \frac{1}{n} \Big(T\left(\max\{k, n - k\} \right) + bn \Big) + \frac{1}{n} \Big(T(n - 1) + bn \Big)
$$

$$
\leq \frac{1}{n} \left(\sum_{k=1}^{n-1} T\left(\max\{k, n - k\} \right) \right) + bn + \frac{1}{n} \Big(T(n - 1) + bn \Big).
$$

If n is odd and we write out $\sum_{k=1}^{n-1} T\left(\max\{k, n - k\} \right)$, we get

$$
T(n - 1) + T(n - 2) + \cdots + T\left(\left\lceil \frac{n}{2} \right\rceil \right) + T\left(\left\lceil \frac{n}{2} \right\rceil \right) + \cdots + T(n - 2) + T(n - 1),
$$

---

*This subsection can be skipped without loss of continuity.

which is $2 \sum_{k=\lceil n/2 \rceil}^{n-1} T(k)$. If n is even and we write out $\sum_{k=1}^{n-1} T(\max\{k, n-k\})$, then we get

$$T(n-1) + T(n-2) + \cdots + T\left(\frac{n}{2}\right) + T\left(1 + \frac{n}{2}\right) + \cdots + T(n-2) + T(n-1),$$

which is at most $2 \sum_{k=n/2}^{n-1} T(k)$. Thus, we can replace our recurrence with

$$T(n) \leq \begin{cases} (2/n)\left(\sum\limits_{k=n/2}^{n-1} T(k)\right) + \dfrac{1}{n} T(n-1) + bn & \text{if } n > 1, \\ d & \text{if } n = 1. \end{cases} \tag{5.39}$$

If n is odd, then the lower limit of the sum is a half-integer, so the possible integer values of the dummy variable k run from $\lceil n/2 \rceil$ to $n-1$. Because this is the natural way to interpret a fractional lower limit, and because it corresponds to what we wrote in both the n even and n odd case above, we adopt this convention.

Exercise 5.6-9 Show that every solution to the recurrence in Recurrence 5.39 has $T(n) = O(n)$.

We can prove this by induction. We try to prove that $T(n) \leq cn$ for some constant c. By the natural inductive hypothesis, we get that

$$T(n) \leq \frac{2}{n}\left(\sum_{k=n/2}^{n-1} ck\right) + \frac{1}{n}c(n-1) + bn$$

$$= \frac{2}{n}\left(\sum_{k=1}^{n-1} ck - \sum_{k=1}^{\lceil n/2 \rceil - 1} ck\right) + \frac{1}{n}c(n-1) + bn$$

$$\leq \frac{2c}{n}\left(\frac{(n-1)n}{2} - \frac{((n/2)-1)n/2}{2}\right) + c + bn$$

$$= \frac{2c}{n}\frac{(3n^2/4) - (n/2)}{2} + c + bn$$

$$= \frac{3}{4}cn + \frac{c}{2} + bn$$

$$= cn - \left(\frac{1}{4}cn - bn - \frac{c}{2}\right).$$

Notice that so far, we have only assumed that there is some constant c such that $T(k) < ck$ for $k < n$. We can choose a larger c than the one given to us by this assumption without changing the inequality $T(k) < ck$. By choosing c so that $cn/4 - bn - c/2$ is nonnegative (for example, $c \geq 8b$ makes this term at least $bn - 4b$, which is nonnegative for $n \geq 4$), we conclude the proof and have another proof of Theorem 5.24.

This kind of careful analysis arises when we are trying to get an estimate of the constant in a big O bound, which we decided not to do in this case.

Important Concepts, Formulas, and Theorems

1. *Expected running time.* When the running time of an algorithm is different for different inputs of the same size, we can think of the running time of the algorithm as a random variable on the sample space of inputs and analyze the expected running time of the algorithm. This gives us a different understanding from studying only the worst-case running time.

2. *Randomized algorithm.* A *randomized algorithm* is an algorithm that depends on choosing something randomly.

3. *Random number generator.* A *random number generator* is a procedure that generates a number that appears to be chosen at random. Usually the designer of a random number generator tries to generate numbers that appear to be uniformly distributed.

4. *Insertion sort.* A recursive description of insertion sort is that to sort $A[1:n]$, first we sort $A[1:n-1]$, and then we insert $A[n]$ by shifting each element greater than $A[n]$ one place to the right and then inserting the original value of $A[n]$ into the place we have opened up. If $n = 1$, we do nothing.

5. *Expected running time of insertion sort.* If $T(n)$ is the expected time to use insertion sort on a list of length n, then there are constants c and c' such that $T(n) \leq T(n-1) + cn$ and $T(n) \geq T(n-1) + c'n$. This means that $T(n) = \Theta(n^2)$. However, the best-case running time of insertion sort is $\Theta(n)$.

6. *Conditional expected value.* We define the *conditional expected value* of X, given F, by $E(X|F) = \sum_{x:x \in F} X(x)P(x)/P(F)$. This is equivalent to $E(X|F) = \sum_{i=1}^{k} x_i P((X = x_i)|F)$.

7. *Randomized selection algorithm.* In the randomized selection algorithm, to select the ith-smallest element of a set A, we randomly choose a pivot element p in A, divide the rest of A into those elements that come before p (in the underlying order of A) and those that come after, put the pivot into the smaller set, and then recursively apply the randomized selection algorithm to find the appropriate element of the appropriate set.

8. *Running time of randomized select.* RandomSelect has expected running time $O(n)$. Because it does less computation than the deterministic selection algorithm, on average, a good implementation will run faster than a good implementation of the deterministic algorithm. However, the worst-case behavior is $\Theta(n^2)$.

9. *QuickSort.* *QuickSort* is a sorting algorithm in which we randomly choose a pivot element p in A, divide the rest of A into those elements that come before p (in the underlying order of A) and those that come after, put the pivot into the smaller set, recursively apply the QuickSort algorithm to sort each of the smaller sets, and concatenate the two sorted lists. We do nothing if a set has size 1.

10. *Running time of QuickSort.* QuickSort has expected running time $O(n \log n)$. It has worst-case running time $\Theta(n^2)$. Good implementations of QuickSort have proved to be faster, on average, than good implementations of other sorting algorithms.

Problems

1. Given an array A of length n (chosen from some set that has an underlying ordering), you can select the largest element of the array by first setting $L = A[1]$ and then comparing L to the remaining elements of the array, one at a time, replacing L with $A[i]$ if $A[i]$ is larger than L. Assume that the elements of A are randomly chosen. For $i > 1$, let $X_i = 1$ if an element i of A is larger than any element of $A[1{:}i - 1]$. Let $X_1 = 1$. What does $X_1 + X_2 + \cdots + X_n$ have to do with the number of times you assign a value to L? What is the expected number of times you assign a value to L? a/h

2. Let $A[i{:}j]$ denote the array of items in Positions i through j of Array A. In one possible implementation of selection sort, you would

    • use the method from Problem 1 to find the largest element of Array A and its Position k in the array,

    • exchange the elements in Positions k and n of Array A, and

    • apply the same procedure recursively to Array $A[1{:}n - 1]$.

    (Actually, this is what you would do if $n > 1$; if $n = 1$, you would do nothing.) What is the expected total number of times you assign a value to L in the selection sort algorithm?

3. Show that if H_n stands for the nth harmonic number, then

$$H_n + H_{n-1} + \cdots + H_2 = \Theta(n \log n) .$$ a/h

4. In a card game, you remove the jacks, queens, kings, and aces from an ordinary deck of cards and shuffle them. You draw a card. If it is an ace, you are paid $1.00, and the game is repeated. If it is a jack, you are paid $2.00, and the game ends. If it is a queen, you are paid $3.00, and the game ends. If it is a king, you are paid $4.00, and the game ends. What is the maximum amount of money a rational person would pay to play this game?

5. Why does every solution to $T(n) \leq T(2n/3) + bn$ have $T(n) = O(n)$? a/h

*6. Show that if in RandomSelect, you remove the instruction
```
If H is empty
       put p in H,
```

---

*This problem depends on material marked with an asterisk.

then if $T(n)$ is the expected running time of the algorithm, there is a constant b such that $T(n)$ satisfies the recurrence

$$T(n) \leq \frac{2}{n-1} \sum_{k=n/2}^{n-1} T(k) + bn .$$

Show that if $T(n)$ satisfies this recurrence, then $T(n) = O(n)$.

7. Suppose you have a recurrence of the form

$$T(n) \leq T(a_n n) + T\big((1 - a_n)n\big) + bn, \text{ if } n > 1 ,$$

where a_n is between $1/5$ and $4/5$. Show that all solutions to this recurrence are of the form $T(n) = O(n \log n)$. a/h

8. Prove Theorem 5.23.

*9. A tighter (up to constant factors) analysis of QuickSort is possible by using ideas very similar to those used for RandomSelect. More precisely, use Theorem 5.23 similarly to the way it was used for select. Write the recurrence you get when you do this. Show that this recurrence has solution $O(n \log n)$. To show this, you will probably want to prove that $T(n) \leq c_1 n \log n - c_2 n$ for some constants c_1 and c_2.

10. It is possible to write a version of RandomSelect analogous to Slower QuickSort. That is, when you pick out the random pivot element, check if it is in the middle half; discard it if it is not. Write this modified selection algorithm, give a recurrence for its running time, and show that this recurrence has solution $O(n)$.

11. One idea often used in selection is that instead of choosing a random pivot element, we choose three random pivot elements and then use the median of these three as the pivot. What is the probability that a randomly chosen pivot element is in the middle half? What is the probability that the median of three randomly chosen pivot elements is in the middle half? Does this justify the choice of using the median of three elements as the pivot? a/h

12. Is the expected running time of QuickSort $\Omega(n \log n)$?

13. (This problem assumes that you understand the construction of a binary search tree.) A random binary search tree on n keys is formed by first randomly ordering the keys and then inserting them in that order. Why is it that in at least half of the random binary search trees, both subtrees of the root have between $n/4$ and $3n/4$ keys? If $T(n)$ is the expected height of a random binary search tree on n keys, explain why

$$T(n) \leq \frac{1}{2} T(n) + \frac{1}{2} T\left(\frac{3}{4}n\right) + 1 .$$

---

*This problem depends on material marked with an asterisk.

(Think about the *definition* of a binary tree. It has a root, and the root has two subtrees. What did we say about the possible sizes of those subtrees?) What is the expected height of a one-node binary search tree? Show that the expected height of a random binary search tree is $O(\log n)$. a/h

14. (This problem assumes you understand the construction of a binary search tree.) The expected time for an unsuccessful search in a random binary search tree on n keys (see Problem 13 for a definition) is the expected depth of a leaf node. Arguing as in Problem 13 and the proof of Theorem 5.24, find a recurrence that gives an upper bound on the expected depth of a leaf node in a binary search tree, and use it to find a big O upper bound on the expected depth of a leaf node.

15. (This problem assumes you understand the construction of a binary search tree.) The expected time for a successful search in a random binary search tree on n nodes (see Problem 13 for a definition) is the expected depth of a node of the tree. With probability $1/n$, the node is the root, which has depth 0; otherwise, the expected depth is 1 plus the expected depth of a node in one of its subtrees. Argue, as in Problem 13 and the proof of Theorem 5.24, that if $T(n)$ is the expected depth of a node in a binary search tree (and if $T(i-1) \le T(i)$ for all $i > 1$), then

$$T(n) \le \frac{n-1}{n}\left(\frac{1}{2}T(n) + \frac{1}{2}T\left(\frac{3}{4}n\right)\right) + 1 .$$

What big O upper bound does this give you on the expected depth of a node in a random binary search tree on n nodes? a/h

5.7 PROBABILITY DISTRIBUTIONS AND VARIANCE

Distributions of Random Variables

We have given meaning to the term *expected value*. For example, if we flip a coin 100 times, the expected number of heads is 50. But to what extent do we expect to see 50 heads? Would it be surprising to see 55, 60, or 65 heads instead? To answer this kind of question, we have to analyze how much we expect a random variable to deviate from its expected value. First, we show how to construct a graph that illustrates how the values of a random variable are distributed around its expected value. The **distribution function** D of a random variable X with finitely many values is the function on the values of X defined by

$$D(x) = P(X = x) .$$

You probably recognize the distribution function from the role it played in the definition of *expected value*. The distribution function of the random variable X assigns to each value x of the random variable the probability that X achieves that value. (Thus, D is a function whose domain is the set of values of X.) When the values of X are integers, it is convenient to visualize the distribution function using a diagram called a **histogram**. Figure 5.8 shows histograms for the distribution of the "number of heads" random variable for ten flips of a coin and the "number of right answers" random variable for someone taking a ten-question test with probability .8 of getting a correct answer. What is a histogram? The histograms in Figure 5.8 are graphs that show, for each integer value x of X, a rectangle of width 1 and centered at x whose height (and thus area) is proportional to the probability $P(X = x)$. Histograms can be drawn with nonunit-width rectangles. When people draw a rectangle with a base ranging from $x = a$ to $x = b$, the area of the rectangle is the probability that X is between a and b.

Figure 5.8: *Examples of histograms*

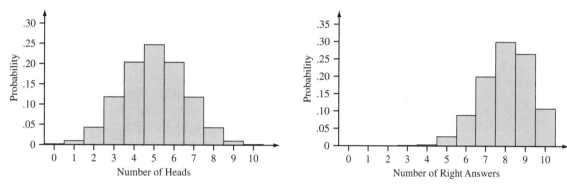

The function D defined by $D(a, b) = P(a \leq X \leq b)$ is often called a **cumulative distribution function**. When sample spaces can be infinite, it doesn't always make sense to assign probability weights to individual members of our sample space, and yet cumulative distribution functions still make sense. Thus, for infinite sample spaces, the treatment of probability is often based on random variables and their cumulative distribution functions. Histograms are a natural way to display information about the cumulative distribution function.

The histograms in Figure 5.8 show the difference between the two distributions. They also show that we can expect the number of heads to be somewhat near the expected number, though as few as two heads or as many as eight are also not out of the question. We see that the number of right answers tends to be clustered between six and ten; so, in this case, we can expect the random variable to be reasonably close to the expected value. With more coin flips or more questions, however, will the results spread out? Relatively speaking, should we expect to be closer to or farther from the expected value? In Figure 5.9, we see the results of 25 coin flips or 25 questions. The expected

Figure 5.9: *Histograms of 25 trials*

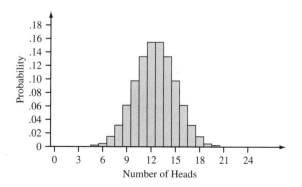 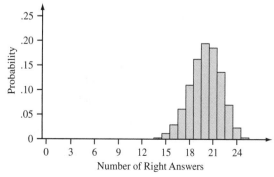

number of heads is 12.5. The histogram makes it clear that we can expect the vast majority of our results to have between 9 and 16 heads. Virtually all the results lie between 5 and 20. Thus, the results are not spread as broadly (relatively speaking) as they were with just 10 flips. As with the coin tossing histogram, the test score histogram with 25 questions seems even more tightly packed around its expected value. Essentially, all the scores lie between 14 and 25. Although we can still tell the difference between the shapes of the histograms, they have become somewhat similar in appearance.

Figure 5.10 shows the 30 most relevant values for 100 flips of a coin and for a 100-question test. Now the two histograms have almost the same shape, though the test histogram is still more tightly packed around its expected value. The number of heads has virtually no chance of deviating by more than 15 from its expected value, and the test score has almost no chance of deviating by more than 11 from the expected value. Thus, the spread has only doubled, even though the number of trials has quadrupled. In both cases, the curve formed by the tops of the rectangles seems quite similar to the bell-shaped curve, called the **normal curve**, that arises in so many areas of science. In the test taking curve, however, you can see a bit of difference between the lower left side and the lower right side.

Figure 5.10: *100 independent trials*

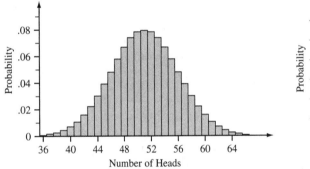
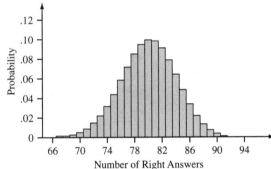

We saw that we need about 30 values to see the most relevant probabilities for 100 trials, whereas we need 15 values to see the most relevant probabilities for 25 independent trials. This might lead us to predict that we need only about 60 values to see essentially all the results in 400 trials. As Figure 5.11 shows, this is indeed the case. Although the

Figure 5.11: *400 independent trials*

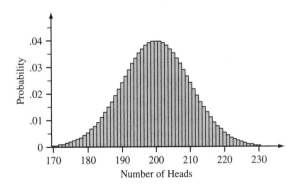
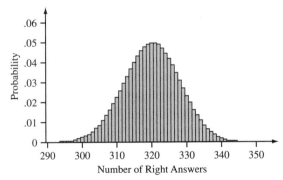

test taking distribution is still more tightly packed than the coin flipping distribution, we have to examine the former closely to find any asymmetry. These experiments suggest that the spread of a distribution (for independent trials) grows as the square root of the number of trials grows, because each time we quadruple the number of elements, we double the spread. They also suggest that there is some common kind of bell-shaped limiting distribution function for at least the distribution of successes in independent trials that have two outcomes. However, without a theoretical foundation, we don't know how far the truth of our observations extends. Thus, we seek an algebraic way to measure the difference between a random variable and its expected value.

Variance

Exercise 5.7-1

Suppose X is the number of heads in four flips of a coin. Let Y be the random variable $X - 2$, or the difference between X and its expected value. Compute $E(Y)$. Does $E(Y)$ effectively measure how much we expect to see X deviate from its expected value? Compute $E(Y^2)$. Try repeating the process with X being the number of heads in ten flips of a coin and Y being $X - 5$.

Before answering these questions, we state a trivial, but useful, lemma (which appeared as Problem 10 in Section 5.4) and a corollary, showing that the expected value of an expectation is that expectation.

Lemma 5.26

If X is a random variable that always takes on the value c, then $E(X) = c$.

Proof: $E(X) = P(X = c) \cdot c = 1 \cdot c = c.$ ∎

We can think of a constant c as a random variable that always takes on the value c, and thus, we can simply write $E(c)$ for the expected value of this random variable. In this case, our lemma says that $E(c) = c$. This lemma has an important corollary.

Corollary 5.27

Let X be a random variable on a sample space. Then $E\big(E(X)\big) = E(X)$.

Proof: When we think of $E(X)$ as a random variable, it has a constant value traditionally denoted by μ. By Lemma 5.26, we have that $E\big(E(x)\big) = E(\mu) = \mu = E(x).$ ∎

Returning to Exercise 5.7-1, we can use linearity of expectation and Corollary 5.27 to show that

$$E\big(X - E(X)\big) = E(X) - E\big(E(X)\big) \tag{5.40}$$
$$= E(X) - E(X)$$
$$= 0 .$$

Thus, this is not a particularly useful measure of how close a random variable is to its expectation. If a random variable is sometimes above its expectation and sometimes below, then we would like these two differences to somehow add together rather than cancel each other out. This suggests that we should try to convert the values of $X - E(X)$ to positive numbers and then take the expectation of these positive numbers as our measure of spread. There are two natural ways to make numbers positive: taking their absolute value and squaring them. It turns out that to prove something that involves the spread of expected values, squaring is more useful. Maybe we could have guessed this because we see that the spread seems to grow with the square root, and the square root isn't related to the absolute value in the way it is related to the squaring function. On the other hand, as we saw in Exercise 5.7-1, computing expected values of these squares from what we now know is time consuming. A bit of theory will make it easier.

We define the **variance** $V(X)$ of a random variable X as the expected value $E\left((X - E(X))^2\right)$. We can also express this as a sum over the individual elements of the sample space S to get that

$$V(X) = E\left((X - E(X))^2\right) = \sum_{s:s \in S} P(s)(X(s) - E(X))^2.$$

Let's apply this definition to compute the variance of the number X of heads in four flips of a coin. We have

$$V(X) = (0-2)^2 \cdot \frac{1}{16} + (1-2)^2 \cdot \frac{1}{4} + (2-2)^2 \cdot \frac{3}{8} + (3-2)^2 \cdot \frac{1}{4} + (4-2)^2 \cdot \frac{1}{16} = 1.$$

Computing the variance for ten flips of a coin involves some very inconvenient arithmetic. It would be nice to have a computational technique that would save us from having to figure out large sums if we want to compute the variance for 10 or even 100 or 400 flips of a coin so that we may check our intuition about how the spread of a distribution grows. We saw before that the expected value of a sum of random variables is the sum of the expected values of the random variables. This was very useful in making computations.

Exercise 5.7-2   What is the variance for the number of heads in one flip of a coin? What is the sum of the variances for four independent trials of one flip of a coin?

Exercise 5.7-3   We have a nickel and a quarter in a cup. We withdraw one coin. What is the expected amount of money we withdraw? What is the variance? We return the coin to the cup and then withdraw two coins, one after the other, without replacement. What is the expected amount of money we withdraw? What is the variance? What is the expected amount of money and variance for the first draw? For the second draw?

Compute the variance for the number of right answers when we answer one question with probability .8 of getting the right answer (note that the number of right answers is either 0 or 1, but the expected value need not be). Compute the variance for the number of right answers when we answer five questions with probability .8 of getting the right answer. Is there a relationship between these two variances?

In Exercise 5.7-2, we can compute the variance

$$V(X) = \left(0 - \frac{1}{2}\right)^2 \cdot \frac{1}{2} + \left(1 - \frac{1}{2}\right)^2 \cdot \frac{1}{2} = \frac{1}{4}.$$

Thus, we see that the variance for one flip is $1/4$ and the sum of the variances for four flips is 1. In Exercise 5.7-4, we see that, for one question, the variance is

$$V(X) = .2(0 - .8)^2 + .8(1 - .8)^2 = .16.$$

For five questions, the variance is

$$4^2 \cdot (.2)^5 + 3^2 \cdot 5 \cdot (.2)^4 \cdot (.8) + 2^2 \cdot 10 \cdot (.2)^3 \cdot (.8)^2$$
$$+ 1^2 \cdot 10 \cdot (.2)^2 \cdot (.8)^3 + 0^2 \cdot 5 \cdot (.2)^1 \cdot (.8)^4 + 1^2 \cdot (.8)^5 = .8.$$

The result is five times the variance for one question.

For Exercise 5.7-3, the expected amount of money for one draw is $0.15. The variance is

$$.5(.05 - .15)^2 + .5(.25 - .15)^2 = .01.$$

For removing both coins, one after the other, the expected amount of money is $0.30, and the variance is 0. Finally, the expected value and variance on the first draw are $0.15 and .01, respectively, and the expected value and variance on the second draw are $0.15 and .01, respectively. Notice that we haven't given units for the variance; had we done so, the units would be "squared dollars." We prefer not to worry about units for variance.

It would be nice if we had a simple method for computing variance by using a rule like "the expected value of a sum is the sum of the expected values." However, Exercise 5.7-3 shows that the variance of a sum is not always the sum of the variances. On the other hand, Exercises 5.7-2 and 5.7-4 suggest that such a result might be true for a sum of variances in independent trials processes. In fact, slightly more than this is true. We say that random variables X and Y are **independent** when the event that X has value x is independent of the event that Y has value y, regardless of the choice of x and y. For example, in n flips of a coin, the number of heads on flip i (which is 0 or 1) is independent of the number of heads on flip j. To show that the variance of a sum of independent random variables is the sum of their variances, we first need to show that the expected value of the product of two *independent* random variables is the product of their expected values.

If X and Y are independent random variables on a sample space S with values $x_1, x_2, \ldots, x_k$ and $y_1, y_2, \ldots, y_m$, respectively, then

$$E(XY) = E(X)E(Y).$$

Proof: We prove the lemma by the following series of equalities. In going from Line 5.41 to Line 5.42, we use the fact that X and Y are independent; the rest of the equation follows from definitions and algebra.

$$E(X)E(Y) = \sum_{i=1}^{k} x_i P(X = x_i) \sum_{j=1}^{m} y_j P(Y = y_j)$$

$$= \sum_{i=1}^{k} \sum_{j=1}^{m} x_i y_j P(X = x_i) P(y = y_j)$$

$$= \sum_{z:\, z \text{ is a value of } XY} z \sum_{(i,j):x_i y_j = z} P(X = x_i) P(Y = y_j) \qquad (5.41)$$

$$= \sum_{z:\, z \text{ is a value of } XY} z \sum_{(i,j):x_i y_j = z} P\big((X = x_i) \wedge (Y = y_j)\big) \qquad (5.42)$$

$$= \sum_{z:\, z \text{ is a value of } XY} z P(XY = z)$$

$$= E(XY). \blacksquare$$

If X and Y are independent random variables, then

$$V(X + Y) = V(X) + V(Y).$$

Proof: Using the definitions of variance, algebra, and linearity of expectation, we have

$$V(X + Y)$$
$$= E\left((X + Y - E(X + Y))^2\right)$$
$$= E\left((X - E(X) + Y - E(Y))^2\right)$$
$$= E\left(\big((X - E(X))^2 + 2(X - E(X))(Y - E(Y)) + (Y - E(Y))^2\big)\right)$$
$$= E\left((X - E(X))^2\right) + 2E\left((X - E(X))(Y - E(Y))\right) + E\left((Y - E(Y))^2\right).$$
$$\qquad (5.43)$$

The first and last terms in Line 5.43 are simply the definitions of $V(X)$ and $V(Y)$, respectively. Note also that if X and Y are independent and b and c are constants, then $X - b$ and $Y - c$ are independent (see Problem 8). Thus, we can apply Lemma 5.28 to the middle term in Line 5.43 to obtain

$$V(X + Y) = V(X) + 2E(X - E(X))E(Y - E(Y)) + V(Y).$$

Now we apply Equation 5.40 to the middle term to show that it is 0, which proves the theorem. ∎

With Theorem 5.29, computing the variance for ten flips of a coin is easy. As usual, we have the random variable X_i, which is 1 or 0, depending on whether the coin comes up heads. We saw that the variance of X_i is 1/4, so the variance for $X_1 + X_2 + \cdots + X_{10}$ is $10/4 = 2.5$.

Exercise 5.7-5

Find the variance for 100 flips of a coin and 400 flips of a coin.

Exercise 5.7-6

The variance in Exercise 5.7-5 grew by a factor of four when the number of trials grew by a factor of four, while the spread we observed in our histograms grew by a factor of two. Can you suggest a natural measure of spread that fixes this problem?

For Exercise 5.7-5, recall that the variance for one flip is 1/4. Therefore, the variance for 100 flips is 25, and the variance for 400 flips is 100. Because this measure grows linearly with the size, we can take its square root to give a measure of spread that grows with the square root of the quiz size—just as our observed "spread" did in the histograms. Taking the square root actually makes intuitive sense, because it "corrects" for the fact that we are measuring expected squared spread rather than expected spread.

The square root of the variance of a random variable is called the **standard deviation** of the random variable and is denoted by σ (or by $\sigma(X)$ when there is a chance for confusion as to what random variable we are discussing). Thus, the standard deviation for 100 flips is 5, and for 400 flips, it is 10. Notice that in both the 100-flip case and the 400-flip case, the "spread" we observed in the histogram was ± 3 standard deviations from the expected value. What about for 25 flips? For 25 flips, the standard deviation is 5/2; so, ± 3 standard deviations from the expected value is a range of 15 points, which is, again, what we observed. For the test scores, the variance is .16 for one question; the standard deviation for 25 questions is 2, giving us a range of 12 points for ± 3 standard deviations. For 100 questions, the standard deviation is 4, and for 400 questions, the standard deviation is 8. Notice again how 3 standard deviations relates to the spread we see in the histograms.

Our observed relationship between the spread and the standard deviation is no accident. A consequence of a theorem of probability known as the central limit theorem is that the percentage of results within 1 standard deviation of the mean in a relatively large number of independent trials with two outcomes is about 68%; the percentage within 2 standard deviations of the mean is about 95.5%; and the percentage within 3 standard deviations of the mean is about 99.7%.

The central limit theorem tells us about the distribution of a sum of independent random variables that have the same distribution function.[15] When the number of random variables we are adding is sufficiently large, the **central limit theorem** tells us the approximate probability of the sum being between a and b standard deviations from its expected value. (For example, if $a = -1.5$ and $b = 2$, then the theorem tells us an approximate probability that the sum is between 1.5 standard deviations less than its expected value and 2 standard deviations more than its expected value.) The central limit theorem tells us that this approximate value[16] is

$$\frac{1}{\sqrt{2\pi}} \int_a^b e^{-\frac{x^2}{2}} \, dx \; .$$

The distribution given by

$$P(a \leq X \leq b) = \frac{1}{\sqrt{2\pi}} \int_a^b e^{-\frac{x^2}{2}} \, dx$$

is called the **normal distribution**. Because many of the things we observe in nature can be thought of as the outcome of multistage processes and the quantities we measure are often the result of adding some quantity at each stage, the central limit theorem "explains" why we should expect to see normal distributions for so many of the things we measure. For example, a person's weight can be thought of as the sum, over all the weeks of his life, of random variables X_i that give his weight change due to food consumption in Week i and random variables Y_i that give his weight change due to exercise in Week i. It is not clear whether this is a natural interpretation for blood pressures. Thus, although we shouldn't be particularly surprised that a person's weights at various times are normally distributed, we don't have the same basis for predicting that blood pressures would be normally distributed, even though they are![17]

---

[15] Actually, the variables can have different distributions, as long as no variable contributes a lot more to the sum than any other, and the variables can be dependent, as long as not too many of them are too highly related to others.

[16] Still more precisely, if we let μ be the expected value of the random variable X_i and σ be its standard deviation (all X_i have the same expected value and standard deviation because they have the same distribution) and if we scale the sum of our random variables by

$$Z = \frac{X_1 + X_2 + \cdots + X_n - n\mu}{\sigma \sqrt{n}} \; ,$$

then the probability that $a \leq Z \leq b$ is

$$\frac{1}{\sqrt{2\pi}} \int_a^b e^{-\frac{x^2}{2}} \, dx \; .$$

[17] Actually, this is a matter of opinion. One might argue that blood pressures respond to many little additive factors.

Exercise 5.7-7

If we want to be 95% sure that the number of heads in n flips of a coin is within $\pm 1\%$ of the expected value, how big does n have to be?

Exercise 5.7-8

What is the variance and standard deviation of the random variable that gives the number of right answers for someone taking a 100-question short-answer test, assuming that each answer is graded either correct or incorrect, if the person knows 80% of the subject material for the test and that the student answers correctly each question she knows? Should we be surprised if such a student scores 90 or above on the test?

Recall that for one flip of a coin, the variance is $1/4$; so, for n flips, it is $n/4$. Thus, for n flips, the standard deviation is $\sqrt{n}/2$. We expect that 95% of our outcomes will be within 2 standard deviations of the mean (in this context, it is common to round 95.5 to 95), so we are asking, when are 2 standard deviations 1% of $n/2$? In other words, we want an n such that $2\sqrt{n}/2 = .01(.5n)$. This is equivalent to $\sqrt{n} = 5 \cdot 10^{-3}n$. Squaring both sides gives $n = 25 \cdot 10^{-6}n^2$, which gives $n = 10^6/25 = 40000$. Therefore, we need to flip a coin 40,000 times to be 95% sure that the number of heads will be within 1% of the expected value of 20,000.

For Exercise 5.7-8, the expected number of correct answers on any given question is .8. The variance for each answer is $.8(1 - .8)^2 + .2(0 - .8)^2 = .8 \cdot .04 + .2 \cdot .64 = .032 + .128 = .16$. Notice that this is $.8(1 - .8)$. The total score is the sum of the random variables giving the number of points on each question. If the questions are independent of each other, then the variance of their sum is the sum of their variances, or 16. Thus, the standard deviation is 4. Because 90% is 2.5 standard deviations above the expected value, the probability of getting a score that far from the expected value is somewhere between .05 and .003, by the central limit theorem. (In fact, it is just a bit more than .01.) Assuming that someone is just as likely to be 2.5 standard deviations below the expected score as above, which is not exactly right but close, we see that it is quite unlikely that someone who knows 80% of the material would score 90% or above on the test. Thus, we should be surprised by such a score and take the score as evidence that the student likely knows more than 80% of the material.

Coin flipping and test taking are two special cases of Bernoulli trials. With the same kind of computations we used for the test score random variable, we can prove the following theorem.

Theorem 5.30

In Bernoulli trials with probability p of success, the variance for one trial is $p(1 - p)$, and for n trials, it is $np(1 - p)$. The standard deviation for n trials is $\sqrt{np(1 - p)}$.

Proof: You are asked to give the proof in Problem 7. ∎

Important Concepts, Formulas, and Theorems

1. *Histogram. Histograms* are graphs that show, for each integer value x of a random variable X, a rectangle of width 1 and centered at x whose height (and thus area) is proportional to the probability $P(X = x)$. Histograms can be drawn with nonunit-width rectangles. When you draw a rectangle with a base ranging from $x = a$ to $x = b$, the area of the rectangle is the probability that X is between a and b.

2. *Expected value of a constant.* If X is a random variable that always takes on the value c, then $E(X) = c$. In particular, $E(E(X)) = E(X)$.

3. *Variance.* The *variance* $V(X)$ of a random variable X is defined as the expected value of $(X - E(X))^2$. This can also be expressed as a sum over the individual elements of the sample space S, which gives
$$V(X) = E\left((X - E(X))^2\right) = \sum_{s:s \in S} P(s)(X(s) - E(X))^2.$$

4. *Independent random variables.* Random variables X and Y are *independent* when the event that X has value x is independent of the event that Y has value y, regardless of the choice of x and y.

5. *Expected product of independent random variables.* If X and Y are independent random variables on a sample space S, then $E(XY) = E(X)E(Y)$.

6. *Variance of sum of independent random variables.* If X and Y are independent random variables, then $V(X + Y) = V(X) + V(Y)$.

7. *Standard deviation.* The square root of the variance of a random variable is called the *standard deviation* of the random variable and is denoted by σ (or by $\sigma(X)$ when there is a chance for confusion as to what random variable we are discussing).

8. *Variance and standard deviation for Bernoulli trials.* In Bernoulli trials with probability p of success, the variance for one trial is $p(1 - p)$, and for n trials, it is $np(1 - p)$. The standard deviation for n trials is $\sqrt{np(1 - p)}$.

9. *Central limit theorem.* The central limit theorem says that the sum of independent random variables with the same distribution function is approximated well as follows: The probability that the sum is between a and b is

$$\frac{1}{\sqrt{2\pi}} \int_a^b e^{-\frac{x^2}{2}} \, dx$$

when the number of random variables being added is sufficiently large. This implies that the probability that a sum of independent random variables is within 1, 2, or 3 standard deviations of its expected value is approximately .68, .955, and .997, respectively. (The theorem holds more generally when the random variables have different distributions, provided that no one of them "dominates" the rest, or when the random variables are not independent, provided that not too many of them are very similar to others.)

Problems

1. Suppose a student who knows 60% of the material covered in a chapter of a textbook is going to take a five-question objective (each answer is either right or wrong, not multiple choice or true-false) quiz. Let X be the random variable that gives the number of questions the student answers correctly for each quiz in the sample space of all quizzes the instructor could construct. What is the expected value of the random variable $X - 3$? What is the expected value of $(X - 3)^2$? What is the variance of X? a/h

2. In Problem 1, let X_i be the number of correct answers the student gets on Question i, that is, X_i is either 0 or 1. What is the expected value of X_i? What is the variance of X_i? How does the sum of the variances of X_1 through X_5 relate to the variance of X for Problem 1?

3. A dime and a 50-cent piece are in a cup. You withdraw one coin. What is the expected amount of money you withdraw? What is the variance? You then draw a second coin, without replacing the first. What is the expected amount of money you withdraw? What is the variance? Suppose instead that you consider withdrawing two coins from the cup together. What is the expected amount of money you withdraw, and what is the variance? What does this example show about whether the variance of a sum of random variables is the sum of their variances? a/h

4. If the quiz in Problem 1 has 100 questions, what is the expected number of right answers, the variance of the expected number of right answers, and the standard deviation of the number of right answers?

5. Estimate the probability that a person who knows 60% of the material gets a grade strictly between 50 and 70 in the quiz described in Problem 4. a/h

6. What is the variance of the number of right answers for someone who knows 80% of the material on which a 25-question quiz is based? What if the quiz has 100 questions? 400 questions? How can you "correct" these variances for the fact that the "spread" in the histogram for the "number of right answers" random variable only doubled when the number of questions in a test was quadrupled?

7. Prove Theorem 5.30. a/h

8. Show that if X and Y are independent and b and c are constant, then $X - b$ and $Y - c$ are independent.

9. A nickel, a dime, and a quarter are in a cup. Withdraw two coins, first one and then the second, without replacement. What is the expected amount of money and variance for the first draw? For the second draw? For the sum of both draws? a/h

10. What are the expected number of failures, the variance of the number of failures, and the standard deviation of the number of failures in n independent trials with probability p of success? Compare your answers with the corresponding results for successes, and explain any interesting observations.

11. What are the variance and standard deviation for the sum of the tops of n dice that you roll? `a/h`

12. How many questions need to be on a short-answer test for you to be 95% sure that someone who knows 80% of the course material gets a grade between 75% and 85%?

13. Is a score of 70% on a 100-question true-false test consistent with the hypothesis that the test-taker was just guessing? What about a 10-question true-false test? (This is not a "plug and chug" problem; you have to come up with your own definition of "consistent with.") `a/h`

14. Given a random variable X, how does the variance of cX relate to that of X?

15. Draw a graph of the equation $y = x(1 - x)$ for x between 0 and 1. What is the maximum value of y? Why does this show that the variance (see Problems 7 and 10) of the "number of successes" random variable for n independent trials is less than or equal to $n/4$? `a/h`

16. This problem develops an important law of probability known as **Chebyshev's law**. Suppose you are given a real number $r > 0$ and you want to estimate the probability that the difference $|X(x) - E(X)|$ of a random variable from its expected value is more than r.

    a. Let $S = \{x_1, x_2, \ldots, x_n\}$ be the sample space, and let $E = \{x_1, x_2, \ldots, x_k\}$ be the set of all x such that $|X(x) - E(X)| > r$. By using the formula that defines $V(X)$, show that

    $$V(X) > \sum_{i=1}^{k} P(x_i)r^2 = P(E)r^2 .$$

    b. Show that the probability of $|X(x) - E(X)| \geq r$ is no more than $V(X)/r^2$. This is called Chebyshev's law.

17. With the help of Problem 15 (among others), show that in n independent trials with probability p of success, you have that

    $$P\left(\left|\frac{\text{number of successes} - np}{n}\right| \geq r\right) \leq \frac{1}{4nr^2} . \text{ `a/h`}$$

18. This problem derives from Chebyshev's law an intuitive law of probability known as the **law of large numbers**. Informally, the **law of large numbers** says that if you repeat an experiment many times, the fraction of the time that an event occurs is very likely to be close to the probability of the event. The law applies to independent trials with probability p of success. It states that for any positive number s, no matter how small, you can make the probability of the number X of successes being between $np - ns$ and $np + ns$ as close to 1 as

you choose by making the number n of trials large enough. For example, you can make the probability of the number of successes being within 1% (or 0.1%) of the expected number as close to 1 as you wish.

    a. Show that the probability of $|X(x) - np| \geq sn$ is no more than $p(1-p)/s^2n$.

    b. Explain why part a means you can make the probability of $X(x)$ being between $np - sn$ and $np + sn$ as close to 1 as you want by making n large.

19. On a true-false test, the score is often computed by subtracting the number of wrong answers from the number of right ones and converting that number to a percentage of the number of questions. On a true-false test graded this way, what is the expected score of someone who knows 80% of the material in a course? How does this scheme change the standard deviation in comparison with an objective test? What must you do to the number of questions to be able to be a certain percent sure that someone who knows 80% gets a grade within 5 points of the expected percentage score? a/h

20. Another way to bound the deviance from the expectation is known as **Markov's inequality**, which says that if X is a random variable taking only nonnegative values, then

$$P(X > kE(X)) \leq \frac{1}{k}$$

for any $k \geq 1$. Prove this inequality.

6

GRAPHS

6.1 GRAPHS

In this chapter, we study graphs, which are a fundamental topic in discrete mathematics and computer science. As we will see, we can use graphs to model many common situations and to naturally describe many algorithms. Graphs are also an ideal venue for developing a deeper understanding of proof by induction, especially strong induction.

Exercise 6.1-1

Figure 6.1 shows a stylized map of some cities in the eastern United States (Boston, New York, Pittsburgh, Cincinnati, Chicago, Memphis, New Orleans, Atlanta, Washington DC, and Miami). A company has major offices with data-processing centers in each of these cities, and as its operations have grown, it has leased dedicated communication lines between certain pairs of these cities to allow for efficient communication among the computer systems. Each blue dot in the figure stands for a data center, and each line stands for a dedicated communication link. What is the minimum number of links that could be used to send a message from B (Boston) to NO (New Orleans)? Give a route with this number of links.

Exercise 6.1-2

Which city (or cities) has (or have) the most communication links emanating from it (or them)?

Exercise 6.1-3

What is the total number of communication links in the figure?

Figure 6.1: *A stylized map of some eastern U.S. cities*

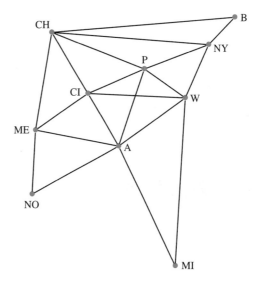

Figure 6.1 is a drawing of what we call a "graph." A **graph** consists of a set of **vertices**[1] and a set of **edges**[2] and has the property that each edge has two (not necessarily different) vertices, called its **endpoints**, associated with it. We say that the edge **joins** the endpoints, and we say that two endpoints are **adjacent** if they are joined by an edge. When a vertex is an endpoint of an edge, we say that the edge and the vertex are **incident**. Several more examples of graphs are given in Figure 6.2. Graphs model situations in which there are relationships among pairs of objects. In Figure 6.1, our objects are the cities, and the relationship is being joined by a communication link. More generally, we represent objects as vertices, and we represent a relationship between two objects as an edge connecting their vertices. Other examples include a graph in which the vertices represent biological species and two vertices are joined by an edge if their species have a common ancestor, or a graph in which the vertices represent people and an edge is drawn between two vertices if the people attended the same school.

Figure 6.2: *Some examples of graphs*

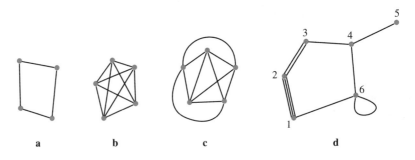

a   b   c   d

---

[1] Another common name for vertex is *node*.

[2] Another common name of edge is *arc,* though some authors restrict this usage to directed graphs.

The relationships we have mentioned are all symmetric; that is, whatever relationship may exist between two vertices a and b also exists between b and a. Graphs model symmetric relationships. One can also study directed graphs, which model relationships that are not necessarily symmetric. Much of what we do for graphs holds for directed graphs as well; we do not pursue the idea of directed graphs in this book.

To **draw** a graph, we draw a point (in our case, a blue circle) in the plane for each vertex; then, for each edge, we draw a (possibly curved) line between the vertices that correspond to the endpoints of the edge. The only vertices that may be touched by the line representing an edge are the endpoints of that edge. Notice that Figure 6.2d has three edges joining vertices 1 and 2, two edges joining vertices 2 and 3, and one edge joining vertex 6 to itself. This last edge has two identical endpoints. Note that in Figure 6.2, sometimes the vertices are labeled, and sometimes they aren't. We label the vertices to give them meaning, as in Figure 6.1, or when we know we will refer to them, as in Figure 6.2d.

Figure 6.1 and the first three graphs in Figure 6.2 are examples of **simple graphs**, which are graphs that have at most one edge joining each pair of distinct vertices and no edges joining a vertex to itself.[3] If there is an edge joining vertex x and vertex y in a simple graph, we denote it by $\{x, y\}$. Thus, in Figure 6.1, $\{P, W\}$ denotes the edge between Pittsburgh and Washington DC. Sometimes it will be helpful to have a symbol to stand for a graph. The phrase "Let $G = (V, E)$" is shorthand for "Let G stand for a graph with vertex set V and edge set E." We say that Figure 6.2d has a loop at vertex 6 and multiple edges joining vertices 1 and 2 and vertices 2 and 3. More precisely, an edge that joins a vertex to itself is called a **loop**, and if there is more than one edge joining x and y, then the graph is said to have **multiple edges** between those two vertices.

Figures 6.2b and 6.2c are different drawings of the same graph, which consists of five vertices and one edge between each pair of distinct vertices. It is called the complete graph on five vertices and is denoted by K_5. In general, a **complete graph** on n vertices is a graph with n vertices that has an edge between each pair of vertices. We use K_n to stand for a complete graph on n vertices. Figures 6.2b and 6.2c illustrate that there are many different ways of drawing a given graph. The two drawings also demonstrate two different ideas: Figure 6.2b shows that each vertex is adjacent to each other vertex and suggests that there is a high degree of symmetry. Figure 6.2c shows that it is possible to draw the graph so that only one pair of edges crosses; other than the one place where two edges cross, the only places where edges touch each other are at their endpoints. In fact, it is impossible to draw K_5 so that no edges cross, a fact that we explain later in this chapter.

In Exercise 6.1-1, the links referred to are edges of the graph, and the cities are the vertices of the graph. It is possible to get from the vertex for Boston to the vertex for New Orleans by using three communication links, namely, the edge from Boston

---

[3]The terminology of graph theory has not yet been standardized, because it is a relatively young subject. The terminology we are using here is the most popular terminology in computer science. However, some graph theorists would reserve the word *graph* for what we have just called a *simple graph* and would use the word *multigraph* for what we have called a *graph*.

to Chicago, the edge from Chicago to Memphis, and the edge from Memphis to New Orleans. A **path** in a graph is an alternating sequence of vertices and edges such that

- it starts and ends with a vertex,

- each edge joins the vertex before it in the sequence to the vertex after it in the sequence, and

- no vertex appears more than once in the sequence.

If a is the first vertex in the path and b is the last vertex in the path, then we say the path is from a to b. Thus, the path from Boston to New Orleans is

$$B\{B, CH\}CH\{CH, ME\}ME\{ME, NO\}NO \, .$$

Because the graph is simple, there is exactly one edge between successive vertices in this list. Therefore, we can also use the shorter notation B, CH, ME, NO to describe the same path. The **length of a path** is the number of edges it has; so our path from Boston to New Orleans has length 3. By inspecting the map, we see that there is no shorter path from Boston to New Orleans. The length of a shortest path between two vertices in a graph is called the **distance** between them. Thus, the distance from Boston to New Orleans in the graph of Figure 6.1 is three.

Some applications lead us to pathlike sequences in which vertices can be repeated. A **walk** satisfies the first two conditions for a path but need not satisfy the third.[4] The **length of a walk** is the number of edges it has.

The following lemma will prove useful later.

| Lemma 6.1 | If there is a walk between two distinct vertices x and y of a graph G, then there is a path between x and y in G. |

Proof: If the walk is a path, then we are done. If not, let z be a vertex that appears more than once in the walk from x to y. We create a shorter walk by removing the part of the walk between the first and last occurrences of z in the walk, including the last z but not the first. Then z will appear only once in the new walk. This process can be repeated until there are no vertices that appear more than once. At that point, the walk is a path. ∎

The Degree of a Vertex

In Exercise 6.1-2, the city with the most communication links is Atlanta (A). We say the vertex A has degree 6, because six edges are incident to it. More generally, the **degree** of a vertex in a graph is the number of times it is incident with edges of the graph; that is, the degree of a vertex x is the number of edges between x and

---

[4]Some texts use the word "path" for what we have just defined as a walk and use the phrase "simple path" for what we have defined as a path.

other vertices plus twice the number of loops at vertex x. In Figure 6.2d, vertex 2 has degree 5, and vertex 6 has degree 4.

Exercise 6.1-4 In a graph like the one in Figure 6.1, it is somewhat difficult to count the edges, because you might forget which ones you've counted and which ones you haven't. Is there a relationship between the number of edges in a graph and the degrees of the vertices? If so, find it. (*Hint:* Computing degrees of vertices and number of edges in some relatively small examples of graphs should help you discover a formula.)

In Exercise 6.1-4, examples such as those in Figure 6.2 convince us that the sum of the degrees of the vertices is twice the number of edges. How can we prove this? One way is to count the total number of incidences between vertices and edges. Each edge has exactly two incidences, so the total number of incidences is twice the number of edges. But the degree of a vertex is the number of incidences it has, so the sum of the degrees of the vertices is also the total number of incidences. Therefore, the sum of the degrees of the vertices of a graph is twice the number of edges. Thus, to compute the number of edges of a graph, we can sum the degrees of the vertices and divide by 2. There is another proof of this result that uses induction.

Theorem 6.2 Suppose a graph has a finite number of edges. Then the sum of the degrees of the vertices is twice the number of edges.

Proof: The proof proceeds by induction on the number of edges in the graph. If a graph has no edges, then each vertex has degree 0 and the sum of the degrees is 0, which is twice the number of edges. Now suppose that $e > 0$ and that the theorem is true whenever a graph has fewer than e edges. Let G be a graph with e edges and let ϵ be an edge of G.[5] Let G' be the graph (on the same vertex set as G) that we get by deleting ϵ from the edge set E of G. Then G' has $e - 1$ edges, and so, by our inductive hypothesis, the sum of the degrees of the vertices of G' is $2(e - 1)$. Now there are two possible cases. Either e was a loop, in which case one vertex of G' has degree two less in G' than it has in G, or e has two distinct endpoints, in which case exactly two vertices of G' have degree one less than their degree in G. Thus, in both cases, the sum of the degrees of the vertices in G' is two less than the sum of the degrees of the vertices in G. Therefore, the sum of the degrees of the vertices in G is $(2e - 2) + 2 = 2e$. Thus, the truth of the theorem for graphs with $e - 1$ edges implies the truth of the theorem for graphs with e edges. Therefore, by the principle of mathematical induction, the theorem is true for a graph with any finite number of edges. ∎

---

[5]Because it is very handy to have e stand for the number of edges of a graph, we will use Greek letters such as epsilon (ϵ) to stand for the edges of a graph. It is also handy to use v to stand for the number of vertices of a graph, so we use other letters near the end of the alphabet, such as w, x, y, and z, to stand for vertices.

There are several instructive points in the proof of Theorem 6.2. First, because it wasn't clear from the outset whether we would need to use strong or weak induction, we made the inductive hypothesis that we would normally make for strong induction. However, in the course of the proof, we saw that we only needed to use weak induction; so that is how we wrote our conclusion. This is not a mistake. We used our inductive hypothesis correctly; we just didn't need to use it for every possible value it covered.

Second, instead of saying that we would take a graph with $e - 1$ edges and add an edge to get a graph with e edges, we said that we would take a graph with e edges and remove an edge to get a graph with $e - 1$ edges. This is because we need to prove that the result holds for *every* graph with e edges. By using the second approach, we avoided the need to say that "every graph with e edges may be built up from a graph with $e - 1$ edges by adding an edge," because in the second approach, we started with an arbitrary graph on e edges. In the first approach, we would have proved that the theorem was true for all graphs that could be built from an $(e - 1)$-edge graph by adding an edge, and we would have had to say explicitly that every graph with e edges could be built in this way.

In Exercise 6.1-3, the sum of the degrees of the vertices (working from left to right) is

$$2 + 4 + 5 + 5 + 6 + 5 + 2 + 5 + 4 + 2 = 40,$$

and so the graph has 20 edges.

Connectivity

All of the examples we have seen so far have a property that is not common to all graphs—namely, that for every pair of vertices, there is a path between them.

Exercise 6.1-5
The company with the computer network in Figure 6.1 needs to reduce its expenses. It is currently leasing each of the communication lines shown in the figure. Because it can send information from one city to another through one or more intermediate cities in the graph, it decides to lease only the minimum number of communication lines it needs to be able to send a message from any city to any other city by using any number of intermediate cities. What is the minimum number of lines it needs to lease? Give two examples of subsets of the edge set with this number of edges (lines) that will allow communication between any two cities. Then give two examples of a subset of the edge set with this number of edges (lines) that will not allow communication between any two cities.

Some experimentation with the graph in Figure 6.1 convinces us that if we keep eight or fewer edges, there is no way to communicate among the cities (we explain this more precisely later on). However, we also see that there are quite a few sets of nine edges that suffice for communication among all the cities. Figure 6.3 shows two sets of nine edges each that allow communication among all the cities and two sets of nine edges each that do not allow communication among all the cities.

Notice that in Figures 6.3a and 6.3b, it is possible to get from any vertex to any other vertex by a path. A graph is called **connected** when, for each pair of vertices of the

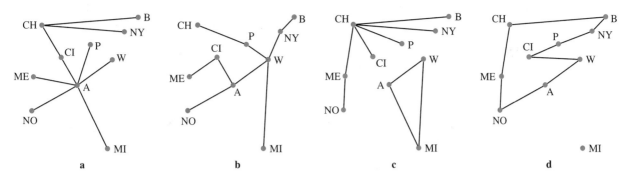

Figure 6.3: *Selecting nine edges from the stylized map of some eastern U.S. cities*

graph, there is a path between these two vertices. Notice that in Figure 6.3c, it is not possible to find a path from Atlanta to Boston, for example, and in Figure 6.3d, it is not possible to find a path from Miami to any of the other vertices. Thus, these last two graphs are not connected, and we call them **disconnected**. In Figure 6.3d, we say that Miami is an **isolated vertex**. We say two vertices are **connected** if there is a path between them. Thus, in Figure 6.3c, the vertices for Boston and New Orleans are connected.

The relationship of being connected divides the set of vertices into mutually exclusive classes;[6] that is, it partitions the vertices of the graph. How do we know this? The class containing Boston, for example, is all vertices connected to Boston. If two vertices are in that set, then they both have paths to Boston, and there is a walk between them using Boston as an intermediate vertex. By Lemma 6.1, there must be a path between the vertices, so they are connected. If a vertex x is in the set containing Boston and another vertex y is not, then they cannot be connected. If they were connected, the walk from y to x and then on to Boston would connect y to Boston, which would mean y was in the class containing Boston after all. Thus, the relation of being connected partitions the vertex set of the graph into disjoint classes. Though we made this argument with respect to the vertex Boston in the specific case of Figure 6.3c, it is a perfectly general argument that applies to arbitrary vertices in arbitrary graphs.

We call the relationship of "being connected to" the **connectivity relation**. We call the blocks into which this relationship partitions the graph **connectivity classes**. There is no edge of a graph between two vertices in different connectivity classes, because if there were, then everything in one class would be connected to everything in the other class and the two classes would have to be the same. Thus, we also end up with a partition of the edges into disjoint sets. If a graph has edge set E, and C is a connectivity class, then $E(C)$ denotes the set of edges whose endpoints are both in C. Because no edge connects vertices in different connectivity classes, each edge must be in some set $E(C)$. The graph consisting of a connectivity class C, together with the edges $E(C)$, is called a **connected component** of our original graph. From now on, our emphasis will be on connected components rather than on connectivity classes, and we will describe a connected component by listing its vertices. Note that Figures 6.3c and 6.3d each have two connected components. In Figure 6.3c, the vertex sets of the connected

---

[6]For those who have studied Section 1.4, this means that the relationship is an equivalence relation.

components are {NO, ME, CH, CI, P, NY, B} and {A, W, MI}. In Figure 6.3d, the connected components are {NO, ME, CH, B, NY, P, CI, W, A} and {MI}. Two other examples of graphs with multiple connected components are shown in Figure 6.4.

Figure 6.4: *A simple graph G_1 with three connected components and a graph G_2 with four connected components*

Cycles

In Figures 6.3c and 6.3d, we see a feature that we don't see in Figures 6.3a and 6.3b, namely, a walk that leads from a vertex back to itself. A walk with at least one edge that starts and ends at the same vertex, but that has no other repeated vertices or edges, is called a **cycle**. Similarly, a walk that starts and ends with the same vertex is called a **closed walk**. The closed walks in Figures 6.3c and 6.3d are cycles A, W, M, A and NO, ME, CH, B, NY, P, CI, W, A, NO, respectively. We don't normally distinguish which point on a cycle is the starting point; for example, we consider the cycle A, W, MI, A to be the same as the cycle W, MI, A, W.

Let's compare Figures 6.3d and 6.1. In both graphs, NO, ME, CH, B, NY, P, CI, W, A, NO is a cycle. In Figure 6.3d, the only edges on the set of vertices in the cycle are the edges of the cycle. In contrast, some vertices in the cycle of Figure 6.1 are joined by other edges, too. We wish to be able to distinguish between these two cases.

In general, a graph H is called a **subgraph** of the graph G if all the vertices and edges of H are vertices and edges of G. In other words, $H = (V', E')$ is a subgraph of $G = (V, E)$ if $V' \subseteq V$ and $E' \subseteq E$. A graph H is called an **induced subgraph** of G if H is a subgraph of G and every edge of G connecting vertices of H is an edge of H. Thus, the graph G_1 in Figure 6.4 has an induced K_4 (complete graph on four vertices) and an induced cycle on three vertices (which also happens to be an induced K_3). It has a subgraph that is a cycle on four vertices, but it does not have an induced subgraph that is a cycle on four vertices. It has some induced paths on three vertices as well. Can you find one?

Notice that in graph G_2 of Figure 6.4, there are cycles with one edge and cycles with two edges. We call a graph a **cycle on n vertices**, or an **n-cycle**, and denote it by C_n if its vertex set is the vertex set of a cycle and its edge set is the edge set of that cycle. We say that a graph is a **path on n vertices** and denote it by P_n if its vertex set is the vertex set of a path and its edge set is the edge set of that path. Thus, Figure 6.2a is a drawing of C_4. Graph G_2 of Figure 6.4 has an induced P_3 and an induced C_2 as subgraphs.

Trees

The graphs in Figures 6.3a and 6.3b are called trees. We have redrawn them slightly in Figure 6.5 to clarify why they are called trees. Note that the graphs drawn in Figures 6.3a and 6.3b and in Figures 6.5a and 6.5b are connected and have no cycles.

Figure 6.5: *A visual explanation of the name* tree

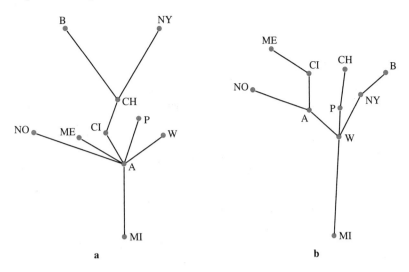

a   b

Definition 6.1   A connected graph with no cycles is called a tree.[7]

Note that the graph with one vertex and no edges is, by this definition, a tree.

Other Properties of Trees

Our definition of a tree left out several other properties of trees that we could have discovered by a further analysis of Figure 6.3.

Exercise 6.1-6   Given any two vertices in a tree, how many distinct paths are there between these two vertices?

Exercise 6.1-7   Is it possible to delete an edge from a tree and have it remain connected?

Exercise 6.1-8   If $G = (V, E)$ is a graph, and we add an edge that joins vertices of V, what can happen to the number of connected components?

Exercise 6.1-9   How many edges does a tree with v vertices have?

Exercise 6.1-10   Does every tree have a vertex of degree 1? If the answer is "yes," explain why. If the answer is "no," try to find additional conditions that will guarantee that a tree satisfying these conditions has a vertex of degree 1.

---

[7]The student who has experience with rooted trees, binary trees, or binary search trees should note that we are not talking about these kinds of trees in this section. They are the subject of the next section.

Figure 6.6: *A graph with multiple paths from x to y*

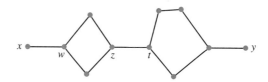

For Exercise 6.1-6, suppose we have two distinct paths from a vertex x to a vertex y. The paths begin with the same vertex x and might have some more edges in common, as in Figure 6.6. Let w be the last vertex after (or including) x that the paths share before they become different. For visualizing the argument, let us focus on the path that goes upward at the vertices marked w and t and the path that goes downward at these two vertices. The paths must come together again at y, though they might come together earlier. Let z be the first vertex the paths have in common after w. Then there are two paths from w to z that have only w and z in common. Taking one of these paths from w to z and the other from z to w gives us a cycle, and so the graph is not a tree. We have shown that if a graph has two distinct paths from x to y, then it is not a tree. By contrapositive inference, then, if a graph is a tree, it does not have two distinct paths between two vertices x and y. We state this result as a theorem.

Theorem 6.3	There is exactly one path between each pair of vertices in a tree.

> **Proof:** By the definition of a tree, there is at least one path between each pair of vertices. By our argument above, there is at most one path between each pair of vertices. Thus, there is exactly one path. ∎

For Exercise 6.1-7, note that if ϵ is an edge from x to y, then x, ϵ, y is the unique path from x to y in the tree. Suppose we delete ϵ from the edge set of the tree. If there were still a path from x to y in the resulting graph, then it would also be a path from x to y in the tree, which would contradict Theorem 6.3. Thus, the only possibility is that there is no path between x and y in the resulting graph; thus, it is not connected and is therefore not a tree.

For Exercise 6.1-8, if the endpoints are in the same connected component, then the number of connected components will not change. If the endpoints of the edge are in different connected components, then the number of connected components can decrease by one. Because an edge has two endpoints, it is impossible for the number of connected components to decrease by more than one when we add an edge. This paragraph and the previous one lead us to the following useful lemma.

Lemma 6.4	Removing one edge from the edge set of a tree gives a graph with two connected components, each of which is a tree.

> **Proof:** Suppose that ϵ is an edge from x to y in a tree. We have seen that the graph that we get by deleting ϵ from the edge set of the tree is not connected,

so the graph has at least two connected components. But adding the edge back in can only reduce the number of connected components by one. Therefore, the graph has exactly two connected components. Because neither has any cycles, both are trees. ■

Figure 6.7: *Two trees on four vertices*

a   b

In Exercise 6.1-9, our trees with ten vertices had nine edges. If we draw a tree on two vertices, it will have one edge; if we draw a tree on three vertices, it will have two edges. There are two different-looking trees on four vertices, as shown in Figure 6.7, and each has three edges. On the basis of these examples, we conjecture that a tree on n vertices has $n - 1$ edges. One approach to proving this is to try to use induction. To do so, we have to see how to build every tree from smaller trees or how to take a tree and break it into smaller ones. In either case, we then have to figure out how to use the truth of our conjecture for the smaller trees to imply its truth for the larger trees. A mistake that people often make at this stage is to assume that every tree can be built from smaller ones by adding a vertex of degree 1. Although that is true for finite trees with more than one vertex (which is the point of Exercise 6.1-10), we haven't proved it yet, so we can't yet use it in proofs of other theorems. Another approach to using induction is to ask whether there is a natural way to break a tree into two smaller trees. There is a way; as we just showed in Lemma 6.4, if you remove an edge ϵ from the edge set of a tree, you get two connected components that are trees. We may assume inductively that the number of edges of each of these trees is one less than its number of vertices. Thus, if the graph with these two connected components has v vertices, then it has $v - 2$ edges. Adding ϵ back in gives us a graph with $v - 1$ edges—so, except for the fact that we have not done a base case, we have proved the following theorem.[8]

| **Theorem 6.5** | For all integers $v \geq 1$, a tree with v vertices has $v - 1$ edges. |

Proof: If a tree has one vertex, it can have no edges, as any edge would have to connect that vertex to itself and would thus give a cycle. A tree with two or more vertices must have an edge in order to be connected. Before the statement of the theorem, we showed how to use the deletion of an edge to complete an inductive proof that a tree with v vertices has $v - 1$ edges. Therefore, for all $v \geq 1$, a tree with v vertices has $v - 1$ edges. ■

---

[8] In Section 4.1, we mentioned that in certain applications of induction, it makes our proofs simpler if we try to understand how to break large instances of our problems into smaller ones, rather than trying to understand how to build smaller instances to get larger ones. This is one example where the approach is useful.

Finally, for Exercise 6.1-10, we can now give a contrapositive argument to show that a finite tree with more than one vertex has a vertex of degree 1. Suppose that G is a graph that is connected and that all vertices of G have degree 2 or more. Then the sum of the degrees of the vertices is at least $2v$, and so, by Theorem 6.2, the number of edges is at least v. Therefore, by Theorem 6.5, G is not a tree. Then, by contrapositive inference, if T is a tree, then T must have at least one vertex of degree 1. This corollary to Theorem 6.5 is so useful that we state it formally.

Corollary 6.6   A finite tree with more than one vertex has at least one vertex of degree 1.

Important Concepts, Formulas, and Theorems

1. *Graph.* A *graph* consists of a set of *vertices* and a set of *edges* and has the property that each edge has two (not necessarily different) vertices, called its *endpoints,* associated with it.

2. *Edge/Adjacent.* We say that an edge in a graph *joins* its endpoints, and we say that two endpoints are *adjacent* if they are joined by an edge.

3. *Incident.* When a vertex is an endpoint of an edge, we say that the edge and the vertex are *incident.*

4. *Drawing a graph.* To *draw* a graph, we draw a point in the plane for each vertex. For each edge, we draw a (possibly curved) line between the points that correspond to the endpoints of the edge. Lines that correspond to edges may only touch the vertices that are their endpoints.

5. *Simple graph.* A *simple graph* is one that has, at most, one edge joining each pair of distinct vertices and no edges joining a vertex to itself.

6. *Loop/Multiple edges.* An edge that joins a vertex to itself is called a *loop,* and we say that we have *multiple edges* between vertices x and y if there is more than one edge joining x and y.

7. *Notation for a graph.* The phrase "Let $G = (V, E)$" is shorthand for "Let G stand for a graph with vertex set V and edge set E."

8. *Notation for edges.* In a simple graph, we use the notation $\{x, y\}$ for an edge from x to y. In any graph, when we want to use a letter to denote an edge, we use a Greek letter like ϵ so that we can save e to stand for the number of edges of the graph.

9. *Complete graph on n vertices.* A *complete graph* on n vertices is a graph with n vertices that has an edge between each pair of vertices. We use K_n to stand for a complete graph on n vertices.

10. *Walk.* We call an alternating sequence of vertices and edges in a graph a *walk* if it starts and ends with a vertex and each edge joins the vertex before it (in the sequence) to the vertex after it (in the sequence).

11. *Path.* A walk is called a *path* if it has no repeated vertices or edges.

12. *Length/Distance.* The *length* of a path is the number of edges. The *distance* between two vertices in a graph is the length of a shortest path between them.

13. *Degree of a vertex.* The *degree* of a vertex in a graph is the number of times it is incident with edges of the graph; that is, the degree of a vertex x is the number of edges from x to other vertices plus twice the number of loops at vertex x.

14. *Sum of degrees of vertices.* The sum of the degrees of the vertices in a graph with a finite number of edges is twice the number of edges.

15. *Connected.* A graph is *connected* if, for each pair of vertices of the graph, there is a path between them. We say that two vertices are *connected* if there is a path between them; so, a graph is connected if each pair of its vertices are connected. The relationship of being connected partitions the vertices of a graph into sets called *connectivity classes.*

16. *Connected component.* If C is a subset of the vertex set of a graph, then we use $E(C)$ to stand for the set of all edges *both* of whose endpoints are in C. The graph consisting of a connectivity class C of the connectivity relation, together with the edges $E(C)$, is called a *connected component* of our original graph.

17. *Closed walk.* A walk that starts and ends at the same vertex is called a *closed walk.*

18. *Cycle.* A walk whose first and last vertices are the same is called a *cycle* if it has at least one edge and all vertices of the walk, except the first and last, are distinct.

19. *Tree.* A connected graph with no cycles is called a tree.

20. *Important properties of trees.*

   a. There is a unique path between each pair of vertices in a tree.

   b. A tree on v vertices has $v - 1$ edges.

   c. Every finite tree with at least two vertices has a vertex of degree 1.

Problems

1. Find a shortest path from vertex 1 to vertex 5 in Figure 6.8. a/h

Figure 6.8

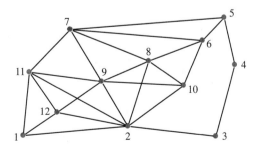

2. Find the longest path possible from vertex 1 to vertex 5 in Figure 6.8.

3. Find the vertex of largest degree in Figure 6.8. What is its degree? a/h

4. How many connected components does the graph in Figure 6.9 have?

Figure 6.9: *A graph with a number of connected components*

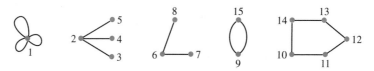

5. Find all induced cycles in Figure 6.9. a/h

6. What is the size of the largest induced K_n in Figure 6.9?

7. Find a largest induced K_n (in words, a largest complete subgraph) in Figure 6.8. a/h

8. Find the size of a largest induced P_n in Figure 6.9.

9. A graph with no cycles is called a **forest**. Show that if a forest has v vertices, e edges, and c connected components, then $v = e + c$. a/h

10. What can you say about a five-vertex simple graph in which every vertex has degree 4?

11. Draw a drawing of K_6 in which only three pairs of edges cross. a/h

12. Either prove true or find a counterexample: A graph is a tree if there is one and only one path between each pair of vertices.

13. Are there connected graphs with v vertices and $v - 1$ edges that are not trees? a/h

14. Are there graphs with v vertices and $v - 1$ edges and no cycles that are not trees? Give a proof to justify your answer.

15. Suppose that a graph G is connected, but, for each edge, deleting that edge leaves a disconnected graph. What can you say about G? Prove it. a/h

16. Show that each tree with four vertices can be drawn with one of the two drawings in Figure 6.7.

17. Draw the minimum number of drawings of trees possible so that each tree with five vertices has one of those drawings. Explain why you have drawn all possible trees. a/h

18. Draw the minimum number of drawings of trees possible so that each tree with six vertices is represented by exactly one of those drawings. Explaining why you have drawn all possible drawings is optional.

19. Find a longest induced cycle in Figure 6.8. a/h

6.2 SPANNING TREES AND ROOTED TREES

Spanning Trees

We introduced our discussion of trees with the example of choosing a minimum-sized set of edges that would connect all the vertices in the graph of Figure 6.1. The kinds of trees we used to solve our original problem have a special name: A tree whose edge set is a subset of the edge set of the graph G is called a **spanning tree** of G if the tree has exactly the same vertex set as G. Thus, Figures 6.3a and 6.3b are spanning trees of the graph of Figure 6.1.

Exercise 6.2-1 Does every connected graph have a spanning tree? Give either a proof or a counterexample.

Exercise 6.2-2 Give an algorithm that determines whether a graph has a spanning tree, finds such a tree if it exists, and takes time bounded above by a polynomial in v and e, where v is the number of vertices and e is the number of edges of the graph.

For Exercise 6.2-1, if the graph has no cycles but is connected, then it is a tree, and thus it has a spanning tree. This makes a good base step for an inductive proof that every connected graph has a spanning tree. Let c be an integer greater than 0, and suppose inductively that when a connected graph has fewer than c cycles, the graph has a spanning tree. Suppose that G is a graph with c cycles. Choose a cycle of G and then choose an edge of that cycle. Deleting that edge (but not its endpoints) reduces the number of cycles by at least one, and so our inductive hypothesis implies that the resulting graph has a spanning tree. But then that spanning tree is also a spanning tree of G. Therefore, by the principle of mathematical induction, every finite connected graph has a spanning tree. We have proved the following theorem.

Theorem 6.7 Each finite connected graph has a spanning tree.

Proof: The proof is given before the statement of the theorem. ∎

In Exercise 6.2-2, we want an algorithm for determining whether a graph has a spanning tree. One natural approach would be to convert the inductive proof of Theorem 6.7 into a recursive algorithm. Doing it in the obvious way, however, would mean having to search for cycles in our graph. A natural way to look for a cycle is to look at each subset of the vertex set to see if that subset is a cycle of the graph. Because there are 2^v subsets of the vertex set, we cannot guarantee that an algorithm that works in this way would find a spanning tree in time that is bounded by a polynomial in v and e. Instead, we use another approach, describing a quite general algorithm, which we can then specialize in different ways for different purposes.

The idea of the algorithm is to build, one vertex at a time, a tree that is a subgraph (not necessarily an induced subgraph) of the graph $G = (V, E)$. (A subgraph of G that is a tree is called a **subtree** of G.) We start with some vertex, say x_0. If there are no edges leaving the vertex and the graph has more than one vertex, then the graph is not connected, and hence does not have a spanning tree. Otherwise, we can choose an edge ϵ_1 that connects x_0 to another vertex x_1. Thus, $\{x_0, x_1\}$ is the vertex set of a subtree of G. If there are no edges that connect some vertex in the set $\{x_0, x_1\}$ to a vertex not in that set, then $\{x_0, x_1\}$ is a connected component of G. In this case, either G is not connected and has no spanning tree or G has just two vertices and we have a spanning tree. However, if there is an edge that connects some vertex in the set $\{x_0, x_1\}$ to a vertex not in that set, then we can use this edge to continue building a tree. This suggests an iterative approach to building the vertex set S of a subtree of our graph one vertex at a time. For the base case of the algorithm, we let $S = \{x_0\}$. For the inductive step, given S, we choose an edge ϵ that leads from a vertex in S to a vertex in $V - S$ (provided such an edge exists) and add it to the edge set E' of the subtree. If no such edge exists, then we stop. If $V = S$ when we stop, then E' is the edge set of a spanning tree. (We can prove inductively that E' is the edge set of a tree on S because adding a vertex of degree 1 to a tree gives a tree.) If $V \neq S$ when we stop, then G is not connected and does not have a spanning tree.

To describe the algorithm a bit more precisely, we give the following pseudocode.

Spantree(V, E)

```
// Assume G is a graph with vertex set V and edge set E.
// This algorithm will find a spanning tree with edge set E' if one exists.
// The sets S ⊆ V and E' ⊆ E are initially empty.
(1)   Choose a vertex x₀ in V
(2)   S = {x₀}
(3)   While there is an edge ε from a vertex y in S to a vertex x not in S
(4)        S = S ∪ {x}
(5)        E' = E' ∪ {ε}
(6)   If (|S| == |V|)
(7)        Print "The edge set of a spanning tree is"
(8)        Print the elements of E'
(9)   Else
(10)       Print "The graph is not connected."
```

Notice that Spantree will continue as long as a vertex in S is connected to a vertex not in S. Thus, when the algorithm stops, S will be the vertex set of a connected component of the graph and E' will be the edge set of a spanning tree of this connected component. This suggests that one use of Spantree is to find connected components of graphs. If we want the connected component containing a specific vertex x, then we make this choice of x_0 in Line 1.

In the algorithm, we deliberately left vague the way in which the vertex x and the edge ϵ are chosen, because there are several different ways to specify x or y and ϵ, each accomplishing a different purpose. Suppose, however, that in Line 3 we are willing to choose any edge from a vertex y in S to a vertex x not in S. We could examine each edge to see if it connected a vertex in S to a vertex not in S. As we shall see, there is a way to keep track of S so that we can test whether a vertex is in S in at most a constant amount of time. Thus, we would need time at most a constant times e to complete the test in the "while" loop. The other steps in the "while" loop each take at most a constant amount of time. Because we repeat the "while" loop at most v times, all executions of that loop should take at most $O(ve)$ time. In Line 6, we need to know $|V|$ and $|S|$. We are likely to know v, which is the number of vertices, before we start; if not, we can compute v before we get started in time no more than a constant times v. We can compute the size of S as we build it. Thus, with the assumptions we have made, we conclude that the algorithm takes $O(v + ve + v) = O(ve)$ time. However, we will see that by being more specific about how we carry out our choices, we can reduce the running time.

Breadth-First Search

One way to guarantee a faster running time would be to arrange our choice of ϵ so that we examined each edge no more than some constant number of times between the start and end of the algorithm. Suppose we look for edges from vertices in S to vertices not in S as follows: We first consider all edges incident with x_0 as possible choices for ϵ; we then consider all edges incident with vertices at distance 1 from x_0 as possible choices for ϵ; and then continue with distances 2, 3, and so on. In this way, if an edge can be used to connect a vertex in S with a vertex not in S, then we will discover this fact the first time we look at the edge. If we later consider this edge from its other endpoint, it would already connect two vertices in S. Because each edge has two endpoints, each edge would be considered at most twice. One carefully organized special case of this idea is called **breadth-first search (BFS)**.

To give a simple description of breadth-first search, we use a data structure called a **queue**, which models customers standing in line for service at a cash register or bank teller. As customers arrive, they go to the end of the line. When the server is free, the first person in line leaves the line and is served.

We can think of a queue as a list of items to which we can do exactly two things—we can add an item x to the end of the queue and we can remove an item from the front of the queue. We say that we **enqueue** x onto Q when we add it to the end of Q, and we say that we **dequeue** an item from Q when we remove it from the front of the queue. There are a number of ways to implement queues so that each operation takes constant time.[9] We can use a queue to keep the elements of S in the order in which they were added to S. Now we use the idea of a queue to describe more precisely the process of breadth-first search. We begin by putting x_0, our starting vertex, at the end of the queue and into S. Then we do the following until we run out of vertices on our queue:

1. Dequeue a vertex w from the queue.

2. For each edge ϵ incident with w, if the edge ϵ joins w to a vertex z not in S, add ϵ to E', add z to S, and enqueue z.

---

[9]Cormen et al. [13] (pp. 200–203) show how to implement a queue so that the enqueue and dequeue operations both take constant time.

To give a pseudocode description of this algorithm, we assume that the vertices are numbered $1, 2, \ldots, v$. This lets us keep track of what vertices are in the set S by using an array Intree of trues and falses. Intree[x] is true if and only if vertex x is in S. By looking in Intree we can test in constant time whether a vertex is in S.[10]

There are a number of ways to represent the edge set of a graph in a computer. One way is to give a list, called an **adjacency list**, for each vertex, which lists all vertices adjacent to that vertex. If there are two edges from x to y, we list y twice in the adjacency list for x and x twice in the adjacency list for y. In the general case of multiple edges, we list each adjacency as many times as there are edges that give the adjacency. We assume in our pseudocode that the edges are given in this way. That is, E is an array whose ith element is a *list* of the vertices adjacent to vertex i.

In our pseudocode, we use S and E' as we did in Spantree. We also assume that we are given a vertex x_0 from which we are to start the search.

```
BFSpantree(x_0, V, E)

// Assume V contains vertices numbered 1, 2,..., v.
// Assume E is an array with v entries, and entry i of E is a list of the
// vertices adjacent to vertex i.
// Assume the parameter x_0 is the starting vertex for the BFS.
// The output of the algorithm is either the edge set of a spanning tree of
// the graph or the edge set of a spanning tree of the connected component
// that contains x_0.
// Assume Q is a queue that will be initialized to an empty queue.
(1)   Intree = an array of length v with each entry initialized to "false"
(2)   S = {x_0}
(3)   n = 1
(4)   E' = Ø
(5)   Q = Ø
(6)   Intree[x_0] = "true"
(7)   Enqueue x onto Q
(8)   While there is at least one vertex on Q
(9)       Dequeue the first element from Q and assign it to y
(10)      For each element x of the list E[y]
(11)          If (!Intree[x])
(12)              Enqueue x onto Q
(13)              S = S ∪ {x}
(14)              Intree[x]= "true"
(15)              E' = E' ∪ {{x,y}}
(16)              n = n + 1
(17) If (n == v)
(18)      Print "The edge set of a spanning tree of the graph is"
(19)      Print the elements of E'
```

---

[10]It just involves a bit more bookkeeping (that the authors didn't want to burden you with) to do the test in constant time if you have a different vertex set.

```
(20) Else
(21)      Print "The vertex set of the connected component containing" $x_0$ "is"
(22)      Print the elements of $S$
(23)      Print "The edge set of a spanning tree of the connected component"
(24)      Print "containing" $x_0$ "is"
(25)      Print the elements of $E'$
```

How long does it take to run this algorithm? Note that the "while" loop in Line 8 runs (at most) once for each vertex. When this selected vertex is y, the number of times that the "for" loop in Line 10 runs is the degree of y. It takes a constant time to dequeue an element from Q and assign it to y. The steps in the "for" loop each take at most a constant amount of time. Thus, the total time for the "for" loop is at most a constant times the degree of y. The total time for the "while" loop is the sum of the times for each of its iterations. This sum is no more than a constant times the sum of the degrees of the vertices of the graph—that is, no more than a different constant times the number of edges. The initialization of the array and the printing of the vertex set take $O(v)$ time. The printing of the edge set of the tree also takes $O(v)$ time. Therefore, the time required to carry out the algorithm is $O(v + e)$.

We said that our method would first consider edges incident with x_0, and then edges incident with vertices of distance 1 from x_0, and continuing with distances 2, 3, and so on. Let's show why.

Lemma 6.8   For each nonnegative integer d, all vertices of distance d from the starting vertex x_0 of a breadth-first search tree are added to the vertex set S of the tree before those of distance $d + 1$ or more from x_0.

Proof: We add a vertex to S when we add it to the queue. When we add a vertex x other than x_0 to the queue, we are adding it because it is adjacent to some other vertex z already in the tree. (We say that we are adding x from z.) Because such a vertex is added from an adjacent vertex, its distance from x_0 is at most one more than the distance from x_0 of the vertex from which it was added. With this in mind, we prove our lemma by induction.

Because we first add x_0 to the queue, our lemma is true for $d = 0$. Suppose inductively that all vertices of distance $d - 1$ from x_0 are added to the queue (and thus to S) before any vertices of distance d from x_0. Let x be a vertex of distance d from x_0, and let y be a vertex of distance $d + 1$ or more from x_0. (See Figure 6.10.) Then x is adjacent to a vertex of distance $d - 1$ (but no smaller distance) from x_0. By the inductive hypothesis, all vertices of distance $d - 1$ from x_0 are added to the tree before any vertices of distance d from x_0. (In Figure 6.10, vertices of distance d from x_0 are on the circle and those of distance less than d are inside the circle.) From this, we can conclude that all vertices of distance $d - 1$ from x_0 are added to the tree before x. At least one of these vertices is adjacent to x, and so x is added to the queue from one of these vertices, which we shall call x_{d-1}.

Figure 6.10: *Vertices closer to x_0 are added to the tree sooner.*

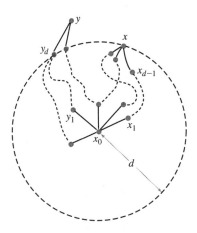

If y were added from a vertex of distance less than or equal to $d - 1$ from x_0, the vertex y would be of distance at most d from x_0. Therefore, y is added from a vertex y_d of distance d or more from x_0. By the inductive hypothesis, x_{d-1} is added to the queue before y_d. Thus, vertices added from x_{d-1} are added to the queue before vertices added from y_d. Therefore, x is added to the tree before y. Thus, all vertices of distance d from x_0 are added to the tree before any vertices of distance $d + 1$ or more. Hence, by the principle of mathematical induction, for every integer $d \geq 0$, all vertices of distance d are added to the vertex set of the tree before any vertices of distance $d + 1$ or more. ∎

Although we introduced breadth-first search to get an algorithm that quickly determines a spanning tree of a graph or a spanning tree of the connected component of a graph containing a given vertex, the algorithm does more for us.

Exercise 6.2-3 How does the distance from x_0 to y in a breadth-first search tree, centered at x_0, in a graph G, relate to the distance from x_0 to y in G?

In fact, the unique path from x_0 to y in a breadth-first search spanning tree of a graph G is a shortest path from x_0 to y in G; thus, the distance from x_0 to another vertex in G is the same as the distance in a breadth-first search spanning tree centered at x_0. This makes it easy to compute the distance between a vertex x_0 and all of the other vertices in a graph.

Theorem 6.9 The unique path from x_0 in a breadth-first search spanning tree, centered at the vertex x_0, in a graph G, to a vertex y is a shortest path from x_0 to y in G. Thus, the distances from x_0 to y in G are the same as distances in a breadth-first search spanning tree of G.

Proof: We prove this theorem by induction on the distance d of a vertex from x_0. Clearly the theorem is true if $d = 0$. Suppose now that whenever x has distance $d - 1$ from x_0 in G, it has distance $d - 1$ from x_0 in the tree. Let y be a vertex of distance d from x_0 in G. On a shortest path from x_0 to y, there is a vertex x' of distance $d - 1$ from x_0. By Lemma 6.8, y is added to the tree after all vertices of distance $d - 1$, and because there is at least one vertex of distance $d - 1$ adjacent to y, the vertex y must be added from a vertex of distance $d - 1$ or less. However, y cannot be adjacent to a vertex of distance less than $d - 1$ to x_0 (or else its distance from x_0 would be less than d). For this reason, when y is added to the tree, y can only be adjacent in T to vertices of distance $d - 1$ (in G and, thus, by the inductive hypothesis, in the tree) from x_0. Thus, the unique path from x_0 to y in the tree must have length d. Therefore, by the principle of mathematical induction, the theorem holds for all nonnegative distances. ∎

Rooted Trees

A breadth-first search spanning tree of a graph is not simply a tree. It is actually a tree with a selected vertex—namely, x_0—and is one example of what we call a rooted tree. A **rooted tree** consists of a tree with a selected vertex, called a **root**, in the tree. Another kind of rooted tree you have likely seen is a binary search tree. It is fascinating how much additional structure is provided to a tree when we select a vertex and call it a root. Figure 6.11 shows a tree with a chosen vertex and the result of redrawing the tree in a more standard way. The standard way computer scientists draw rooted trees is with the root at the top and all the edges sloping down (as you might expect to see with a family tree).

We adopt the language of family trees—ancestor, descendant, parent, and child—to describe rooted trees in general. In Figure 6.11, we say that vertex j is a child of vertex i and a descendant of vertex r, as well as a descendant of vertices f and i. We say that vertex f is an ancestor of vertex i. Vertex r is the parent of vertices a, b, c, and f. Each of those four vertices is a child of vertex r. Vertex r is an ancestor of all of the other vertices in the tree. In general, in a rooted tree with root r, a vertex x is an **ancestor** of a vertex y and a vertex y is a **descendant** of a vertex x if x is on the unique

Figure 6.11: *Two different drawings of the same rooted tree*

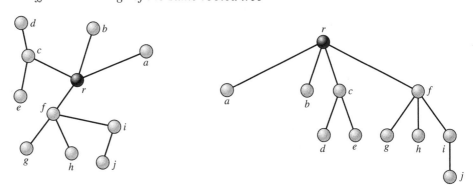

path from the root to y. Vertex x is a **parent** of vertex y and y is a **child** of vertex x in a rooted tree if x is the unique vertex adjacent to y on the unique path from r to y. A vertex can have only one parent but many ancestors. A vertex is its own ancestor or descendant, but it cannot be its own parent or child. A vertex with no children is called a **leaf vertex** or an **external vertex**; other vertices are called **internal vertices**.

Exercise 6.2-4 The definition of a parent implies that a vertex in a rooted tree can have at most one parent. Explain why. Does every vertex in a rooted tree have a parent?

In Exercise 6.2-4, suppose that x is not the root. Then, because there is a unique path between a vertex x and the root of a rooted tree and because there is a unique vertex on that path adjacent to x, each vertex other than the root has a unique parent. The root, however, has no parent.

Exercise 6.2-5 A binary tree is a special kind of rooted tree that has some additional structure that makes it tremendously useful as a data structure. To describe the idea of a binary tree, it is useful to think of a tree with no vertices, which we call the **null tree** or **empty tree**. We can then recursively describe a **binary tree** as

- an empty tree (a tree with no vertices), or

- a structure T consisting of a root vertex, a binary tree called the left subtree of the root, and a binary tree called the right subtree of the root. If the left or right subtree is nonempty, then its root vertex is joined by an edge to the root of T.

Thus, a single vertex is a binary tree with an empty right subtree and an empty left subtree. A rooted tree with two vertices can occur in two ways as a binary tree, either with a root and a left subtree consisting of one vertex or as a root and a right subtree consisting of one vertex. Draw all binary trees on four vertices in which the root vertex has an empty right child. Draw all binary trees on four vertices in which the root has a nonempty left child and a nonempty right child.

Exercise 6.2-6 A binary tree is a **full binary tree** if it is not empty and each vertex has either two nonempty children or two empty children (recall that a vertex with no children is called a leaf or external vertex). Are there any full binary trees with an even number of vertices? Prove that your answer is correct.

Exercise 6.2-7 What is the relationship between the number of internal vertices and the number of external vertices in a full binary tree?

For Exercise 6.2-5, we have the five binary trees shown in Figure 6.12 as our answer to the first question. Then, in Figure 6.13, we have four more trees that answer the second question.

Figure 6.12: *The four-vertex binary trees whose root has an empty right child*

Figure 6.13: *The four-vertex binary trees whose root has both a left and a right child*

For Exercise 6.2-6, because a full binary tree is not empty, it must have an odd number of vertices. We can prove this inductively. A full binary tree with one vertex has an odd number of vertices. Suppose inductively that $n > 1$ and that any full binary tree with fewer than n vertices has an odd number of vertices. For a full binary tree with $n > 1$ vertices, the root must have two nonempty children. Thus, removing the root gives us two binary trees rooted at the children of the original root and each with fewer than n vertices. By the definition of "full," each subtree rooted in the two children must be a full binary tree. The number of vertices of the original tree is one more than the total number of vertices of these two trees. Because this is a sum of three odd numbers, it must be odd. Thus, by the principle of mathematical induction, a full binary tree must have an odd number of vertices.

For Exercise 6.2-7, we give drawings of some full binary trees in Figure 6.14. The drawings suggest that the number of internal vertices is one less than the number of external vertices, though more pictures—or better yet, a proof—would be needed to be really convincing. Let's try a proof by induction that the number of internal vertices is one less than the number of external vertices. Clearly this is true for a full binary tree with one vertex, because that vertex is an external vertex. Thus, assume that in a full binary tree with fewer than n vertices, the number of internal vertices is one less than the number of external vertices. Take a full binary tree T on $n > 1$ vertices and remove the root vertex, giving two binary trees T_1 and T_2 on fewer than n vertices. Because T is a full binary tree, each of its vertices has either zero or two children. Then each vertex of T_1 or T_2 has either zero or two children, so they are full binary trees. If T_1 has v_1 internal vertices and T_2 has v_2 internal vertices, then by the inductive hypothesis, they have $v_1 + 1$ and $v_2 + 1$ external vertices, respectively. But the external vertices of T are exactly those of T_1 and T_2, so T has $v_1 + v_2 + 2$ external vertices. The internal vertices of T are the root and the internal vertices of T_1 and T_2, which means T has $v_1 + v_2 + 1$ internal vertices. Therefore, the number of internal vertices of T is one less than the number of external vertices of T. Thus, by the principle of mathematical

Figure 6.14: *Some full binary trees*

induction, for all full binary trees, the number of internal vertices is equal to one less than the number of external vertices.

Recall that in Section 4.1, we said that there are circumstances in which trying to build an example from smaller examples is not as good a way of finding an inductive proof as trying to see how to decompose a larger example into smaller ones. Here is a case in point. Removing a root vertex gives us an immediate inductive proof; however, it is not immediately clear what the various ways of pasting together smaller full binary trees to give larger ones are and whether all full binary trees on n vertices can be given in this way. Another example in which induction works in this way occurs in Exercise 6.2-6. For instance, a possible way to attempt to grow a full binary tree to a larger full binary tree is to add a new leaf node to some vertex. However, this is doomed to failure, because adding a vertex of degree 1 to a full binary tree never gives a full binary tree.

The definition we gave for "binary tree" was inductive because that type of definition makes it easy for us to prove things about binary trees. We remove the root, apply the inductive hypothesis to the binary tree or trees that result, and then use that information to prove our result for the original tree. We could have defined a binary tree as a special kind of rooted tree, such that

- each vertex has at most two children,

- each child is specified to be a left or right child, and

- a vertex has at most one of each kind of child.

Although this definition works, it is less convenient than the inductive definition.

There is a similar inductive definition of a rooted tree. Because we have already defined rooted trees, we will pretend that we are now defining a new object called an r-tree. The recursive definition states that an r-tree is either a single vertex, called a root, or a graph consisting of a vertex called a root and a set of disjoint r-trees, each of which has its root attached by an edge to the original root. We can then prove, as a theorem, that a graph is an r-tree if and only if it is a rooted tree. Thus, by replacing "r-tree" with "rooted tree" in our inductive definition, we have another definition of a rooted tree. Sometimes inductive proofs for rooted trees are easier if we use the method of removing the root and applying the inductive hypothesis to the rooted trees that result, as we did for binary trees in our solution of Exercise 6.2-6.

Important Concepts, Formulas, and Theorems

1. *Spanning tree.* A tree whose edge set is a subset of the edge set of the graph G is called a *spanning tree* of G if the tree has exactly the same vertex set as G.

2. *Queue.* We can think of a queue as a list of items to which we can do exactly two things: We can add an item x to the end of the queue, and we can remove an item from the front of the queue. We say that we *enqueue x* onto Q when we add it to the end of Q, and we say that we *dequeue* an item from Q when we remove it from the front of the queue.

3. *Breadth-first search.* We create a *breadth-first search (BFS)* tree centered at x_0 in the following way: We begin by enqueueing x_0 at the end of a queue and putting x_0 into S, which becomes the vertex set of the proposed BFS tree. Then we do the following until we run out of vertices on our queue:

   a. Dequeue a vertex w from the queue.

   b. For all edges ϵ incident with w, if ϵ joins w to a vertex z not in S, then add ϵ to E', add z to S, and put z on the end of the queue.

   Now S is the vertex set of the connected component containing x_0, and E' is the edge set of a breadth-first search spanning tree of that component.

4. *Breadth-first search and distances.* You may compute the distance from a vertex y to a vertex x by doing a breadth-first search centered at x and then computing the distance from x to y in the breadth-first search tree. In particular, the path from x to y in a breadth-first search tree of G centered at x is a shortest path from x to y in G.

5. *Rooted tree.* A *rooted tree* consists of a tree with a selected vertex, called a *root,* in the tree.

6. *Ancestor/Descendant.* In a rooted tree with root r, a vertex x is an *ancestor* of a vertex y, and vertex y is a *descendant* of vertex x if x is on the unique path from the root to y.

7. *Parent/Child.* In a rooted tree with root r, a vertex x is a *parent* of a vertex y and y is a *child* of vertex x if x is the unique vertex adjacent to y on the unique path from r to y.

8. *Leaf vertex/External vertex.* A vertex with no children in a rooted tree is called a *leaf vertex,* a *leaf,* or an *external vertex.*

9. *Internal vertex.* A vertex of a rooted tree that is not a leaf vertex is called an *internal vertex.*

10. *Binary tree.* We recursively describe a *binary tree* as

    - an empty tree (a tree with no vertices), or
    - a structure T consisting of a root vertex, a binary tree called the left subtree of the root, and a binary tree called the right subtree of the root. If the left or right subtree is nonempty, then its root vertex is joined by an edge to the root of T.

11. *Full binary tree.* A binary tree is a *full binary tree* if it is nonempty and each vertex has either two nonempty children or two empty children.

12. *Recursive definition of a rooted tree.* The recursive definition of a rooted tree states that it is either a single vertex, called a root, or a graph consisting of a vertex called a root and a set of disjoint rooted trees, each of which has its root attached by an edge to the original root.

Problems

1. Find all spanning trees (list their edge sets) of the graph in Figure 6.15. a/h

Figure 6.15

2. Show that a finite graph is connected if and only if it has a spanning tree.

3. Draw all rooted trees on five vertices. The order and the place in which you write the vertices on the page is unimportant. If you would like to label the vertices (as in Figure 6.11), that is fine, but don't give two different ways of labeling or drawing the same tree. a/h

4. Draw all rooted trees on six vertices with four leaf vertices. If you would like to label the vertices (as in Figure 6.11), that is fine, but don't give two different ways of labeling or drawing the same tree.

5. Find a tree with more than one vertex and with the property that all the rooted trees you get by picking different vertices as roots are different as rooted trees. (Two rooted trees are the same (*isomorphic*), if they each have one vertex or if you can label them so that they have the same labeled root and the same labeled subtrees.) a/h

6. Create a breadth-first search tree centered at vertex 12 for the graph in Figure 6.8, and use your tree to compute the distance of each vertex from vertex 12.

7. Draw all full binary trees on seven vertices. a/h

8. The *depth* of a vertex in a rooted tree is defined to be the number of edges on the unique path to the root. The *height* of a rooted tree is the maximum of the depths of its vertices. A binary tree is *complete* if it is full and all its leaves have the same depth. How many vertices does a complete binary tree of height 1 have? Height 2? Height d? (Proof required for height d.)

9. Based on Problem 8, what is the minimum height of *any* binary tree on v vertices? (Please prove this.) a/h

10. As defined in Problem 8, a binary tree is complete if it is full and all its leaves have the same depth. A vertex that is not a leaf vertex is called an *internal vertex*. What is the relationship between the number I of internal vertices and the number L of leaf vertices in a complete binary tree?

11. The *internal path length* of a binary tree is the sum, taken over all internal vertices of the tree, of the depth of the vertex. The *external path length* of a binary tree is the sum, taken over all leaf vertices of the tree, of the depth of the

vertex (see Problem 8 for a definition of "depth"). Show that in a nonempty full binary tree with n internal vertices, internal path length i, and external path length e, you have $e = i + 2n$. a/h

12. Prove that a graph is an r-tree, as defined at the end of this section, if and only if it is a rooted tree.

13. Use the inductive definition of a rooted tree (r-tree) given at the end of this section to give another proof that a rooted tree with n vertices has $n - 1$ edges if $n \geq 1$. a/h

14. Figure 6.16 has numbers added to the edges of the graph of Figure 6.1 to give what is usually called a **weighted graph**—a graph with numbers, often called **weights**, associated with its edges. We use $w(\epsilon)$ to stand for the weight of the edge ϵ. In this case, these numbers represent the lease fees, in thousands of dollars, for the communication lines that the edges represent. Because the company is choosing a spanning tree from the graph to save money, it is natural that it would want to choose the spanning tree with minimum total cost. To be precise, a **minimum spanning tree** in a weighted graph is a spanning tree of the graph such that the sum of the weights on the edges of the spanning tree is a minimum among all spanning trees of the graph.

Figure 6.16: *A stylized map of some eastern U.S. cities*

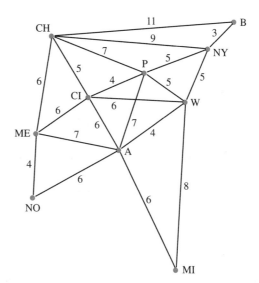

Give an algorithm to select a spanning tree of minimum total weight from a weighted graph, and apply the algorithm to find a minimum spanning tree of the weighted graph in Figure 6.16. Show that your algorithm works, and analyze its running time.

6.3 EULERIAN AND HAMILTONIAN GRAPHS

Eulerian Tours and Trails

Exercise 6.3-1

In an article generally acknowledged to be one of the origins of graph theory, reprinted in Biggs, Lloyd, and Wilson[7], Leonhard Euler (pronounced "Oiler") described a geographic problem that he offered as an elementary example of what he called "the geometry of position." The problem, known as the Königsberg Bridge problem, concerns the town of Königsberg in Prussia (now Kaliningrad in Russia), which is shown in a schematic map (circa 1700) in Figure 6.17. Euler tells us that the citizens of Königsberg amused themselves by trying to find a walk through town that crossed each of the seven bridges once and only once and ended where it started. Is such a walk possible?

Figure 6.17: *A schematic map of Königsberg*

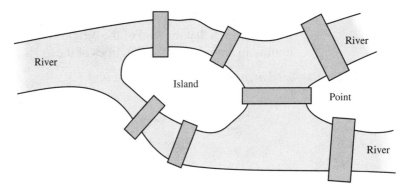

In Exercise 6.3-1, such a walk will enter a landmass on a bridge and leave it on a different bridge. So, except for the starting and ending point, the walk requires two new bridges each time it enters and leaves a landmass. Thus, each landmass must be at the end of an even number of bridges. However, as we see in Figure 6.17, each landmass is at the end of an odd number of bridges. Therefore, no such walk is possible.

We can represent the map in Exercise 6.3-1 more compactly with the graph in Figure 6.18. In graph-theoretic terminology, Euler's question asks whether there is a walk, starting and ending at the same vertex, that uses each edge exactly once.

Figure 6.18: *A graph to replace the schematic map of Königsberg*

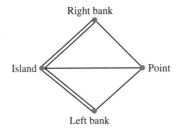

Exercise 6.3-2	Determine whether the graph in Figure 6.1 (in Section 6.1) has a closed walk that includes each edge of the graph exactly once, and find one if it does.

Exercise 6.3-3	Find the strongest condition, or conditions, you can find that must be satisfied by all graphs with a walk that starts and ends at the same place and that includes each vertex at least once and each edge once and only once. Such a walk is known as an **Eulerian tour** or **Eulerian circuit**.

Exercise 6.3-4	Find the strongest condition, or conditions, you can find that must be satisfied by all graphs with a walk that starts and ends at different places and that includes each vertex at least once and each edge once and only once. A walk where no edge appears more than once is called a **trail**, so this kind of walk is known as an **Eulerian trail**.

Exercise 6.3-5	Determine whether the graph in Figure 6.1 has an Eulerian trail, and find one if it does.

The graph in Figure 6.1 cannot have a closed walk that includes each edge exactly once, because if the initial vertex of the walk were W, then the number of edges incident with W would have to be one at the beginning of the walk, plus two for each time W appears before the end of the walk, plus one more for the time W would appear at the end of the walk. Thus, the degree of W would have to be even. But if W were not the initial vertex of a closed walk including all the edges, then each time we entered W on one edge, we would have to leave it on a second edge; so, the number of edges incident with W would have to be even. Thus, in Exercise 6.3-2, there is no closed walk that includes each edge exactly once.

Notice that in any graph with an Eulerian circuit, each vertex, except for the starting-finishing one, will be paired with two new edges (those preceding and following it on the walk) each time it appears on the walk. This is similar to our argument for a walk through Königsberg. Therefore, each of these vertices is incident with an even number of edges. Furthermore, the starting vertex is incident with one edge at the beginning of the walk and with a different edge at the end of the walk. Each other time the starting vertex occurs, it will be paired with two edges. Thus, this vertex is incident with an even number of edges as well. Therefore, a natural condition that a graph must satisfy if it has an Eulerian tour is that each vertex has even degree. But Exercise 6.3-3 asks for the strongest condition or conditions we could find that a graph with an Eulerian tour would satisfy. How do we know whether this is as strong a condition as we could devise? In fact, it isn't—the graph in Figure 6.19 clearly has no Eulerian tour because it is disconnected, even though every vertex has even degree.

The point that Figure 6.19 makes is that to have an Eulerian tour, a graph must be connected as well as having only vertices of even degree. Thus, perhaps the strongest conditions we can find for having an Eulerian tour are that the graph is connected and every vertex has even degree. Again, the question comes up, "How do we show that these conditions are as strong as possible, if indeed they are?" We showed that a condition was not as strong as possible by giving an example of a graph that satisfied the condition but that did not have an Eulerian tour. What if we could show that no such example is possible? If we could prove that there is an Eulerian tour in every connected graph in which each vertex has even degree, then we would show that our condition is as strong as possible.

Figure 6.19: *This graph has no Eulerian tour, even though each vertex has even degree.*

| Theorem 6.10 | A finite graph has an Eulerian tour if and only if it is connected and each vertex has even degree. |

Proof: A graph must be connected to have an Eulerian tour because there must be a walk that includes each vertex and therefore each pair of vertices must be connected by a path. Similarly, as explained earlier, each vertex must have even degree for a graph to have an Eulerian tour. Therefore, we need only show that if a graph is connected and each vertex has even degree, then it has an Eulerian tour. We do so with a recursive construction. The base of our recursive construction is a procedure that forms a closed walk that starts and ends at x_0, but may not include all the edges. We simply start by taking an edge out of x_0 to a neighboring vertex, say x_1. We add the edge $\{x_0, x_1\}$ to the walk, delete it from E, and continue by taking another edge incident to x_1. Because each vertex of G has even degree, whenever there is one vertex incident to x_1, there must also be another; when we remove both these edges, every vertex of G still has even degree. Thus, we continue in this manner with vertices x_2, x_3, and so on until we reach x_0 again. Observe that because each vertex has even degree, we will eventually reach x_0 again. This gives us a closed walk C. If C contains all the edges of G, we stop. As we constructed the walk, we deleted the edges of this closed walk from the edge set of G, giving us a graph $G' = (V, E')$ in which each vertex has even degree because we have removed two edges incident with each vertex of the closed walk (or else we have removed a loop).

However, G' need not be connected. Each connected component of G' is a connected graph in which each vertex has even degree. Further, each connected component of G' contains at least one element x_i of the closed walk whose edges we deleted. (Suppose that a connected component K contained no x_i. Because G is connected, there is a path in G for each i from each vertex in K to each vertex x_i. Choose the shortest such path, and suppose that it connects a vertex y in K to x_j. Then no edge in the path can be in the closed walk or else we would have a shorter path from y to a different vertex x_i. Therefore, removing the edges of the closed walk leaves y connected to x_j in K, so that K contains an x_i after all, which is a contradiction.) Each connected component has fewer edges than G, so we may assume inductively that each connected component has an Eulerian tour. Now we may begin to construct an Eulerian tour of G recursively by starting at x_0 and taking an Eulerian tour of the connected component containing x_0. Now suppose we have constructed a walk that contains the vertices $x_1, x_2, \ldots, x_k$, as well as all the vertices and edges of each connected component of G' containing at least one of these vertices. If this is not already an Eulerian tour, there is an edge ϵ_{k+1} to a vertex x_{k+1} in our original closed walk. Add this edge and vertex to the walk we are constructing. If the vertices and edges of the connected component of G' containing x_{k+1} are not already in our tour, we add an Eulerian tour of the connected component of G' containing x_{k+1} to the walk we are constructing. Every vertex is in some connected component of G', and every edge is either an edge of the first closed walk or an edge of some connected component of G'. Therefore, when we add the last edge and vertex of our original closed walk to the walk that we have been constructing, every vertex and edge of the graph will have to be in the walk we have constructed. Further, by the way we constructed this walk, no edge appears more than once. Thus, if G is connected and each vertex of G has even degree, then G has an Eulerian tour. ∎

A graph with an Eulerian tour is called an **Eulerian graph**.

In Exercise 6.3-4, each vertex other than the initial and final vertices of the walk must have even degree by the same reasoning that we used for Eulerian tours. But the initial vertex must have odd degree. This is because the first time we encounter this vertex in our Eulerian trail, it is incident with one edge in the walk, but each succeeding time, it is incident with two edges in the walk. Similarly, the final vertex must have odd degree. This makes it natural to guess the following theorem.

Theorem 6.11   A graph G has an Eulerian trail if and only if G is connected and all but two of the vertices of G have even degree.

Proof: We have already shown that if G has an Eulerian trail, then all but two vertices of G have even degree, and these two vertices have odd degree.

Suppose that G is a connected graph in which all but two vertices have even degree. Suppose the two vertices of odd degree are x and y. Add an edge ϵ' joining x and y to the edge set of G to get G'. Then G' has an Eulerian tour by Theorem 6.10. One of the edges of the tour is the added edge. We may traverse the tour starting with any vertex and any edge following that vertex in the tour; thus, we may begin the tour with either $x\epsilon'y$ or $y\epsilon'x$. By removing the first vertex and ϵ' from the tour, we get an Eulerian trail in G. ∎

By Theorem 6.11, there is no Eulerian trail in Exercise 6.3-5.

Euler made a big deal in his paper of explaining why it is necessary for each landmass to have an even number of bridges, but he seemed to consider the process of constructing the walk rather self-evident, as if it were hardly worthy of comment. For us, however, proving that the construction is possible if each landmass has an even number of bridges (that is, showing that the condition that each landmass has an even number of bridges is a sufficient condition for the existence of an Eulerian tour) was a much more significant effort than proving that having an Eulerian tour requires each landmass to have an even number of bridges. The standards of what is required to back up a mathematical claim have changed over the years!

Finding Eulerian Tours

Notice that our proof of Theorem 6.10 gives a recursive algorithm for constructing a tour: We find a closed walk W starting and ending at a vertex we choose, create the graph $G - W$ that results from removing the closed walk, and then follow our closed walk, pausing each time we enter a new connected component of $G - W$ to construct recursively an Eulerian tour of the component and traverse it before returning to following our closed walk.

We will soon give pseudocode for this algorithm. Note that when we recursively find a walk through a connected component of $G - W$, the algorithm will remove all the edges of that connected component. Therefore, even if several vertices in W are in the connected component, we will construct the entire walk when the first of these vertices is encountered, and all edges from the component will be removed from the other vertices in this process.

We need to do the following three operations on walks:

- CreateWalk(x,y): Creates and returns a walk with a single edge that starts at vertex x and ends at vertex y.

- AppendToWalk(W,x): Adds vertex x to the end of walk W, adding an edge from the current end of the walk to x.

- SpliceWalks(W_1,x,W_2): Assumes that x is a vertex in walk W_1 and that walk W_2 begins and ends at x. Changes walk W_1 so that it goes from its beginning to x, follows W_2 to its end at x, and then continues from x to the end of walk W_1.

We also assume the existence of a procedure RemoveEdge(x,y,E), which removes one edge connecting x and y from the edge set E. Finally, Degree(x,E) should be the degree of x in the current edge set E.

FindEulerianTour(V,E,x_0)

```
// Assume every vertex of V has even degree.
// Assume x_0 is a vertex in V of degree > 0.
// Returns a walk that begins and ends at x_0 containing all edges in the
// connected component containing x_0.
// The algorithm first finds a closed walk starting and ending at x_0.
(1)    y = a vertex adjacent to x_0
(2)    W = CreateWalk(x_0,y)
(3)    RemoveEdge(x_0,y,E)
(4)    While (y ≠ x_0)
(5)        x = y
```

```
(6)        y = a vertex adjacent to x
(7)        AppendToWalk(W,y)
(8)        RemoveEdge(x,y,E)
(9)  W₁ = W
(10) For each vertex x in W
(11)       While (Degree(x,E) > 0)
(12)            W₂ = FindEulerianTour(V,E,x)
(13)            SpliceWalks(W₁,x,W₂)
(14) Return W₁
```

It is possible to use linked structures to implement this algorithm so that each operation on walks takes $O(1)$ time. The operations on the graph can also be implemented in $O(1)$ time. Because each time we find an adjacency, we remove an edge from E, the total time spent in the loop in Lines 4–8 to find a walk is proportional to the length of the walk. The time needed to copy W to W_1 in Line 9 is no more than some constant times the amount of time spent in the loop in Lines 4–8. The same is true in the recursive calls, and eventually every edge is removed from E and added to W. Thus, our algorithm's time is proportional to the number of edges, or $\Theta(e)$.

Hamiltonian Paths and Cycles

A natural question to ask in light of our work on Eulerian tours is whether we can state necessary and sufficient conditions for a graph to have a closed walk that includes each *vertex* exactly once (except for the beginning and end). An answer to this question could be quite useful. For example, a salesperson might have to plan a trip through a number of cities connected by a network of airline routes. Planning the trip so the salesperson would travel through a city only when stopping there for a sales call would minimize the number of flights needed. This question came up in a game called "Around the World," designed by William Rowan Hamilton. In this game, the vertices of the graph were the vertices of a dodecahedron (a twelve-sided solid in which each side is a pentagon), and the edges were the edges of the dodecahedron. The object of the game was to design a trip that started at one vertex, visited each vertex once, and then returned to the starting vertex along an edge. Hamilton suggested that two players could take turns, one choosing the first five cities on a tour, and the other trying to complete the tour. It is because of this game that a cycle that includes each vertex of the graph exactly once (thinking of the first and last vertex of the cycle as the same) is called a **Hamiltonian cycle**. A graph is called Hamiltonian if it has a Hamiltonian cycle. A **Hamiltonian path** is a path that includes each vertex of the graph exactly once.

It turns out that nobody yet knows (and as we explain briefly at the end of this section, it may be reasonable to expect that nobody will find) useful necessary and sufficient conditions for a graph to have a Hamiltonian cycle or a Hamiltonian path. What would make necessary and sufficient conditions useful? Useful conditions would be significantly easier to verify than trying all permutations of the vertices to see if taking the vertices in the order of that permutation defines a Hamiltonian cycle or path. Because people have been unable to find useful necessary and sufficient conditions, this branch of graph theory has evolved into theorems that give sufficient conditions for a graph to have a Hamiltonian cycle or path. Such theorems say that all graphs of a certain type have Hamiltonian cycles or paths, but they do not characterize all graphs that have Hamiltonian cycles or paths.

Exercise 6.3-6   Describe all values of n such that a complete graph on n vertices has a Hamiltonian path. Describe all values of n such that a complete graph on n vertices has a Hamiltonian cycle.

Exercise 6.3-7   Determine whether the graph in Figure 6.1 has a Hamiltonian cycle or path. If it does, determine one.

Exercise 6.3-8   Try to find an interesting condition involving the degrees of the vertices of a simple graph that guarantees the graph will have a Hamiltonian cycle. Does your condition apply to graphs that are not simple? (There is more than one condition to try and therefore more than one reasonable answer to this exercise. For example, you might ask if a graph in which each vertex has degree $n - 2$ has a Hamiltonian cycle.)

In Exercise 6.3-6, the path consisting of one vertex and no edges is a Hamiltonian path but not a Hamiltonian cycle in the complete graph on one vertex. (Recall that a path consisting of one vertex and no edges is not a cycle.) Similarly, the path with one edge in the complete graph K_2 is a Hamiltonian path but not a Hamiltonian cycle, and because K_2 has only one edge, there is no Hamiltonian cycle in K_2. In the complete graph K_n, any permutation of the vertices is a list of the vertices of a Hamiltonian path. If $n \geq 3$, such a Hamiltonian path from x_1 to x_n can be converted to a Hamiltonian cycle by adding the edge from x_n to x_1, followed by the vertex x_1. (This gives a cycle starting and ending at x_1 and including each vertex other than x_1 exactly once.) Thus, each complete graph has a Hamiltonian path, and each complete graph with more than three vertices has a Hamiltonian cycle.

In Exercise 6.3-7, the path with vertices NO, A, MI, W, P, NY, B, CH, CI, and ME is a Hamiltonian path. Adding the edge from ME to NO and the vertex NO gives the Hamiltonian cycle NO, A, MI, W, P, NY, B, CH, CI, ME, NO.

Now consider Exercise 6.3-8. Based on our observation that the complete graph on n vertices has a Hamiltonian cycle if $n > 2$, we might let our condition be that the degree of each vertex is one less than the number of vertices. This would be uninteresting, however, because it would simply restate what we already know for complete graphs. The reason we could say that K_n has a Hamiltonian cycle when $n > 3$ is that when we enter a vertex, there is always a remaining edge on which we could leave the vertex. However, the condition that each vertex has degree $n - 1$ is stronger than we need for the entering-leaving condition, because until we are at the second-to-last vertex of the cycle, we have more choices than we need for edges on which to leave the vertex. On the other hand, it might seem that even if n were rather large, the condition that each vertex should have degree $n - 2$ would not be sufficient to guarantee a Hamiltonian cycle. It might be possible that, as illustrated in Figure 6.20, when we

Figure 6.20: *The path 1, 2, 3, 4, 5 cannot be extended to a Hamiltonian cycle.*

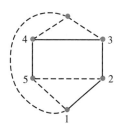

get to the second-to-last vertex that we hoped to have on the cycle, all of the $n - 2$ vertices to which the vertex is adjacent might already be on the cycle and different from the first vertex. Thus, we would not have an edge on which we could leave that vertex. However, there is the possibility that, as in Figure 6.21, when we had an earlier choice, we might have made a different choice to include this vertex earlier on the cycle, giving a different set of choices at the second-to-last vertex. In fact, if $n > 3$ and each vertex has degree at least $n - 2$, then we could choose vertices for a path more or less as we did for the complete graph until we arrive at vertex $n - 1$. At that point, we could complete a Hamiltonian path, unless x_{n-1} is adjacent only to the first $n - 2$ vertices on the path, as in Figure 6.22. In this last case, the first $n - 1$ vertices would form a cycle, because x_{n-1} would be adjacent to x_1. Suppose y is the vertex not yet on the path (vertex 6 in Figure 6.22). Because y has degree $n - 2$ and is not adjacent to x_{n-1}, we find that y would have to be adjacent to the first $n - 2$ vertices on the path. Then, because $n > 3$, we could take the walk $x_1 y x_2 \ldots x_{n-1} x_1$ (which is 1, 6, 2, 3, 4, 5, 1 in Figure 6.22), and we would have a Hamiltonian cycle. Of course, unless n were 4, we could also insert y between x_2 and x_3 (or any x_{i-1} and x_i such that $i < n - 1$), so we would still have a great deal of flexibility. To push this kind of reasoning further, in our next theorem we will introduce a new technique that appears often in graph theory. We discuss our use of the technique after the proof.

Figure 6.21: *Making a better choice early on lets us find a Hamiltonian cycle.*

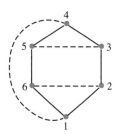

Figure 6.22: *The path 1, 2, 3, 4, 5 cannot be extended to a Hamiltonian cycle.*

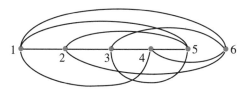

Dirac's Theorem If every vertex of a v-vertex simple graph G with at least three vertices has degree at least $v/2$, then G has a Hamiltonian cycle.

Proof: Suppose, for the sake of contradiction, that there is a graph G_1 with no Hamiltonian cycle in which each vertex has degree at least $v/2$. If we add to the edge set of G_1 an edge joining two existing vertices in G_1, then each vertex will still have degree at least $v/2$. If we add all possible edges to G_1, we will get a complete graph, and it will have a Hamiltonian cycle. Thus, if we continue adding edges one at a time to G_1, then we will at some point reach a graph that does have a Hamiltonian cycle. Instead, suppose we add edges to G_1 until we reach a graph G_2 with no Hamiltonian cycle but with the property that if we add any edge to G_2, then we get a Hamiltonian cycle. We say that G_2 is **maximal** with respect to not having a Hamiltonian cycle. Suppose that x and y are not adjacent in G_2. Adding an edge to G_2 between x and y gives a graph with a Hamiltonian cycle, and x and y must be joined by the added edge in this Hamiltonian cycle. (Otherwise G_2 would have a Hamiltonian cycle.) Thus, G_2 has a Hamiltonian path $x_1x_2 \ldots x_v$ that starts at $x = x_1$ and ends at $y = x_v$. Furthermore, x and y are not adjacent.

Before we stated our theorem, we considered a case in which we had a cycle on $f - 1$ vertices and in which we were going to add an extra vertex between two adjacent vertices. This resulted in a path on v vertices from $x = x_1$ to $y = x_v$ that we then wanted to convert to a cycle. If we had that y is adjacent to some vertex x_i on the path while x is adjacent to x_{i+1}, then we could construct the Hamiltonian cycle $x_1x_{i+1}x_{i+2} \ldots x_v x_i x_{i-1} \ldots x_2 x_1$. But for this proof, we are assuming that our graph does not have a Hamiltonian cycle. Thus, for each x_i that x is adjacent to on the path $x_1x_2 \ldots x_v$, we know that y is not adjacent to x_{i-1}. Because all vertices are on the path, x is adjacent to at least $v/2$ vertices among x_2 through x_v. Thus, y is not adjacent to at least $v/2$ vertices among x_1 through x_{v-1}. But there are only $v - 1$ vertices—namely, x_1 through x_{v-1}—to which y could be adjacent because it is not adjacent to itself. Thus, y is adjacent to at most $v - 1 - v/2 = v/2 - 1$ vertices. This is a contradiction. Therefore, if each vertex of a simple graph has degree at least $v/2$, then the graph has a Hamiltonian cycle. ∎

The new technique used in our proof was that of assuming we had a maximal graph (G_2) that did not have our desired property and then using this maximal graph in a proof by contradiction.

Exercise 6.3-9

Suppose $v = 2k$. Consider a graph G that consists of two complete graphs, one with k vertices $x_1, \ldots, x_k$ and one with $k + 1$ vertices $x_k, \ldots, x_{2k}$. Notice that we get a graph with exactly $2k$ vertices, because the two complete graphs have one vertex in common. How do the degrees of the vertices relate to v? Does the resulting graph have a Hamiltonian cycle? What does this say about whether we can reduce the lower bound on the degree in Theorem 6.12?

Exercise 6.3-10 In Exercise 6.3-9, is there a similar example in the case $v = 2k + 1$?

In Exercise 6.3-9, the vertices that lie in the complete graph with k vertices, with the exception of x_k, have degree $k - 1$. Because $v/2 = k$, this graph does not satisfy the hypothesis of Dirac's theorem, which assumes that every vertex of the graph has degree at least $v/2$. Figure 6.23 shows the case in which $k = 3$.

Figure 6.23: *The vertices of K_4 are white or blue; those of K_3 are black or blue.*

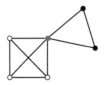

The graph in Figure 6.23 has no Hamiltonian cycle. If an attempt at a Hamiltonian cycle begins at a white vertex in this figure, then after crossing the blue vertex to include the black ones, we can never return to a white vertex without using the blue one a second time. The situation is similar if we try to begin a Hamiltonian cycle at a black vertex. If we try to begin a Hamiltonian cycle at the blue vertex, we would next have to include all white vertices or all black vertices; we would then be stymied because we would have to take our path through the blue vertex a second time to change colors between white and black. As long as $k \geq 2$, the same argument shows that our graph has no Hamiltonian cycle. Thus, the lower bound of $v/2$ in Dirac's theorem is **tight**; that is, we have a way to construct a graph with minimum degree $v/2 - 1$ (when v is even) for which there is no Hamiltonian cycle. If $v = 2k + 1$, then we might consider two complete graphs of size $k + 1$ joined at a single vertex. Each vertex other than the one at which the graphs are joined would have degree k, and we would have $k < k + 1/2 = v/2$. So again, the minimum degree would be less than $v/2$. The same kind of argument that we used with the graph in Figure 6.23 would show that as long as $k \geq 1$, we have no Hamiltonian cycle.

If you analyze our proof of Dirac's theorem, you will see that we really used only a consequence of the condition that all vertices have degree at least $v/2$—namely, that for any two vertices, the sum of their degrees is at least v.

Theorem 6.13 **(Ore's Theorem)** If G is a v-vertex simple graph with $v \geq 3$ such that for each two nonadjacent vertices x and y the sum of the degrees of x and y is at least v, then G has a Hamiltonian cycle.

Proof: See Problem 13. ∎

NP-Complete Problems

At the beginning of our discussion of Hamiltonian cycles, we mentioned that the problem of determining whether a graph has a Hamiltonian cycle seems significantly more difficult than the problem of determining whether a graph has an Eulerian tour. On the surface, however, these two problems have significant similarities.

- Both problems ask whether a graph has a particular property. (Does this graph have a Hamiltonian cycle/Eulerian tour?) The answer is simply "yes" or "no."

- For both problems, there is additional information we can provide that makes it relatively easy to check a "yes" answer if there is one. (The additional information is a closed walk. We simply check whether the closed walk includes each vertex or each edge exactly once.)

But there is also a striking difference between the two problems. It is reasonably easy to find an Eulerian tour in a graph that has one (we saw that the time to use the algorithm implicit in the proof of Theorem 6.10 is $O(e)$, where e is the number of edges of the graph). However, nobody has found a polynomial time algorithm for solving the Hamiltonian cycle problem. This puts us in an interesting position. If someone gets lucky and guesses a permutation of the vertices that is a Hamiltonian path, then we can quickly verify the person's claim to have a Hamiltonian path. However, in a graph of reasonably large size, we have no practical method for finding a Hamiltonian path.

This is the essential difference between the class **P** of problems said to be solvable in polynomial time and the class **NP** of problems said to be solvable in nondeterministic polynomial time. We are not going to describe these problem classes in their full generality; a course in the theory of computation or algorithms is a more appropriate place for such a discussion. However, to give a sense of the difference between these kinds of problems, we will talk about them in the context of graph theory. A question about whether a graph has a certain property is called a **graph decision problem**. Two examples are "Does this graph have an Eulerian tour?" and "Does this graph have a Hamiltonian cycle?" A graph decision problem has a "yes" or "no" answer. Thus, the question "What is the length of the longest path in G?" is not a decision problem, but the question "Is there a path of length k in G?" is.

A **P-algorithm**, or polynomial-time algorithm, for a property takes a graph as input and, in time $O(n^k)$ (where k is a positive integer independent of the input graph and n is a measure of the amount of information needed to specify the input graph), outputs the answer "yes" if and only if the graph does have the property. We say that the algorithm **accepts** the graph if it answers "yes." (Notice that we don't specify what the algorithm does if the graph does not have the property, except that it doesn't output "yes.") We say that a property of graphs is **in the class P** if there is a P-algorithm that accepts exactly the graphs with the property.

Many decision problems for which no P-algorithm is known seem to be hard in the same way that the Hamiltonian cycle problem is hard—namely, there is a P-algorithm for checking a single possible solution to determine if it is a solution, but there are an exponential (or worse) number of possible solutions that we might have to check before finding a "yes" answer. These problems seem different from problems in which even verifying that a proposed solution is indeed a solution takes more than polynomial time. Is there a way to characterize "polynomial-time checkable" problems? How can we specify a "possible solution" in a way that would work for any problem?

An **NP**-algorithm (nondeterministic polynomial-time algorithm) for a graph property takes a graph G whose representation has size n and $O(n^j)$ additional information for some integer j independent of G.[11] You can think of this additional information as a possible solution, though the algorithm can use it any way it chooses. If the algorithm, perhaps using the additional information, can determine that G has the desired property in $O(n^k)$ time, where k is an integer independent of G, then it outputs "yes." If it cannot determine that the graph has the desired property, even if it uses the additional information, then it can do anything except answer "yes." For example, for the property of being Hamiltonian, the extra information might consist of a permutation of the vertex set of the graph. It would then check the permutation to see if the vertices, in the order given, form a Hamiltonian cycle. It would output "yes" if they do.

We call such an algorithm nondeterministic because whether it outputs "yes" for a given input graph is determined not merely by the graph but also by the additional information. In particular, the algorithm might or might not answer "yes" for a graph that has the given property. It depends on the additional information.

We say the algorithm **accepts** a graph if there is some choice of additional information that will cause the algorithm to output "yes." There may be many other choices of additional information that do not lead to the algorithm outputting "yes," but that does not matter. As long as there is any choice of additional information that causes the algorithm to output "yes," we say that the algorithm accepts the graph.

We say that a property is **in the class NP** if there is an **NP**-algorithm that accepts exactly the graphs with the property. Because graph decision problems ask us to decide whether a graph has a given property, we adopt the notation **P** and **NP** to describe problems as well. We say that a decision problem is in **P** or **NP** if the graph property it asks us to decide is in **P** or **NP**, respectively.

When we say that a nondeterministic algorithm *uses* the additional information, we are thinking of "use" in a very loose way. In particular, for a graph decision problem in **P**, the algorithm could simply ignore the additional information and use the polynomial-time algorithm to determine whether the answer should be "yes." Thus, every graph property in **P** is also in **NP**.

Some problems in **NP**, like the Hamiltonian path problem, have an exciting feature: if they can be solved in polynomial time, then every problem in **NP** can be solved in polynomial time. Such problems, called **NP-complete**, are the hardest problems in **NP**. If an **NP**-complete problem is in **P**, then **P** and **NP** are the same class. This result would be very surprising, because it would mean that being told a possible answer never makes it significantly easier to solve a problem. However, the question as to whether **P** and **NP** are the same class of problems has vexed computer scientists since it was introduced in

---

[11]The size of the problem is the number of bits needed to write down the problem using a reasonable representation. We are not going to formally define "reasonable." When we ask whether a graph has an Eulerian tour, we could measure the size of the problem by the number of vertices or by the number of edges. When we ask whether a weighted graph has a spanning tree of weight w or less, not only is the number of vertices or edges of the graph important, but so is the number of digits in the numbers—and perhaps the way in which we represent the numbers. For our purposes, this intuitive idea of the size of a problem should suffice.

1971. (See the end of Section 6.4 for a discussion of the context in which this question arose.) It is one of the most important unsolved problems in computer science.

Thus, knowing that a problem, for example the Hamiltonian cycle problem, is **NP**-complete does not prove that there is no polynomial-time algorithm for it. It does mean that a polynomial-time algorithm for the problem would also give polynomial-time algorithms for thousands of other problems for which people have been unable to find polynomial-time algorithms. Some of these problems have been studied for hundreds of years. If a problem is **NP**-complete, then it is very unlikely that we will find a polynomial-time algorithm to solve it. Our time is probably better spent trying to do something else.

Proving That Problems Are NP-Complete*

It is natural to ask how we might prove that a problem is **NP**-complete. In 1971, Stephen Cook[12] and (independently) Leonid Levin[25] introduced the concept of **NP**-completeness. Cook showed that the "satisfiability" problem is **NP**-complete. This problem is as follows: Given a Boolean expression with variables and the logical connectives "and," "or," and "not," is there a way to assign true and false to the variables that makes the entire expression true?[12] Cook showed that the satisfiability problem is **NP**-complete by showing that it was possible to use a very complex Boolean expression to model the steps of a simple computer called a nondeterministic Turing machine. Any assignment of variables that satisfied the expression would define a sequence of valid steps of a computation that ended in the computer saying "yes."

Once we know that a particular problem is **NP**-complete, we can use it to prove that other problems are **NP**-complete using a technique called a **reduction**, which is a type of transformation between problems. We next look at examples of such tranformations and then abstract the general principle from the examples.

We have claimed that the Hamiltonian cycle problem is **NP**-complete. We can use this claim to show that the following problem, called the **k-cycle problem**, is **NP**-complete: Given a graph G and an integer k, does G have a cycle of length exactly k?

The idea of reduction is to show that if we had an algorithm to solve the k-cycle problem in polynomial time, then we could use that algorithm to solve the Hamiltonian cycle problem in polynomial time. But solving the Hamiltonian cycle problem in polynomial time would mean that any problem in **NP** could be solved in polynomial time, because the Hamiltonian cycle problem is **NP**-complete. Thus, solving the k-cycle problem in polynomial time would mean that any problem in **NP** could be solved in polynomial time. Therefore, by definition, the k-cycle problem is **NP**-complete.

How can we use the k-cycle problem to solve the Hamiltonian cycle problem? This is quite easy, because the Hamiltonian cycle problem is a special case of the k-cycle problem. Given a graph G with v vertices and an algorithm to solve the k-cycle problem, we could find out if G has a Hamiltonian cycle by asking the algorithm if G has a cycle of length v. We have transformed an instance of the Hamiltonian cycle

---

*The material in this section is not used later in the book except for problems marked with an asterisk.

[12]Boolean expression is another name for what we called a symbolic compound statement in Chapter 3.

problem (Does graph G have a Hamiltonian cycle?) into an instance of the k-cycle problem (Does graph G have a cycle of length v?). The first question and the second question always give the same answer for any graph G.

Let's look at a more complicated case. The clique problem asks, Given a graph G and an integer n, does G have K_n as a subgraph? (That is, does G have n vertices such that there is an edge between every pair of vertices?) It is known that the clique problem is **NP**-complete, though we are not in a position to explain why here. The independent set problem asks, Given a graph G and an integer n, is there a set of n vertices such that there are no edges between any pair of vertices? We wish to show that the independent set problem is **NP**-complete.

We assume that we have an algorithm for solving the independent set problem in polynomial time, and we show that we can use this algorithm to solve the clique problem in polynomial time. The transformation is as follows: Suppose we want to know if a graph G has a clique of size n. We could create a new graph G', called the **complement** of G. The graph G' has the same vertex set as G, but there is an edge between a pair of vertices in G' if and only if there is not an edge between those vertices in G. Constructing G' would take $O(v^2)$ time. We could then use our algorithm for solving the independent set problem to determine whether G' has an independent set of size n. Because we have reversed edges and nonedges when constructing G' from G, an independent set in G' is a clique in G. Thus, the question "Does G have a clique of size n?" always has the same answer that "Does G' have an independent set of size n?" has. Constructing G' can be done in polynomial time, and we assumed that running the algorithm to solve the independent set problem can be done in polynomial time. Thus, by using this transformation, we can solve the clique problem in polynomial time, which implies that we can solve any problem in **NP** in polynomial time. Thus, the independent set problem is **NP**-complete.

By doing transformations like these (and more complicated ones), computer scientists and mathematicians have created a long list [17] of **NP**-complete problems, which continues to evolve. This list is useful because if we need to solve a problem that we have never seen before, we can find out if it is on the list before we spend months trying to solve it. But what if the problem is not on the list, and we try everything we can think of to solve the problem with no luck? The next natural thing to do is determine if we can find an **NP**-complete problem that we can transform into our difficult problem. If we can, then even though we don't have a solution, we know that a solution, if there is one, is likely to be difficult for anybody to find.

The general technique is as follows: To prove that a problem Q is **NP**-complete, assume you have an algorithm that solves that problem in polynomial time. Pick another problem Q' that is known to be **NP**-complete. Show how to take any instance of the **NP**-complete problem Q' and transform it into an instance of problem Q such that the answer to the original problem is "yes" if and only if the answer to the transformed problem is "yes."[13] Show that the transformation that you specified takes polynomial time. If problem Q can be solved in polynomial time, then so can Q',

---

[13] An *instance* of a problem is a case of the problem in which all parameters are specified; for example, a particular instance of the Hamiltonian cycle problem is a case of the problem for a particular graph.

because to solve an instance of Q', we first transform it into an equivalent instance of Q and then run the algorithm for solving Q and return the answer. Both the transformation and the run of the algorithm are polynomial, so the whole process is. Because Q' is **NP**-complete and can be solved in polynomial time, any problem in **NP** can be solved in polynomial time. Therefore, by definition, Q is **NP**-complete.

This brief discussion of **NP**-completeness is intended to give a sense of the nature and importance of the subject. We restricted ourselves to graph problems for two reasons. First, we expect you to have a sense of what a graph problem is. Second, no treatment of graph theory is complete without at least some explanation of how some problems seem to be much more intractable than others. However, there are **NP**-complete problems throughout mathematics and computer science. There are even **NP**-complete problems that arise in areas as diverse as biology, physics, and social sciences such as economics. Providing a real understanding of the subject would require much more time than is available in an introductory course in discrete mathematics.

Important Concepts, Formulas, and Theorems

1. *Eulerian graphs and tours.* A graph that has a walk that starts and ends at the same place and that includes each vertex at least once and each edge once and only once is called an *Eulerian graph.* Such a walk is known as an *Eulerian tour* or *Eulerian circuit.*

2. *Characterizing Eulerian graphs.* A graph has an Eulerian tour if and only if it is connected and each vertex has even degree.

3. *Eulerian trail.* A walk that includes each vertex of the graph at least once and each edge of the graph exactly once but that has different first and last endpoints is an *Eulerian trail.*

4. *Characterizing graphs with Eulerian trails.* A graph G has an Eulerian trail if and only if G is connected and all but two of the vertices of G have even degree.

5. *Hamiltonian graphs and cycles.* A cycle that includes each vertex of a graph exactly once (thinking of the first and last vertex of the cycle as the same) is called a *Hamiltonian cycle.* A graph is called *Hamiltonian* if it has a Hamiltonian cycle.

6. *Hamiltonian path.* A *Hamiltonian path* is a path that includes each vertex of the graph exactly once.

7. *Dirac's theorem.* If every vertex of a v-vertex simple graph G with at least three vertices has degree at least $v/2$, then G has a Hamiltonian cycle.

8. *Ore's theorem.* If G is a v-vertex simple graph with $v \geq 3$ such that for each two nonadjacent vertices x and y the sum of the degrees of x and y is at least v, then G has a Hamiltonian cycle.

9. *Graph decision problem.* A question about whether a graph has a certain property is called a *graph decision problem.*

10. **P**-*algorithm/Polynomial-time algorithm/Accepts.* A **P**-*algorithm,* or polynomial-time algorithm, for a property takes a graph as input and, in time

$O(n^k)$ (where k is a positive integer independent of the input graph and n is a measure of the amount of information needed to specify the input graph), outputs the answer "yes" if and only if the graph does have the property. We say that the algorithm *accepts* the graph if it answers "yes."

11. *Problem class* **P**. We say that a property of graphs is *in the class* **P** if there is a **P**-algorithm that accepts exactly the graphs with the property.

12. **NP**-*algorithm/Nondeterministic polynomial-time algorithm.* An **NP**-*algorithm* (nondeterministic polynomial-time algorithm) for a graph property takes a graph G whose representation is size n and also takes $O(n^j)$ additional information for some integer j independent of G. If the algorithm can determine from G and perhaps the additional information that G has the desired property in $O(n^k)$ time, where k is an integer independent of G, then it outputs "yes." If it cannot determine from G and the additional information that the graph has the desired property, then it can do anything except answer "yes."

13. **NP**-*complete.* A graph decision problem in **NP** is called **NP**-complete if a polynomial-time algorithm for that problem implies a polynomial-time algorithm for every problem in **NP**.

Problems

1. For each graph in Figure 6.24, either explain why the graph does not have an Eulerian circuit or find an Eulerian circuit. a/h

Figure 6.24

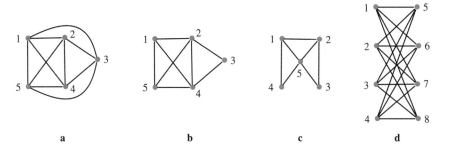

2. For each graph in Figure 6.25, either explain why the graph does not have an Eulerian trail or find an Eulerian trail.

Figure 6.25

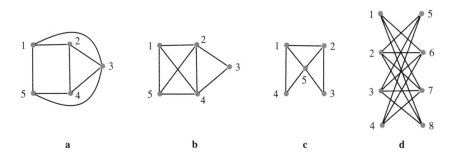

3. What is the minimum number of new bridges that would have to be built in Königsberg and where could they be built to give a graph with an Eulerian circuit? a/h

4. If a new bridge were built in Königsberg between the island and the top bank of the river and another between the island and the bottom bank of the river, could you take a walk that crosses all of the bridges and uses none twice? Explain either where you could start and end in that case or why you couldn't do it.

5. For which values of n does the complete graph on n vertices have an Eulerian circuit? a/h

6. The hypercube graph Q_n has as its vertex set the n-tuples of zeros and ones. Two of these vertices are adjacent if and only if they are different in one position. The name "hypercube" comes from the fact that Q_3 can be drawn in three-dimensional space as a cube. For what values of n is Q_n Eulerian?

7. For what values of n is the hypercube graph Q_n (see Problem 6) Hamiltonian? a/h

8. Give an example of a graph that has a Hamiltonian cycle but no Eulerian circuit and a graph that has an Eulerian circuit but no Hamiltonian cycle.

9. The complete bipartite graph $K_{m,n}$ is a graph with $m + n$ vertices. These vertices are divided into a set of size m and a set of size n. We call these sets the **parts** of the graph. Within each set, there are *no* edges, but between each pair of vertices in different sets, there is an edge. The graph $K_{4,4}$ is pictured in Figure 6.24d.

    a. For what values of m and n is $K_{m,n}$ Eulerian? a/h

    b. For which values of m and n is $K_{m,n}$ Hamiltonian? a/h

10. Show that the edge set of a graph in which each vertex has even degree may be partitioned into edge sets of cycles of the graph.

11. A cut vertex of a graph is a vertex whose removal (along with all edges incident with it) increases the number of connected components of the graph. Describe any circumstances under which a graph with a cut vertex can be Hamiltonian. a/h

12. Which of the graphs in Figure 6.26 satisfy the hypotheses of Dirac's theorem? Of Ore's theorem? Which have Hamiltonian cycles?

Figure 6.26

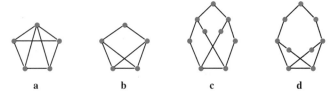

a　　　　b　　　　c　　　　d

13. Prove Theorem 6.13. `a/h`

*14. The Hamiltonian path problem is the problem of determining whether a graph has a Hamiltonian path. Explain why this problem is in **NP**. Explain why the problem of determining whether a graph has a Hamiltonian path is **NP**-complete.

15. We form the Hamiltonian closure of a graph G by constructing a sequence of graphs G_i with $G_0 = G$ and with G_i formed from G_{i-1} by adding an edge between two nonadjacent vertices whose degree-sum is at least v. When we reach a G_i to which we cannot add such an edge, we call it a Hamiltonian closure of G. Prove that a Hamiltonian closure of a simple graph G is Hamiltonian if and only if G is Hamiltonian. `a/h`

16. Show that a simple connected graph has one and only one Hamiltonian closure.

6.4 MATCHING THEORY

The Idea of a Matching

Exercise 6.4-1   Suppose a school board is deciding among applicants for faculty positions. The school board has positions for teachers in a number of different grades, a position for an assistant librarian, two coaching positions, and one position each for high school math and English teachers. The board has many applicants, each of whom can fill more than one of the positions. The board would like to know whether it's possible to fill all the positions with the people who have applied for jobs and have been judged as qualified.

Table 6.1 shows a sample of the kinds of applications that a school district might get for its positions. An x below an applicant's number means that the applicant

Table 6.1: *Some sample job application data*

Job \ Applicant	1	2	3	4	5	6	7	8	9
Assistant librarian	x		x	x					
Second grade	x	x	x	x					
Third grade	x	x		x					
High school math				x	x	x			
High school English				x		x	x		
Asst. baseball coach						x	x	x	x
Asst. football coach					x	x		x	

---

*This problem depends on material in the text marked with an asterisk.

qualifies for the position to the left of the x. Thus, Candidate 1 is qualified to teach second grade and third grade and to be an assistant librarian. The assistant coaches teach physical education when they are not coaching, so a coach can't also hold one of the listed teaching positions. Draw a graph in which the vertices are labeled 1 through 9 for the applicants, and L, S, T, M, E, B, and F for the positions. Draw an edge from an applicant to a position if that applicant can fill that position. Use the graph to help decide if it is possible to fill all the positions from among the applicants deemed suitable. If you can do so, give an assignment of people to jobs. If you cannot, try to explain why not.

Exercise 6.4-2

Table 6.2 shows a second sample of the kinds of applications a school district might get for its positions. Draw a graph as before and use it to help you decide if it is possible to fill all the positions from among the applicants deemed suitable. If you can do so, give an assignment of people to jobs. If you cannot, try to explain why not.

Table 6.2: *Some other sample job application data*

Job \ Applicant	1	2	3	4	5	6	7	8	9
Library assistant				x	x				
Second grade	x	x	x					x	
Third grade	x	x		x		x			x
High school math				x	x	x			
High school English					x	x			
Asst. baseball coach	x		x			x	x	x	x
Asst. football coach				x	x	x			

Figure 6.27a shows a graph of the data from Table 6.1.

Figure 6.27: *A graph of the data from Table 6.1*

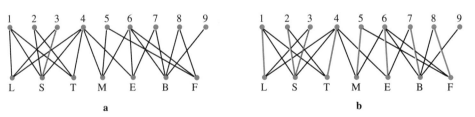

a   b

From the figure, we see that, as shown in blue in Figure 6.27b, L:1, S:2, T:4, M:5, E:6, B:7, and F:8 is one assignment of jobs to people. This assignment picks out a set of edges that share no endpoints. For example, the edge from L to 1 has no endpoint among the endpoints of the edges {S, 2}, {T, 4}, {M, 5}, {E, 6}, {B, 7}, and {F, 8}—namely S, T, M, E, B, F, 2, 4, 5, 6, 7, and 8. A set of edges in a graph that share no endpoints is called a **matching** of the graph.[14] Thus, we have a matching between jobs and people who can fill the jobs. Because we don't want to assign two jobs to one person or two people to one job, a matching is exactly the sort of solution we were looking for. Notice that the edge from L to 1 is a matching all by itself. Thus, we weren't simply looking for a matching; we were looking for a matching that fills all the jobs. A matching is said to **saturate** a set X of vertices if every vertex in X is matched. Thus, Exercise 6.4-1 asked for a matching that saturates the jobs. In this case, a matching that saturates all the jobs is a matching that is as large as possible, so it is also a **maximum** matching—a matching that is at least as large as any other matching.

Figure 6.27 is an example of a bipartite graph. A graph is called **bipartite** whenever its vertex set can be partitioned into two sets X and Y so that each edge connects a vertex in X with a vertex in Y. We can think of the jobs as the set X and the applicants as the set Y. Each of the two sets is called a **part** of the graph. A part of a bipartite graph is an example of an independent set. A subset of the vertex set of a graph is called **independent** if no two of its vertices are joined by an edge. Thus, a graph is bipartite if and only if its vertex set is a union of two independent sets. Notice that a bipartite graph cannot have any loop edges, because a loop would connect a vertex to a vertex in the same set. More generally, a vertex joined to itself by a loop cannot be in an independent set.

In a bipartite graph, it is sometimes easy to pick out a maximum matching simply by staring at a drawing of the graph. However, that is not always the case. Figure 6.28 is a graph of the data in Table 6.2. Staring at this figure gives us many matchings, but no matching that saturates the set of jobs. Staring, though, is not a valid proof technique, unless we can describe very well what we are staring at. Perhaps you tried to construct a matching by matching with something like {L, 4}, {S, 2}, {T, 7}, {M, 5}, {E, 6}, and {B, 8}. If so, then you were probably frustrated when you got to F and found that 4, 5, and 6 were already used. You may then have gone back and tried to redo your earlier choices to keep one of 4, 5, or 6 free, only to find that you couldn't. You couldn't do this because jobs L, M, E, and F are adjacent only to people 4, 5, and 6. Thus, there are only three people qualified for these four jobs, and so there is no way you can fill them all.

Figure 6.28: *A graph of the data from Table 6.2*

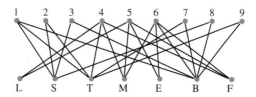

We call the set $N(S)$ of all vertices that are adjacent to at least one vertex of S the **neighborhood** of S or the **neighbors** of S. In these terms, there is no matching that

---

[14]In light of this definition, we should use our standard edge notation for matchings. The matching we described by L:1, S:2, T:4, M:5, E:6, B:7 and F:8 thus becomes {{L, 1}, {S, 2}, {T, 4}, {M, 5}, {E, 6}, {B, 7}, {F, 8}}.

saturates a part X of a bipartite graph if there is some subset S of X such that the set $N(S)$ of the neighbors of S is smaller than S. We can summarize this discussion as follows.

Lemma 6.14 If we can find a subset S of a part X of a bipartite graph G such that $|N(S)| < |S|$, then there is no matching of G that saturates X.

Proof: A matching that saturates X must saturate S. But if there is such a matching, each element of S must be matched to a different vertex, and this vertex cannot be in S because $S \subseteq X$. Therefore, there are edges from vertices in S to at least $|S|$ different vertices not in S; so, $|N(S)| \geq |S|$, which is a contradiction. Thus, there is no such matching. ∎

Applying Lemma 6.14 yields a proof that there is no matching that saturates all the jobs in Exercise 6.4-2, which means the matching {{L, 4}, {S, 2}, {T, 7}, {M, 5}, {E, 6}, {B, 8}} is a maximum matching for the graph in Figure 6.28.

Another possible method for proving that there is no larger matching than the one we originally found is the following: When we matched L to 4, we may have noted that 4 is an endpoint of quite a few edges. Then, when we matched S to 2, we may have noted that S is an endpoint of quite a few edges, and so is T. In fact, 4, S, and T touch 12 edges of the graph, and there are only 23 edges in the graph. If we could find three more vertices that touch the remaining edges of the graph, then we would have six vertices, at least one of which is incident with every edge. A set of vertices such that at least one of them is incident with each edge of a graph G is called a **vertex cover of the edges of G**, or a **vertex cover of G** for short. What does this have to do with a matching? Each matching edge would have to touch one, or perhaps two, of the vertices in a vertex cover of the edges. Thus, the number of edges in a matching is always less than or equal to the number of vertices in a vertex cover of the edges of a graph. Therefore, if we can find a vertex cover of size 6 in Figure 6.28, then we will know that there is no matching that saturates the set of jobs because there are seven jobs. For future reference, we state our result about the size of a matching and the size of a vertex cover as a lemma.

Lemma 6.15 The size of a matching in a graph G is no more than the size of a vertex cover of G.

Proof: The proof is given in the preceding discussion. ∎

We have seen that because 4, S, and T cover more than half of the edges of the graph in Figure 6.28, they are good candidates for being members of a relatively small vertex cover of the graph. Continuing through the edges that we first examined, we see that 5, 6, and B are good candidates for a small vertex cover as well. In fact, {4, S, T, 5, 6, B} form a vertex cover. Because we have a vertex cover of size 6, we know a maximum matching has size no more than 6. Thus, the six-edge matching we already found is a maximum matching. Therefore, with the data in Table 6.2, it is not possible to fill all of the jobs.

Making Matchings Bigger

Practical problems involving matchings will usually lead us to search for the largest possible matching in a graph. To see how to use a matching to create a larger one, we will assume that we have two matchings of the same graph and see how they differ, especially how a larger one differs from a smaller one.

Exercise 6.4-3

In the graph G of Figure 6.27, let M_1 be the matching

$$\{\{L, 1\}, \{S, 2\}, \{T, 4\}, \{M, 5\}, \{E, 6\}, \{B, 9\}, \{F, 8\}\},$$

and let M_2 be the matching

$$\{\{L, 4\}, \{S, 2\}, \{T, 1\}, \{M, 6\}, \{E, 7\}, \{B, 8\}\}.$$

For sets S_1 and S_2, the **symmetric difference** of S_1 and S_2, denoted by $S_1 \triangle S_2$, is $(S_1 \cup S_2) - (S_1 \cap S_2)$. Compute the set $M_1 \triangle M_2$, and draw the graph with the same vertex set as G and edge set $M_1 \triangle M_2$. Use different colors or textures for the edges from M_1 and M_2 so that you can see their interaction. Describe as succinctly as possible the kinds of graphs you see as connected components.

Exercise 6.4-4

In Exercise 6.4-3, one of the connected components suggests a way to modify M_2 by removing one or more edges and substituting one or more edges from M_1 that will give you a larger matching M_2' related to M_2. In particular, this larger matching should saturate everything M_2 saturates and more. What is M_2', and what else does it saturate?

Exercise 6.4-5

Consider the matching $M = \{\{S, 1\}, \{T, 4\}, \{M, 6\}, \{B, 8\}\}$ in the graph of Figure 6.28. How does it relate to the path 3, S, 1, T, 4, M, 6, F? Say as much as you can about the set M' that you obtain from M by deleting the edges of M that are in the path and adding to the result the edges of the path that are not in M.

In Exercise 6.4-3,

$$M_1 \triangle M_2 = \{\{L, 1\}, \{L, 4\}, \{T, 4\}, \{T, 1\}, \{M, 5\}, \{M, 6\}, \{E, 6\}, \{E, 7\}, \\ \{B, 8\}, \{F, 8\}, \{B, 9\}\}.$$

The graph for the edge set $M_1 \triangle M_2$ is shown in Figure 6.29. The edges of M_2 are dashed. As you can see, the graph consists of a cycle with four edges alternating between edges of M_1 and M_2, a path with four edges alternating between edges of M_1 and M_2, and a path with three edges alternating between edges of M_1 and M_2. We call a path or cycle an **alternating path** or **alternating cycle** for a matching M of a graph

G if its edges alternate between edges in M and edges not in M. We call a path or cycle an alternating path or alternating cycle for M_1 and M_2 if its edges alternate between M_1 and M_2. Thus, our connected components are alternating paths and cycles for M_1 and M_2. The graph we drew in Figure 6.29 shows all the ways in which two matchings can differ, as summarized in the following lemma.

Figure 6.29

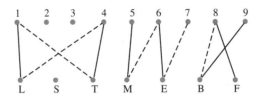

Lemma 6.16   **(Berge's Lemma)** If M_1 and M_2 are matchings of a graph $G = (V, E)$, then each connected component of $M_1 \triangle M_2$ is either a cycle with an even number of vertices or a path. Further, the cycles and paths are alternating cycles and paths for M_1 and M_2.

Proof: Figure 6.29 illustrates this proof. Each vertex of the graph $(V, M_1 \triangle M_2)$ has degree 0, 1, or 2. If a component has no cycles, then it is a tree, and the only kind of tree that has only vertices of degree 1 and 2 is a path. If a component has a cycle, then it cannot have any edges other than the edges of the cycle incident with its vertices, because the graph would then have a vertex of degree 3 or more. Thus, the component must be a cycle. If two edges of a path or cycle in $(V, M_1 \triangle M_2)$ share a vertex, then they cannot come from the same matching, because two edges in the same matching do not share a vertex. Therefore, alternating edges of a path or cycle of $(V, M_1 \triangle M_2)$ must come from different matchings. In particular, this implies that a cycle in the symmetric difference has an even number of vertices. ∎

Corollary 6.17   If M_1 and M_2 are matchings of a graph $G = (V, E)$ and if $|M_2| < |M_1|$, then there is an alternating path for M_1 and M_2 that starts and ends with vertices saturated by M_1 but not by M_2.

Proof: Because an even alternating cycle and an even alternating path in $(V, M_1 \triangle M_2)$ have equal numbers of edges from M_1 and M_2, then we have that at least one component must be an alternating path with more edges from M_1 than M_2, as in Figure 6.29 (where the component in question is $\{8, 9, B, F\}$). Otherwise $|M_2| \geq |M_1|$. Because this is a component of $(V, M_1 \triangle M_2)$, all of its edges must come from M_1 or M_2. Because the edges alternate between the two matchings, the only way for the path to have more edges from M_1 than M_2 is for it to have its endpoints lie only in edges of M_1, so they are saturated by M_1 but not by M_2. ∎

The path with three edges in Exercise 6.4-3 has two edges of M_1 and one edge of M_2. We see that if we remove {B, 8} from M_2 and add {B, 9} and {F, 8}, then we get the matching

$$M_2' = \{\{L, 4\}, \{S, 2\}, \{T, 1\}, \{M, 6\}, \{E, 7\}, \{B, 9\}, \{F, 8\}\} \, .$$

This answers the question of Exercise 6.4-4. Notice that this matching saturates everything M_2 does and also saturates vertices F and 9.

Figure 6.30: *The path and matching of Exercise 6.4-5*

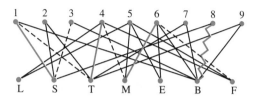

Figure 6.30 shows the matching edges of the path in Exercise 6.4-5 in blue and the nonmatching edges of the path as dashed. The edge of the matching not in the path is shown as zigzag. Notice that the dashed edges and the zigzag edge form a matching that is larger than M and that saturates all the vertices that M does, in addition to 3 and F. The path begins and ends with unmatched vertices for M, namely, 3 and F, and alternates between matching edges and nonmatching edges. All but the first and last vertices of such a path lie on matching edges of the path, and the endpoints of the path do not lie on matching edges. Thus, no edges of the matching that are not path edges will be incident with vertices on the path. We now delete all the matching edges of the path from M and add all the other edges of the path to M. This gives us a new matching, because by taking every second edge of a path, we get edges that do not have endpoints in common. An alternating path is called an **augmenting path** for a matching M if it begins and ends with M-unsaturated vertices. That is, it is an alternating path that begins and ends with unmatched vertices. Our preceding discussion suggests the proof of the following theorem.

| Theorem 6.18 | **(Berge's Theorem)** A matching M in a graph is of maximum size if and only if M has no augmenting path. Further, if a matching M has an augmenting path P with edge set $E(P)$, then we can create a larger matching by deleting the edges in $M \cap E(P)$ from M and adding the edges of $E(P) - M$. |

Proof: First, if there is a matching M_1 larger than M, then, by Corollary 6.17, there is an augmenting path for M. Thus, if a matching does not have maximum size, then it has an augmenting path. Further, as in our discussion of Exercise 6.4-5, if there is an augmenting path for M, then there is a larger matching than M. In particular, our discussion of that exercise showed that if P is an augmenting path, then we can get such a larger matching by deleting the edges in $M \cap E(P)$ and adding the edges of $E(P) - M$. ■

Although the larger matching of Theorem 6.18 may not contain M as a subset, it does saturate all the vertices that M saturates and two additional vertices.

Proof: Every vertex incident with an edge in M is also incident with some edge of the larger matching. Also, each of the two endpoints of the augmenting path is incident with a matching edge. Because we may have removed edges of M to get the larger matching, it may not contain M. ∎

Matching in Bipartite Graphs

Our examples and exercises have all been bipartite, yet all of our lemmas, corollaries, and theorems about matchings have been about general graphs. In fact, some of these results can be strengthened in bipartite graphs. For example, Lemma 6.15 tells us that the size of a matching is no more than the size of a vertex cover. We shall soon see that in a bipartite graph, the size of a maximum matching actually equals the size of a minimum vertex cover.

Searching for Augmenting Paths in Bipartite Graphs

We have seen that if we can find an augmenting path for a matching M in a graph G, then we can create a bigger matching. Because our goal from the outset has been to create the largest matching possible, this helps us achieve that goal. You may ask, however, how do we find an augmenting path? Recall that a breadth-first search tree centered at a vertex x in a graph contains a path—in fact, a shortest path—from x to every vertex y to which it is connected. Thus, it seems that if we could alternate between matching edges and nonmatching edges when doing a breadth-first search, then we would find alternating paths. In particular, if we add a vertex i to our tree by using a matching edge, then any edge we use to add a vertex *from* vertex i should be a nonmatching edge. And if we add a vertex i to our tree by using a nonmatching edge, then any edge we use to add a vertex *from* vertex i should be a matching edge. (Thus, there is at most one such edge.) Because not all edges are available for use in adding vertices to the tree, the tree we get will not necessarily be a spanning tree of our original graph. However, we can hope that if there is an augmenting path starting at vertex x and ending at vertex y, then we will find it by using breadth-first search starting from x in this alternating manner.

Given the matching {{S, 2}, {T, 4}, {B, 7}, {F, 8}} of the graph in Figure 6.27, use breadth-first search starting at vertex 1 in an alternating way to search for an augmenting path starting at vertex 1. Use the augmenting path that you get to create a larger matching.

Continue using the method of Exercise 6.4-6 until you find a matching of maximum size.

Apply breadth-first search from vertex 0 in an alternating way to Figure 6.31a. Does this method find an augmenting path? Is there an augmenting path?

Figure 6.31: *Matching edges are shown in blue.*

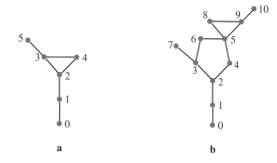

a   b

For Exercise 6.4-6, if we begin at vertex 1, then we add vertices L, S, and T to our queue and our tree. See Figure 6.32a, in which the blue lines, included the dashed one, are the edges of the matching and the dashed lines will be explained later. Vertex 1 is labeled T_0 to show that it is the first vertex in the tree, and L, S, and T are labeled T_1 to indicate that they entered the tree in this first stage. Because L is not incident with a matching edge, we cannot continue the search from there. Also, because S is incident with matching edge {S, 2}, we can use this edge to add vertex 2 to the queue and tree. This is the only vertex we can add from S because we can only use matching edges to add vertices from S. Similarly, from T we can add vertex 4 by using the matching edge {T, 4}. We marked vertices 2 and 4 with T_2 to indicate that they were added to the queue and tree at this stage. All vertices adjacent to vertex 2 have already been added to the queue and tree, but from vertex 4 we can use nonmatching edges to add vertices M and E to our queue and tree. We mark those vertices with T_3 to indicate that they were added to the tree at this stage. Now we can only use matching edges to add vertices to the queue and tree from M or E, but there are no matching edges incident with them, so our alternating search tree stops here. Because M and E are unmatched, we know that we have a path in our tree from vertex 1 to vertex M and a path from vertex 1 to vertex E. The vertex sequence of the path from 1 to M is 1, T, 4, M. The dashed edges in Figure 6.32a indicate the path. Our matching then becomes {{1, T}, {2, S}, {4, M}, {B, 7}, {F, 8}} (see Figure 6.32b, where the matching edges are blue).

Figure 6.32: *Illustrating the process of enlarging a matching*

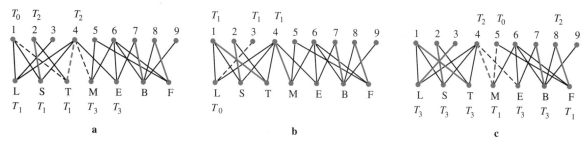

a   b   c

For Exercise 6.4-7, we find another unmatched vertex and repeat the search. For example, working from vertex L, we start a tree by using the edges {L, 1}, {L, 3}, and

{L, 4} to add vertices 1, 3, and 4 to our queue and tree. We could continue working on the tree, but because we see that L{L, 3}3 is an augmenting path, we use it to add the edge {L, 3} to the matching, thus short-circuiting the tree-construction process. Our matching becomes {{1, T}, {2, S}, {L, 3}, {4, M}, {B, 7}, {F, 8}} (see the blue edges in Figure 6.32c). The next unmatched vertex that we see might be vertex 5. Starting from there, we add M and F to our queue and tree. From M, we have the matching edge {M, 4}, and from F, we have the matching edge {F, 8}, so we use them to add the vertices 4 and 8 to the queue and tree. From vertex 4, we add L, S, T, and E to the queue and tree, and from vertex 8, we add vertex B to the queue and tree. All these vertices except E are in matching edges. Because E is in the tree but not incident with a matching edge, it is connected by an augmenting path to vertex 5. The path in the tree from vertex 5 to vertex E has vertex sequence 5, M, 4, E, shown in the tree as dashed. This augmenting path gives us the matching {{1, T}, {2, S}, {L, 3}, {5, M}, {4, E}, {B, 7}, {F, 8}}. You should be able to see this matching in Figure 6.32c; it consists of the two black dashed edges and the blue edges except for the blue dashed edge. Because we now have a matching whose size is the same as the size of a vertex cover, namely, the bottom part of the graph in Figure 6.27, we have a matching of maximum size.

For Exercise 6.4-8, we start at vertex 0 and add vertex 1. You may want to follow along in Figure 6.31, marking the graph in pencil as we did in our solution to Exercise 6.4-7. From vertex 1, we use our matching edge to add vertex 2. From vertex 2, we use our two nonmatching edges to add vertices 3 and 4. However, vertices 3 and 4 are incident with the same matching edge, so we cannot use that matching edge to add any vertices to the tree, and we must stop without finding an augmenting path. From staring at the picture, we see that there is an augmenting path, namely, 0, 1, 2, 4, 3, 5, that gives us the matching {{0, 1}, {2, 4}, {3, 5}}. We would have similar difficulties in discovering either of the augmenting paths in Figure 6.31b.

It turns out to be the odd cycles in Figure 6.31 that prevent us from finding augmenting paths by our modification of breadth-first search. We demonstrate this by describing an algorithm that is a variation on the alternating breadth-first search that we just used in solving our exercises. This algorithm takes a bipartite graph and a matching and either gives us an augmenting path or constructs a vertex cover whose size is the same as the size of the matching. A graph is bipartite if and only if it has no odd cycles (see Problems 13 and 14 for a proof); therefore, this algorithm will prove that a graph must have odd cycles in order to defeat our search strategy.

The Augmentation-Cover Algorithm

We begin with a bipartite graph with parts X and Y and a matching M. (In Figure 6.33, the matching edges are blue.) We label the unmatched vertices in X with a, which stands for alternating. (Figure 6.33a shows these labels, and more.) We number the vertices in sequence as we label them.[15] (Figure 6.33 shows these numbers as subscripts on a.) Starting with $i = 1$ and taking labeled vertices in the order of the numbers we have assigned to them, we use vertex i to do additional labeling as follows, stopping when we have labeled an unmatched vertex in Y or when it is impossible to continue

---

[15]The numbers we assign to the vertices tell when they would be put into a queue in this modified version of breadth-first search.

Figure 6.33: *The augmentation-cover algorithm*

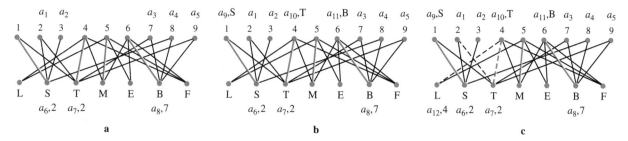

a

b

c

labeling (the first stopping condition is illustrated in Figure 6.33c, and the second in Figure 6.34).

1. If vertex i is in X, then we label all unlabeled vertices adjacent to it with the label a and the name of vertex i. Then we number these newly labeled vertices, continuing our sequence of numbers without interruption. (We show the first iteration of this stage in Figure 6.33a. We show the second iteration of this step (and more) in Figure 6.33c.)

2. If vertex i is in Y and it is incident with an edge of M, then its neighbor in the matching edge cannot yet be labeled. (Matched vertices in X can only be labeled in this step, and because M is a matching, each vertex can be labeled at most once.) We label this neighbor with the label a and the name of vertex i. (We show the first iteration of this stage in Figure 6.33b. We show the second iteration (and more) in Figure 6.33c.)

Figure 6.34: *We can't augment this matching because {4, 5, 6, S, T, B} is a vertex cover.*

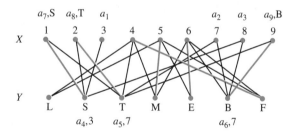

If vertex i is labeled and in Y and it is not incident with an edge of M, then we have discovered an augmenting path. This path starts at vertex i, then goes to the vertex we used to add it (and recorded at vertex i), and so on, back to one of the unlabeled vertices in X. The path is alternating by our labeling method, and it starts and ends with unsaturated vertices, so it is augmenting. (In the case of Figure 6.33, the labeled vertex L in Y is not in a matching edge. In Figure 6.33c, we show the path starting at L as dashed.)

If we continue the labeling process until no more labeling is possible and we do not find an augmenting path, then we let A be the set of labeled vertices. As we shall prove shortly, the set $C = (X - A) \cup (Y \cap A)$ turns out to be a vertex cover whose size is the size of M. (This second case is illustrated in Figure 6.34, in which the blue edges are a matching and the set {4, 5, 6, S, T, B} turns out to be a minimum vertex cover.) We call this algorithm the **augmentation-cover algorithm**.

We now develop pseudocode for the augmentation-cover algorithm. It has four input parameters: the two parts of V called X and Y, an edge set E (each of whose edges connects a vertex in X to a vertex in Y), and a matching M. It also has two output parameters. The first is a set P, which is the edge set of an augmenting path for M if one exists and the empty set otherwise. The second is a set C, which is a minimum vertex cover in the event that there is no augmenting path and which is empty otherwise. Putting a vertex onto the queue is equivalent to labeling it with a and assigning a number to it. The number we assign to it, then, is its position on the queue. Thus, taking the vertices in the order of their numbers is the same as taking them in the order of the queue. The vertex name that we use to label vertex x when we are labeling by hand corresponds to `Pred[x]` in the pseudocode.

In the pseudocode, we assume that if x is a vertex, then it is possible to use x as a subscript of an array. Thus, as in breadth-first search, we assume that the names of our v vertices are the integers 1 through v. (As we pointed out for breadth-first search, changing this assumption is not difficult but involves details we choose not to go into.)

The pseudocode assumes we have a procedure `IsSaturated`. `IsSaturated(x,M)` returns `true` if and only if the vertex x is saturated by some edge in the matching M. This can naively be implemented in $O(|M|)$ time by running through the endpoints of the edges in M and seeing if x is one of them. But we can be more clever. We can preprocess M by creating a Boolean array (an array of `true`s and `false`s) `saturated` of size v such that `saturated[x]` is `true` if and only if an edge in M saturates the vertex x. After $O(v)$ preprocessing time, a call to `IsSaturated` would then take constant time: `IsSaturated(x)` simply looks in the array `saturated` to see if `saturated[x]` is `true`. Then, v calls to `IsSaturated` would take $O(v)$ time rather than $O(v^2)$ time.

Note that the algorithm does not need to keep track of the subscripts on the vertices added to A. These subscripts were used so that we would process the vertices in A in the order that they were added to the queue. The subscripts amount to a "by hand" implementation of a queue.

```
Augmentation-Cover(X, Y, E, M, P, C)

// Assume that V = X ∪ Y contains vertices numbered 1, 2,..., v.
// Assume that E is an array with v entries, and entry i of E is a
// list of the vertices adjacent to vertex i.
// Assume that M is a set of edges in a matching.
// If the graph has an augmenting path, then when the algorithm
// returns, P will contain the edges of an augmenting path and C will
// be empty. If there is no augmenting path, then P will be empty and
// C will contain a vertex cover.
// Assume Q is a queue that will be initialized to an empty queue.
// When the algorithm returns, A will consist of the vertices added
// to Q during the course of the algorithm.
// If an alternating path is found, then Pred(x) will precede x on
// that path if Pred(x) ≠ 0.
(1)   InA = an array of length v with each entry initialized to "false"
(2)   Pred = an array of length v with each entry initialized to 0
```

```
(3)    P = ∅
(4)    A = ∅
(5)    Q = ∅
(6)    C = ∅
(7)  For each element x of X
(8)        If (!IsSaturated(x,M))
(9)              Enqueue x onto Q
(10)             A = A ∪ {x}
(11)             InA[x] = "true"
(12) While there is at least one vertex in Q
(13)       Dequeue z from Q
(14)       If (z ∈ X)
(15)             For each vertex w in the list E[z]
(16)                  If (InA[w] == "false")
(17)                        Enqueue w onto Q
(18)                        A = A ∪ {w}
(19)                        InA[w] = "true"
(20)                        Pred[w] = z // Remember which vertex we came from.
(21)       Else If (IsSaturated(z, M))  // z must be in Y because this "Else"
                                        // corresponds to the "if" in Line 14.
(22)             x = z's neighbor in M
(23)             Enqueue x onto Q
(24)             A = A ∪ {x}
(25)             InA[x] = "true"
(26)             Pred[x] = z // Remember which vertex we came from.
(27)       Else // Have discovered augmenting path
(28)             While (Pred[w] ≠ 0) // Trace back the path
(29)                  P = P ∪ {{w,Pred[w]}}
(30)                  w = Pred[w]
(31)             Return
(32) C = (X − A) ∪ (Y ∩ A)
(33) Return
```

We can use Augmentation-Cover in the algorithm FindMaximumMatching, which finds a maximum matching in a bipartite graph. This procedure takes as input parameters the two parts X and Y of a bipartite graph and its edge set E. It also has two output parameters, the maximum matching M and a vertex cover C of the same size as the maximum matching. As before, we assume that the vertices in $V = X \cup Y$ are the integers 1 to v.

```
FindMaximumMatching(X, Y, E, M, C)

// Assume that V = X ∪ Y contains vertices numbered 1, 2,..., v.
// Assume that E is an array with v entries, and entry i of E is
// a list of the vertices adjacent to vertex i.
// M will contain the edges in a maximum matching when the algorithm
// returns.
// C will contain the vertices of a vertex cover of size |M| when the
// algorithm returns.
```

```
(1)    M = Ø
(2)    Augmentation-Cover(X, Y, E, M, P, C)
(3)    While (P ≠ Ø)
(4)        M = (M − P) ∪ (P − M)
(5)        Augmentation-Cover(X, Y, E, M, P, C)
(6)    Print "The edges of a maximum matching are:" M"."
(7)    Print "A minimum vertex cover is:" C"."
```

Theorem 6.20 **(König-Egerváry Theorem)** In a bipartite graph with parts X and Y, the size of a maximum-sized matching equals the size of a minimum-sized vertex cover.

Proof: By Theorem 6.18 (Berge's theorem), if the augmentation-cover algorithm gives us an augmenting path, then the matching is not maximum sized. By Lemma 6.15, if we can prove that when there is no augmenting path, the set C that the algorithm gives us is a vertex cover whose size is the size of the matching, then we will have proved the theorem. To see that C is a vertex cover, note that every edge incident with a vertex in $X \cap A$ is covered, because its endpoint in Y has been marked with an a (that is, placed in A in Augmentation-Cover); thus, it is in $Y \cap A$. But every other edge must have one vertex in X, so it must be covered by $X − A$. Therefore, C is a vertex cover. If an element of $Y \cap A$ were not matched, it would be an endpoint of an augmenting path, and so all elements of $Y \cap A$ are incident with matching edges. But every vertex of $X − A$ is matched, because A includes all unmatched vertices of X. By step 2 of the augmentation-cover algorithm, which is Lines 21–25 of the pseudocode for Augmentation-Cover, if ϵ is a matching edge with an endpoint in $Y \cap A$, then the other endpoint must be in A. Thus, each matching edge contains only one member of C. Therefore, the size of a maximum matching is the size of C. ∎

Corollary 6.21 When Augmentation-Cover is applied to a bipartite graph and a matching of that graph, it returns either an augmenting path for the matching or a minimum vertex cover whose size equals the size of the matching.

Before we proved the König-Egerváry theorem, we knew that if we could find a matching and a vertex cover of the same size, then we had a maximum-sized matching and a minimum-sized vertex cover. However, in some graphs, we might not be able to test whether a matching is as large as possible by comparing its size with that of a vertex cover, because a maximum-sized matching might be smaller than a minimum-sized vertex cover. The König-Egerváry theorem tells us that in bipartite graphs, this problem never arises, so the test always works for bipartite graphs.

In Exercise 6.4-2, we used a second technique to show that a matching could not saturate the set X of all jobs. In Lemma 6.14, we showed that if we can find a subset S of a part X of a bipartite graph G such that $|N(S)| < |S|$, then there is no matching of G that saturates X. In other words, to have a matching that saturates X in a bipartite graph on parts X and Y, it is necessary that $|N(S)| \geq |S|$ for *every* subset S of X. (When

$S = \emptyset$, then so does $N(S)$.) This necessary condition is called **Hall's condition**, and Hall's theorem says that this necessary condition is sufficient for bipartite graphs.

| Theorem 6.22 | **(Hall's Theorem)** If G is a bipartite graph with parts X and Y, then there is a matching of G that saturates X if and only if $|N(S)| \geq |S|$ for every $S \subseteq X$. |

Proof: In Lemma 6.14, we showed (the contrapositive of the statement) that if there is a matching of G, then $|N(S)| \geq |S|$ for every subset of X. There is no reason to use a contrapositive argument, though; if there is a matching that saturates X, then, because matching edges have no endpoints in common, the elements of each subset S of X will be matched to at least $|S|$ different elements, and these will all be in $N(S)$.

Thus, we need only show that if the graph satisfies Hall's condition, then there is a matching that saturates X. We will do this by showing that X is a minimum-sized vertex cover. Let C be some vertex cover of G. Let $S = X - C$. If ϵ is an edge from a vertex in S to a vertex $y \in Y$, then ϵ cannot be covered by a vertex in $C \cap X$. Therefore, ϵ must be covered by a vertex in $C \cap Y$. This means that $N(S) \subseteq C \cap Y$, so $|C \cap Y| \geq |N(S)|$. By Hall's condition, $|N(S)| \geq |S|$. Therefore, $|C \cap Y| \geq |S|$. Because $C \cap X$ and $C \cap Y$ are disjoint sets whose union is C, we can summarize our remarks with the equation

$$\begin{aligned} |C| &= |C \cap X| + |C \cap Y| \\ &\geq |C \cap X| + |N(S)| \\ &\geq |C \cap X| + |S| \\ &= |C \cap X| + |X - C| \\ &= |X| \,. \end{aligned}$$

We have that X is a vertex cover, and we have just shown that it is a vertex cover of minimum size. Therefore, a matching of maximum size has size $|X|$. Thus, there is a matching that saturates X. ∎

Efficient Algorithms

Although Hall's theorem is quite elegant, applying it requires us to look at every subset of X, which would take us $\Omega\left(2^{|X|}\right)$ time. Similarly, actually finding a minimum vertex cover could involve looking at all (or nearly all) subsets of $X \cup Y$, which would also take us exponential time. However, the augmentation-cover algorithm requires that we examine each edge at most some fixed number of times and then do a little extra work; certainly, no more than $O(e)$ work. We need to repeat the algorithm at most $|X|$ times to find a maximum matching and minimum vertex cover. Thus, in time $O(ev)$, not only can we find out whether we have a matching that saturates X, but we can also find such a matching if it exists and a vertex cover that proves it doesn't exist if it doesn't. However, this algorithm only applies to bipartite graphs. The situation is much more complicated in nonbipartite graphs. In a paper that introduced the idea that an efficient algorithm is one that runs in time $O(n^c)$, where n is the amount of information needed to specify the input and c is a constant, Jack Edmonds[16] developed a more complicated algorithm that extended the idea of a search tree to a more complicated structure, which he called a *flower*. He showed that this algorithm was good in his sense of the word. In a wry twist of fate, the problem of finding a minimum vertex cover (actually, the problem of determining whether there is a vertex cover of size k, where k can be a function of v) is, in fact, **NP**-complete in arbitrary graphs. It is fascinating that the matching problem for general graphs turned out to be solvable in polynomial time, while determining the "natural" upper bound on the size of a matching, an upper bound that originally seemed quite useful, remains out of our reach.

Important Concepts, Formulas, and Theorems

1. *Matching.* A set of edges in a graph that share no endpoints is called a *matching* of the graph.

2. *Saturate.* A matching is said to *saturate* a set X of vertices if every vertex in X is matched.

3. *Maximum matching.* A matching in a graph is a *maximum matching* if it is at least as big as any other matching.

4. *Bipartite graph.* A graph is called *bipartite* whenever its vertex set can be partitioned into two sets X and Y so that each edge connects a vertex in X with a vertex in Y. Each of the two sets is called a *part* of the graph.

5. *Independent set.* A subset of the vertex set of a graph is called *independent* if no two of its vertices are connected by an edge. (In particular, a vertex connected to itself by a loop is not an independent set.) A part of a bipartite graph is an example of an independent set.

6. *Neighborhood.* We call the set $N(S)$ of all vertices that are adjacent to at least one vertex of S the *neighborhood* of S or the *neighbors* of S.

7. *Hall's theorem for a matching in a bipartite graph.* If we can find a subset S of a part X of a bipartite graph G such that $|N(S)| < |S|$, then there is no matching of G that saturates X. If there is no subset $S \subseteq X$ such that $|N(S)| < |S|$, then there is a matching that saturates X.

8. *Vertex cover.* A set of vertices such that at least one of them is incident with each edge of a graph G is called a *vertex cover of the edges* of G, or a *vertex cover* of G for short. In any graph, the size of a matching is less than or equal to the size of any vertex cover.

9. *Alternating path/Augmenting path.* A path is called an *alternating path* for a matching M if, as we move along the path, the edges alternate between edges in M and edges not in M. An *augmenting path* is an alternating path that begins and ends at unmatched vertices. An alternating path for M_1 and M_2 is a path whose edges alternate between edges in M_1 and edges in M_2.

10. *Alternating cycle.* A cycle is called an *alternating cycle* for a matching M if, as we move along the cycle, the edges alternate between edges in M and edges

not in M. A cycle is an alternating cycle for M_1 and M_2 if, as we move along the cycle, the edges alternate between edges in M_1 and edges in M_2.

11. *Berge's lemma.* If M_1 and M_2 are matchings of a graph G, then the connected components of $M_1 \triangle M_2$ are cycles with an even number of vertices and paths. Further, the cycles and paths are alternating cycles and paths for M_1 and M_2.

12. *Berge's corollary.* If M_1 and M_2 are matchings of a graph $G = (V, E)$ and $|M_1| > |M_2|$, then there is an alternating path for M_1 and M_2 that starts and ends with vertices saturated by M_1 but not by M_2.

13. *Berge's theorem.* A matching M in a graph is of maximum size if and only if M has no augmenting path. Further, if a matching M has an augmenting path P with edge set $E(P)$, then we can create a larger matching by deleting the edges in $M \cap E(P)$ from M and adding in the edges of $E(P) - M$.

14. *Augmentation-cover algorithm.* The augmentation-cover algorithm begins with a bipartite graph and a matching of that graph and produces either an augmenting path or a vertex cover whose size equals that of the matching, thus proving that the matching is a maximum matching.

15. *König-Egerváry theorem.* In a bipartite graph with parts X and Y, the size of a maximum-sized matching equals the size of a minimum-sized vertex cover.

Problems

1. In Figure 6.35, find either a maximum matching or a subset S of the set $X = \{a, b, c, d, e\}$ such that $|S| > |N(S)|$. a/h

Figure 6.35: *A bipartite graph*

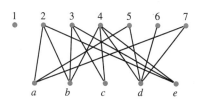

2. Find a maximum matching and a minimum vertex cover in Figure 6.35.

3. In Figure 6.36, find either a matching that saturates the set $X = \{a, b, c, d, e, f\}$ or a set S such that $|N(S)| < |S|$. a/h

Figure 6.36: *A bipartite graph*

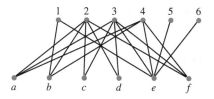

4. Find a maximum matching and a minimum vertex cover in Figure 6.36.

5. In Problems 1–4, when you were able to find a set S with $|S| > |N(S)|$, how did $N(S)$ relate to the vertex cover? Why did this work out as it did? a/h

6. A star is a another name for a tree with one vertex connected to each of n other vertices. (So a star has $n + 1$ vertices.) What are the size of a maximum matching and a minimum vertex cover in a star with $n + 1$ vertices?

7. In Theorem 6.18, is it true that if there is an augmenting path P with edge set $E(P)$ for a matching M, then $M \triangle E(P)$ is a larger matching than M? a/h

8. Find a maximum matching and a minimum vertex cover in Figure 6.31b.

9. In a bipartite graph, is one of the parts always a maximum-sized independent set? What if the graph is connected? a/h

10. Find infinitely many examples of graphs in which a maximum-sized matching is smaller than a minimum-sized vertex cover.

11. Find an example of a graph in which the maximum size of a matching is at least 3 and is half of the size of a minimum vertex cover. a/h

12. Prove or give a counterexample: Every tree is a bipartite graph. (*Note:* A single vertex with no edges is a bipartite graph; one of the two parts is empty.)

13. Prove or give a counterexample: A bipartite graph has no odd cycles. a/h

14. Let G be a connected graph with no odd cycles. Let x be a vertex of G. Let X be all vertices at an even distance from x, and let Y be all vertices at an odd distance from x. Prove that G is bipartite with parts X and Y.

15. What is the sum of the maximum size of an independent set and the minimum size of a vertex cover in a graph G? (*Hint:* It is useful to think both about the independent set and its complement relative to the vertex set.) a/h

6.5 COLORING AND PLANARITY

The Idea of Coloring

Graph coloring was one of the original problems in graph theory. Coloring arose from a question from Francis Guthrie, who noticed that four colors were enough to color the map of the counties of England so that if two counties shared a common boundary line, they received different colors. Guthrie wondered whether this was the case for all maps. His brother Fredrick Guthrie passed this question to Augustus DeMorgan, which is how it seeped into the consciousness of the mathematical community. By thinking of the counties as vertices and drawing an edge between two vertices if their counties share some boundary line, we get a representation of the problem that is independent of such things as the shape of the counties, the amount of boundary line they share, and so on. This representation captures the part of the problem on which we need to focus. We now color the vertices of the graph. For Guthrie's problem, we want to color in such a way that adjacent vertices get different colors. We will return to this problem later in the section; for now, we begin our study with another application of coloring.

The executive committee of the board of trustees of a small college has eight members: Kim, Smith, Jones, Gupta, Ramirez, Wang, Harper, and Chernov. There are six subcommittees with the following membership:

- Investments: K, J, H

- Operations: K, W, G

- Academic affairs: W, S, G

- Development (fund raising): W, C, K

- Budget: S, R, C

- Enrollment: R, C, J, H

Each time the executive committee has a meeting, the following occurs: each subcommittee meets with appropriate college officers, and then the executive committee gets together as a whole to go over subcommittee recommendations and to make decisions. Two subcommittees cannot meet at the same time if they have a member in common, but subcommittees that don't have a member in common can meet at the same time. What is the minimum number of time slots needed to schedule all the subcommittee meetings? Draw a graph in which the vertices are named by the initials of the subcommittee names and in which two vertices are adjacent if their subcommittees have a member in common. Then label the vertices with numbers in such a way that two adjacent vertices get different labels. The numbers represent time slots, so they need not be distinct unless they are on adjacent vertices. What is the minimum possible number of labels you need?

Because map coloring motivated much of graph theory, it is traditional to refer to the process of assigning labels to a graph's vertices as coloring the graph. An assignment of labels to vertices, which is a function from the vertices to the set of labels, is called a **coloring**. The set of possible labels, which is the range of the coloring function, is often referred to as a set of **colors**. Thus, Exercise 6.5-1 asks for a coloring of the graph. However, as with the map problem, the adjacent vertices should have different colors in our coloring. A coloring of a graph is called a **proper coloring** if it assigns different colors to adjacent vertices.

Figure 6.37 shows the graph of Exercise 6.5-1. We call this kind of graph an **intersection graph**, which means that its vertices correspond to sets and that it has an edge between two vertices if and only if the corresponding sets intersect.

Figure 6.37: *The intersection graph of the committees*

The exercise asks us to color the graph with as few colors as possible, regarding the colors as 1, 2, 3, and so on. We represent 1 as a white vertex, 2 as a blue vertex, 3 as a gray vertex, and 4 as a black vertex. The triangle at the bottom of the figure requires three colors simply because all three vertices are adjacent. Because it doesn't matter which three colors we use, our choices of white, blue, and gray are arbitrary. We know that we need at least three colors to color the graph, so it makes sense to try to finish off a coloring using just three colors. Vertex I must be colored differently from E and D; if we use the same three colors, vertex I must have the same color as B. Similarly, vertex A would have to be the same color as E if we use the same three colors. But now none of the colors can be used on vertex O because it is adjacent to three vertices of different colors. Thus, we need at least four colors. We show a proper four-coloring in Figure 6.38.

Figure 6.38: *A proper coloring of the committee intersection graph*

Exercise 6.5-2

How many colors are needed to give a proper coloring of the complete graph K_n?

Exercise 6.5-3

How many colors are needed for a proper coloring of a cycle C_n on $n = 3, 4, 5$, and 6 vertices?

In Exercise 6.5-2, we need n colors to color K_n properly, because each pair of vertices is adjacent and thus must have two different colors. In Exercise 6.5-3, if n is even, we can simply alternate two colors as we go around the cycle. However, if n is odd, using two colors would require that they alternate as we go around the cycle, and when we colored our last vertex, it would be the same color as the first. Thus, we need at least three colors. By alternating two colors as we go around the cycle until we get to the last vertex and coloring it the third color, we get a proper coloring with three colors.

The **chromatic number** of a graph G, traditionally denoted $\chi(G)$, is the minimum number of colors needed to color G properly. Thus, we have shown that the chromatic number of the complete graph K_n is n, the chromatic number of a cycle on an even number of vertices is 2, and the chromatic number of a cycle on an odd number of vertices is 3. We have also shown that the chromatic number of our committee graph is 4.

From Exercise 6.5-2, we see that if a graph G has a subgraph that is a complete graph on n vertices, then we need at least n colors to color those vertices. Thus, we need at least n colors to color G. This is useful enough that we will state it as a lemma.

Lemma 6.23	If a graph G contains a subgraph that is a complete graph on n vertices, then the chromatic number of G is at least n.

Proof: The proof for this lemma is given immediately before the statement. ∎

More generally, if G contains a subgraph that requires at least n colors in a proper coloring, then G itself has chromatic number at least n.

Interval Graphs

An interesting application of coloring arises in the design of optimizing compilers for computer languages. In addition to the usual random access memory (RAM), a computer typically has some memory locations called *registers,* which can be accessed at very high speeds. Thus, values of variables that are going to be used again in the program are kept in registers, if possible, so that they will be quickly available when needed. An optimizing compiler will attempt to decide the time interval in which a given variable may be used during a run of a program and arrange for that variable to be stored in a register for that entire interval of time. Although the time interval is not determined in absolute terms of seconds, the relative endpoints of the intervals can be determined according to when variables first appear and last appear as one steps through the computer code. This is the information needed to set aside registers to use for the variables. We can formulate the problem of assigning variables to registers as a coloring problem. To do so, we draw a graph in which the vertices are labeled with the variable names, and associated to each variable is the interval during which it is used. Two variables can use the same register if they are needed during nonoverlapping time intervals. We can think of our graph on the variables as the intersection graph of the intervals, which means there will be an edge between two vertices (variables) whose time intervals overlap. We want to color the graph properly with a minimum number of registers; we hope that this will be no more than the number of registers that our computer has available. (If it is more than the number of registers, then some of our variables will not be able to fit into registers. This is why we want to use the minimum number of colors.) The problem of assigning variables to registers is called the **register assignment problem**.

An intersection graph of a set of intervals of real numbers is called an **interval graph**. The assignment of intervals to the vertices is called an **interval representation**. Notice that so far in our discussion of coloring, we have not given an algorithm for properly coloring a graph efficiently. This is because the problem of whether a graph has a proper coloring with k colors for any fixed k greater than 2 is another example of an **NP**-complete problem. However, for interval graphs, there is a very simple algorithm for properly coloring the graph in a minimum number of colors.

Exercise 6.5-4	Consider the closed intervals $[1, 4]$, $[2, 5]$, $[3, 8]$, $[5, 12]$, $[6, 12]$, $[7, 14]$, and $[13, 14]$. Draw the interval graph determined by these intervals and find its chromatic number.

Figure 6.39: *The graph of Exercise 6.5-4*

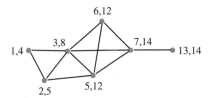

The graph of Exercise 6.5-4 is shown in Figure 6.39. (To avoid cluttering the figure, the graph does not include the square braces around each closed interval.) Because of the way we have drawn this graph, it is easy to see a subgraph that is a complete graph on four vertices. So we know by Lemma 6.23 that the graph has chromatic number at least 4. In fact, Figure 6.40 shows that the chromatic number is exactly 4. This is no accident.

Figure 6.40: *A proper coloring of the graph of Exercise 6.5-4 with four colors*

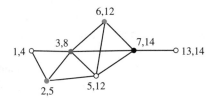

Theorem 6.24   In an interval graph G, the chromatic number is the size of the largest complete subgraph.

Proof: List the intervals of an interval representation of the graph in order of their left endpoints. Color the intervals with the integers 1 through some number n by starting with 1 on the first interval in the list and, for each succeeding interval, using the smallest color not used on any neighbor of the interval earlier in the list. This will clearly give a proper coloring. To see that the number of colors needed is the size of the largest complete subgraph, let n denote the largest color used, and choose an interval I colored with color n. Then, by our coloring algorithm, I must intersect with earlier intervals in the list colored 1 through $n - 1$; otherwise, we could have used a smaller color on I. All of these intervals must contain the left endpoint of I because they intersect I and come earlier in the list. Because they all have a point in common, they form a complete graph on n vertices. Therefore, the minimum number of colors used by *this* coloring algorithm is the size of a complete subgraph of G.

But, by Lemma 6.23, if G contains a complete subgraph on n vertices, then its chromatic number is at least n. Thus, the chromatic number of an interval graph G is the size of the largest complete subgraph of G. ∎

| Corollary 6.25 | An interval graph G may be properly colored, using $\chi(G)$ consecutive integers as colors, by listing the intervals of a representation in order of their left endpoints and going through the list, assigning the smallest color not used on an earlier adjacent interval to each interval in the list. |

Proof: This is the coloring algorithm we used in the proof of Theorem 6.24. ∎

Notice that with the correspondence between numbers and colors that we used before, the coloring in Figure 6.40 is the one given by this algorithm. An algorithm that colors an arbitrary graph G with consecutive integers by listing the graph's vertices in some order, coloring the first vertex in the list 1, and then coloring each succeeding vertex with the least number not used on any adjacent vertices earlier in the list is called a **greedy coloring algorithm**. We have just seen that the greedy coloring algorithm allows us to find the chromatic number of an interval graph. This algorithm takes time $O(n^2)$, because as we go through the list, we might consider every earlier entry when we are considering a given element of the list. It is a good thing that we have a polynomial-time algorithm, because even though we stated in Theorem 6.24 that the chromatic number is the size of the largest complete subgraph, determining whether the size of a largest complete subgraph in a general graph (as opposed to an interval graph) is k (where k may be a function of the number of vertices) is an **NP**-complete problem.

Of course, this assumes that we were given an interval representation of our graph. Suppose we are given a graph that happens to be an interval graph, but we don't know an interval representation. Can we still color the graph quickly? It turns out that there is a polynomial-time algorithm for determining whether a graph is an interval graph and finding an interval representation. This theory is quite beautiful,[16] but it would take us too far afield to pursue it now.

Planarity

We began our discussion of coloring with the map coloring problem. This problem has a special aspect that we did not mention. A map is either drawn on a piece of paper, which is a plane, or on a globe, which is the surface of a sphere. By thinking of the sphere as a completely elastic balloon, we can imagine puncturing it with a pin where nothing is drawn, opening the pinhole a bit by stretching the balloon, and then continuing to stretch the pinhole until we have the surface of the balloon laid out flat on a table. This means that we can think of all maps as drawn in the plane. What does this mean about the graphs we associated with the maps? Let's say, to be specific, that we are talking about the counties of England. In each county, we take an important town in a county and imagine building a road to the boundary of each county with which our first county shares some boundary line (not just a point). These roads, which we can build so that they don't cross each other, are built to the center of the boundary line between two different counties so that the roads join together at that boundary line.

---

[16]See, for example, Golumbic[18].

The towns we choose in each county are the vertices of a graph representing the map, and the roads are the edges. Thus, given a map drawn in the plane, we can draw a graph to represent it in such a way that the edges of the graph do not meet at any point except at their endpoints.[17] A graph is called **planar** if it has a drawing in the plane such that the edges do not meet except at their endpoints. Such a drawing is called a **planar drawing** of the graph. The famous four color problem asked whether all planar graphs have proper four colorings. In 1976, Kenneth Appel and Wolfgang Haken[3], building on some of the early attempts at proving the theorem, used a computer to demonstrate that four colors are sufficient to color any planar graph. Although we do not have time to indicate how their proof went, there is now a book on the subject by Robin Wilson that gives a careful history of the problem, an explanation of what the computer was asked to do, and why, assuming that the computer was correctly programmed, that led to a proof[34]. What we will do here is derive enough information about planar graphs to show that five colors suffice to color a planar graph, as well as give some background on planarity relevant to the design of computer chips.

We start out with two problems that aren't quite realistic but that are suggestive of how planarity enters chip design.

| Exercise 6.5-5 | A circuit is to be laid out on a computer chip in a single layer. The design includes five terminals (think of them as points to which multiple electrical circuits may be connected) that need to be directly connected so that a current can go from any one terminal to any other without sending current to a third terminal. The connections are made with a narrow layer of metal deposited on the surface of the chip, which we will think of as a wire on the surface of the chip. Thus, if one connection crosses another, current in one wire will flow through the other as well. Therefore, the chip must be designed so that each of the $\binom{5}{2}$ pairs of terminals is connected directly by a wire, and no two of these wires cross. Do you think this is possible? |

| Exercise 6.5-6 | As in Exercise 6.5-5, we are laying out a computer circuit. However, we now have six terminals, labeled a, b, c, 1, 2, and 3, such that each of a, b, and c must be connected to each of 1, 2, and 3, but there must be no other connections. As before, the wires cannot touch each other, so we need to design this chip so that no two wires cross. Do you think this is possible? |

The answer to both of these exercises is that it is not possible to design such a chip. One can make compelling geometric arguments to explain why it is not possible, but those arguments require that we simultaneously visualize a large variety of configurations with one picture. Instead, we develop a few equations and inequalities relating to planar graphs that will allow us to give convincing arguments that both these designs are impossible.

---

[17]We are temporarily ignoring a small geographic feature of counties that we will mention when we have the terminology to describe it.

The Faces of a Planar Drawing

If we assume that our graphs are finite, then it is easy to believe that we can draw any edge of a graph as a broken line segment (i.e., a bunch of line segments connected at their ends), such as the edge from f to g in Figure 6.41, rather than a smooth curve. In this way, a cycle in our graph determines a polygon in our drawing. (Typical cycles appear in Figure 6.41.) This polygon may have some of the graph drawn inside it and some of the graph drawn outside it. We say a subset of the plane is **geometrically connected** if between any two points of the region, we can draw a curve without leaving the region.[18] (In our context, you may assume that this curve is a broken line segment, though a careful study of geometric connectivity in general situations is less straightforward.) If we remove all the vertices and edges of the graph from the plane, then we are likely to break the plane into a number of geometrically connected sets. Such a connected set is called a **face** of the drawing.[19] For example, in Figure 6.41,

Figure 6.41: *A typical graph and its faces*

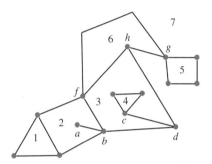

the faces are marked 1 (a triangular face), 2 (a quadrilateral face that has a line segment and point removed for the edge $\{a, b\}$ and the vertex a), 3 (another quadrilateral that has not only a line but also a triangle removed from it), 4 (a triangular face), 5 (a quadrilateral face), 6 (a unique face that is concretely a pentagon but, because the edge from f to h is a broken line segment, is abstractly a triangle since it has three edges on its boundary), and 7 (another unique face whose boundary is concretely a decagon connected at a point to a quadrilateral). Face 7 is called the outside face of the drawing and is the only face with infinite area. Each planar drawing of a graph will have an **outside face**—a face of infinite area in which we can draw a circle that encloses the entire graph. (Remember, we are thinking of our graphs as finite at this point.) Each edge either lies between two faces or has the same face on both of its sides. The edges $\{a, b\}$ and $\{c, d\}$ are the edges of the latter type. Thus, if an edge lies on a cycle, it must divide two faces; otherwise, removing that edge would increase the number of connected components of the graph. An edge whose removal increases the number of

---

[18]The usual thing to say is that it is connected, but we want to distinguish this kind of connectivity from graphical connectivity. (In a more advanced study, we would see how these two apparently different uses of the word "connected" are different aspects of the same idea.) The fine point about counties that we didn't point out earlier is that they are geometrically connected. If they were not, then the graph with a vertex for each county and an edge between two counties that share some boundary line would not necessarily be planar.

[19]More precisely, a connected set in the plane with the vertices and edges removed is a face if it is not a proper subset of any other connected set in the plane with the vertices and edges removed.

---

connected components is called a **cut edge** and cannot lie between two distinct faces. It is straightforward to show that any edge that is not a cut edge lies on a cycle. If an edge lies on only one face, it is a cut edge. To see why, note that we can draw a broken line segment within the face from one side of the edge to the other. See Figure 6.42, in which the broken line segment is shown as dashed. This broken line segment, plus part of the edge, forms a closed curve that encloses part of the graph. Thus, removing the edge disconnects the enclosed part of the graph from the rest of the graph.

Figure 6.42: *A broken line connecting one side of an edge to the other side*

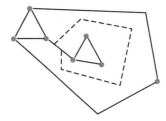

Exercise 6.5-7 Draw some connected planar graphs with at least three faces, and experiment to see if you can find a numerical relationship among v, the number of vertices; e, the number of edges; and f, the number of faces. Check your relationship on the graph in Figure 6.41.

Exercise 6.5-8 In a simple graph, every face has at least three edges. This means that the number of pairs consisting of a face and an edge bordering that face is at least $3f$. Use the fact that an edge borders either one or two faces to get an inequality that relates the number of edges and the number of faces in a connected simple planar graph.

Some playing with planar drawings usually convinces people fairly quickly of the following theorem.

Theorem 6.26 **(Euler's Formula)** In a planar drawing of a connected graph G with v vertices, e edges, and f faces,

$$v - e + f = 2 \, .$$

Proof: We induct on the number of cycles of G. If G has no cycles, then it is a tree, and a tree has one face because all of its edges are cut edges. Then, for a tree, we get $v - e + f = v - (v - 1) + 1 = 2$. Now suppose that G has $n > 0$ cycles. Choose an edge that is between two faces, so it is part of a cycle. Deleting that edge joins the two faces that it was on, so the new graph has $f' = f - 1$ faces. The new graph has the same number of vertices and one less edge. It also has fewer cycles than G, so we have $v - (e - 1) + (f - 1) = 2$ by the inductive hypothesis, which gives $v - e + f = 2$. ■

For Exercise 6.5-8, let's define an edge-face pair to be an edge and a face such that the edge borders the face. According to the exercise, the number of such pairs is at least $3f$ in a simple graph. Because each edge is in either one or two faces, the number of edge-face pairs is also no more than $2e$. This gives

$$3f \leq \text{number of edge-face pairs} \leq 2e \,,$$

or $3f \leq 2e$, so that $f \leq (2/3)e$ in a planar drawing of a graph. We can combine this with Theorem 6.26 to get

$$2 = v - e + f \leq v - e + \frac{2}{3}e = v - \frac{e}{3} \,,$$

which we can rewrite as

$$e \leq 3v - 6$$

in a planar graph.

| Corollary 6.27 | In a connected simple planar graph, $e \leq 3v - 6$.

Proof: The proof of this corollary is given above. ■

In our discussion of Exercise 6.5-5, we said that we would see a simple proof that the circuit layout problem in that exercise was impossible. Notice that the question in that exercise was really the question of whether the complete graph on five vertices, K_5, is planar. If it were, then the inequality $e \leq 3v - 6$ would give us $10 \leq 3 \cdot 5 - 6 = 9$, which is impossible; so, K_5 can't be planar. The inequality of Corollary 6.27 is not strong enough to solve Exercise 6.5-6, which is really asking whether the so-called complete bipartite graph on two parts of size 3, denoted by $K_{3,3}$, is planar. To show that it isn't, we need to refine the inequality of Corollary 6.27 to take into account the special nature of bipartite graphs. In a simple bipartite graph, there are no cycles of size 3, so there are no faces that are bordered by just three edges. Problem 13 asks you to use this fact to prove that in a connected planar simple bipartite graph, $e \leq 2v - 4$.

| Exercise 6.5-9 | Prove or give a counterexample: Every planar graph has at least one vertex of degree 5 or less.

| Exercise 6.5-10 | Prove that every planar graph has a proper coloring with six colors.

In Exercise 6.5-9, suppose that G is a planar graph in which each vertex has degree 6 or more. Then the sum of the degrees of the vertices is at least $6v$ and is also twice the number of edges. Thus, $2e \geq 6v$, or $e \geq 3v$, which is contrary to $e \leq 3v - 6$. This gives us yet another corollary to Euler's formula.

| Corollary 6.28 | Every planar graph has a vertex of degree 5 or less. |

Proof: Each connected component of a planar graph is connected; by the argument before the corollary, each connected component of a planar graph has a vertex of degree 5 or less. Thus, every planar graph has such a vertex. ∎

The Five Color Theorem

We are now in a position to give a proof of the five color theorem, essentially Heawood's proof, which was based on his analysis of an incorrect proof given by Kempe to the four color theorem about ten years earlier in 1879. First, we observe that in Exercise 6.5-10 we can use straightforward induction to show that any planar graph on n vertices can be properly colored in six colors. As a base step, the theorem is clearly true if the graph has six or fewer vertices. So now assume that $n > 6$ and suppose that a graph with fewer than n vertices can be properly colored with six colors. Let x be a vertex of degree 5 or less, as shown in Figure 6.43. We show the edges as dashed because not all the edges we have drawn need to be there. The edges leaving vertices a through e, but going nowhere, are intended to suggest that this configuration sits in some larger graph. Deleting x gives us a planar graph on $n - 1$ vertices. So, by the inductive hypothesis, this graph can be properly colored with six colors. However, only five or fewer of those colors can appear on vertices that were originally neighbors of x, because x had degree 5 or less. In Figure 6.43 these colors are named 1 through 5. Thus, we can put x back into the colored graph, and there is at least one color not used on its neighbors. If we use such a color on x, we have a proper coloring of G. Therefore, by the principle of mathematical induction, every planar graph on $n \geq 1$ vertices has a proper coloring with six colors.

Figure 6.43: *The vertex x has degree at most 5. Edges are dashed because they might not be present.*

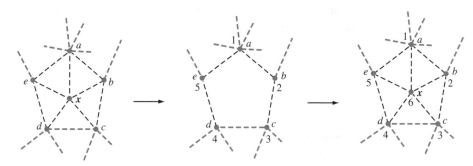

To prove the five color theorem, we make a similar start: We delete a vertex x of degree 5 and properly color the graph that remains. It is possible that when we want to restore x into the graph, five distinct colors are already used on its neighbors. This is where the proof will become interesting.

Theorem 6.29	A planar graph G has a proper coloring with at most five colors.

Proof: We may assume for two reasons that every face, except perhaps the outside face, of our drawing is a triangle. First, if we have a planar drawing with a face that is not a triangle, then we can draw additional edges going through that face until it has been divided into triangles. As we do so, the graph will remain planar. (In Figure 6.43, we would make all the dashed lines in the pentagon containing x into solid lines. If we had a quadrilateral for a face, we would draw a diagonal in it; if we had a pentagon (not the one containing x), we would draw two diagonals; and so on.) Second, if we can prove the theorem for graphs whose faces are all triangles, then we can obtain graphs with nontriangular faces by removing edges from graphs with triangular faces, and a proper coloring remains proper if we remove an edge from our graph. Although this appears to muddy the argument at this point, it makes it possible to give an argument that, at a crucial point, is clearer than it would otherwise be.

Our proof is by induction on the number of vertices of the graph. If G has five or fewer vertices, then it is clearly properly colorable with five or fewer colors. Suppose that G has n vertices and suppose inductively that every planar graph with fewer than n vertices is properly colorable with five colors. We have that G has a vertex x of degree 5 or less. Let G' be the graph obtained by deleting x from G, as in Figure 6.43. By the inductive hypothesis, G' has a coloring with five or fewer colors. Fix such a coloring (as in the second picture in Figure 6.43). If x has degree 4 or less, or if x has degree 5 but is adjacent to vertices colored with only four colors in G', then we may replace x in G' to get G and we have a color available to use on x to get a proper coloring of G. (Can you see how to modify Figure 6.43 to illustrate this?)

Thus, we may assume that x has degree 5 and that in G', five different colors appear on the vertices that are neighbors of x in G. Color all the vertices of G, other than x, as in G'. Let the five vertices adjacent to x be a, b, c, d, and e, in clockwise order, and assume that they are colored with colors 1, 2, 3, 4, and 5, respectively. Further, by our assumption that all faces are triangles, we have that $\{a, b\}, \{b, c\}, \{c, d\}, \{d, e\}$, and $\{e, a\}$ are all edges, so that we have a pentagonal cycle surrounding x. This would be the situation in the third graph of Figure 6.43 if we delete the 6 on vertex x. Consider the subgraph $G_{1,3}$ of G, which has the same vertex set as G but has only edges with endpoints colored 1 and 3. (Some possibilities are shown in Figure 6.44. In this figure, we show only edges connecting vertices colored 1 and 3, as well as dashed lines for the edges from x to its neighbors and the edges between successive neighbors. There may be many more vertices and edges in G.)

The graph $G_{1,3}$ may have a number of connected components. If a and c are not in the same component, then we may exchange the colors on the vertices of the component containing a without affecting the color on c. In this way, we obtain a coloring of G with only four colors—3, 2, 3, 4, and 5 on the vertices a, b, c, d, and e, respectively. We may then use the fifth color (in this case 1) on vertex x, and we have properly colored G with five colors.

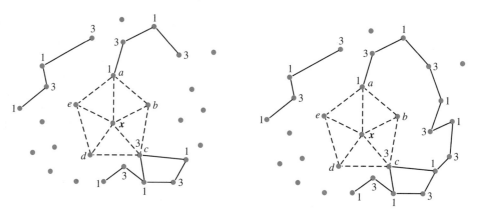

Otherwise, as in the second part of Figure 6.44, because a and c are in the same component of $G_{1,3}$, there is a path from a to c consisting entirely of vertices colored 1 and 3. Temporarily color x with a new color that we call color 6. Then in G, we have a cycle C of vertices colored 1, 3, and 6. This cycle has an inside and an outside. Part of the graph can be on the inside of C, and part can be on the outside. In Figure 6.45, we show two cases for how the cycle could occur: one in which vertex b is inside the cycle C, and one in which it is outside C. (Notice also that in both cases, we have more than one choice for the cycle, because there are two ways in which we could use the quadrilateral at the bottom of the figure.)

Figure 6.45: *Possible cycles in the graph $G_{1,3}$*

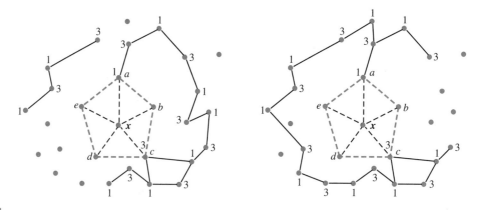

In G, we also have the cycle with vertex sequence a, b, c, d, and e, which is colored with five different colors. This cycle and the cycle C can intersect only in the vertices a and c. Thus, these two cycles divide the plane into four regions: the one inside both cycles, the one outside both cycles, and the two regions inside one cycle but not the other. If b is inside C, then the area inside both cycles is bounded by the cycle $a\{a, b\}b\{b, c\}c\{c, x\}x\{x, a\}a$. Therefore, e and d are not inside the cycle C. If one of d and e is inside C, then both are (because the edge between them cannot cross the cycle), and the boundary of the region inside both cycles is $a\{a, e\}e\{e, d\}d\{d, c\}c\{c, x\}x\{x, a\}a$. In this case, b cannot be inside C. Thus, one of b and d is inside the cycle c, and one is outside it. If we look at the graph $G_{2,4}$ with the same vertex set as G and just the edges connecting vertices colored 2 and 4, the connected component containing b and the connected component containing d must be different—otherwise a path of vertices colored 2 and 4 would have to cross the cycle C colored with 1, 3, and 6. Therefore, in G', we may exchange the colors 2 and 4 in the component containing d. Once we do so, we have only colors 1, 2, 3, and 5 used on vertices a, b, c, d, and e. Thus, we may use this coloring of G' as the coloring for the vertices of G different from x. We may then change the color on x from 6 to 4, and we have a proper five coloring of G. Therefore, by the principle of mathematical induction, every finite planar graph has a proper coloring with five colors. ■

Kempe's argument that seemed to prove the four color theorem was similar to this, though where we had five distinct colors on the neighbors of x and sought to remove one of them, he had four distinct colors on the five neighbors of x and sought to remove one of them. He had a more complicated argument involving two cycles in place of our cycle C, but he missed one of the ways in which these two cycles can interact.[20]

Important Concepts, Formulas, and Theorems

1. *Graph coloring.* An assignment of labels to the vertices of a graph (a function from the vertices to the set of labels) is called a *coloring* of the graph. The set of possible labels (the range of the coloring function) is often referred to as a set of *colors*.

2. *Proper coloring.* A coloring of a graph is called a *proper coloring* if it assigns different colors to adjacent vertices.

3. *Intersection graph.* We call a graph an *intersection graph* if its vertices correspond to sets and it has an edge between two vertices if and only if the corresponding sets intersect.

4. *Chromatic number.* The *chromatic number* of a graph G, traditionally denoted $\chi(G)$, is the minimum number of colors needed to color G properly.

5. *Complete subgraphs and chromatic numbers.* If a graph G contains a subgraph that is a complete graph on n vertices, then the chromatic number of G is at least n.

6. *Interval graph.* An intersection graph of a set of intervals of real numbers is called an *interval graph*. The assignment of intervals to the vertices is called an *interval representation*.

7. *Chromatic number of an interval graph.* In an interval graph G, the chromatic number is the size of the largest complete subgraph.

---

[20]For more history and excerpts from the papers mentioned above, see Biggs, Lloyd, and Wilson[7].

8. *Algorithm to compute the chromatic number and a proper coloring of an interval graph.* An interval graph G may be properly colored using $\chi(G)$ consecutive integers as colors by listing the intervals of a representation in order of their left endpoints and going through the list, assigning the smallest color not used on an earlier adjacent interval to each interval in the list.

9. *Planar graph/Planar drawing.* A graph is called *planar* if it has a drawing in the plane such that edges do not meet except at their endpoints. Such a drawing is called a *planar drawing* of the graph.

10. *Face of a planar drawing.* A geometrically connected set in the plane with the vertices and edges of a planar drawing of a graph removed is a *face* if it is not a proper subset of any other connected set in the plane with the vertices and edges removed.

11. *Cut edge.* An edge whose removal from a graph increases the number of connected components is called a *cut edge* of the graph. A cut edge of a planar graph lies on only one face of a planar drawing.

12. *Euler's formula.* In a planar drawing of a connected graph with v vertices, e edges, and f faces, $v - e + f = 2$. As a consequence, in a connected simple planar graph, $e \leq 3v - 6$.

Problems

1. What is the minimum number of colors needed to color a path on n vertices properly if $n > 1$? a/h

2. What is the minimum number of colors needed to color properly a bipartite graph with parts X and Y?

3. If a graph has chromatic number 2, is it bipartite? Why or why not? a/h

4. Prove that the chromatic number of a graph G is the maximum of the chromatic numbers of its components.

5. A *wheel* on n vertices consists of a cycle on $n - 1$ vertices together with one more vertex, normally drawn inside the cycle, that has an edge (like a spoke) to every vertex of the cycle. What is the chromatic number of a wheel on five vertices? What is the chromatic number of a wheel on an odd number of vertices? a/h

6. A *wheel* on n vertices consists of a cycle on $n - 1$ vertices together with one more vertex, normally drawn inside the cycle, that is connected to every vertex of the cycle. What is the chromatic number of a wheel on six vertices? What is the chromatic number of a wheel on an even number of vertices?

7. The usual symbol for the maximum degree of any vertex in a graph is Δ. Show that the chromatic number of a graph is no more than $\Delta + 1$. (In fact, Brooks proved that if G is not complete or an odd cycle, then $\chi(G) \leq \Delta$. Though there are now many proofs of this fact, none are easy!) a/h

8. Can an interval graph contain an induced cycle with four vertices? Remember that a subgraph of G is an induced subgraph if every edge of G joining two vertices of the subgraph is also an edge of the subgraph.

9. What is the chromatic number of the Petersen graph (see Figure 6.46)? a/h

Figure 6.46: *The Petersen graph*

10. Let G consist of a five-cycle (a cycle on five vertices) and a complete graph on four vertices, with all vertices of the five-cycle joined to all vertices of the complete graph. What is the chromatic number of G?

11. In how many ways can you properly color a tree on n vertices with t colors? a/h

12. In how many ways can you properly color a complete graph on n vertices with t colors?

13. Show that in a simple planar graph with no triangles, $e \leq 2v - 4$. a/h

14. Show that in a simple bipartite planar graph, $e \leq 2v - 4$. Use this fact to prove that $K_{3,3}$ is not planar.

15. Show that in a simple planar graph with no triangles, there is a vertex of degree 3 or less. a/h

16. Show that if a simple planar graph has fewer than 12 vertices, then it has at least 1 vertex of degree 4 or less.

17. In the Petersen graph (Figure 6.46), what is the size of the smallest cycle? Is the Petersen graph planar? a/h

18. Prove the following theorem of Welsh and Powell: If a graph G has degree sequence $d_1 \geq d_2 \geq \cdots \geq d_n$, then $\chi(G) \leq 1 + \max_i[\min(d_i, i - 1)]$ (that is, the maximum over all i of the minimum of d_i and $i - 1$).

19. What upper bounds do Problem 18, the bound you were asked to prove in Problem 7, and the Brooks bound in Problem 7 give you for the chromatic number in Problem 10? Which comes closest to the right value? How close? a/h

Appendix

Equivalence Relations and Congruence Relations

A.1 EQUIVALENCE RELATIONS

Relations

In Section 1.4, we said, "A relationship that divides a set into mutually exclusive classes is called an *equivalence relation.*" There are at least two questions one might ask about this sentence. First, what is a relationship, and second, how does a relationship divide a set into classes? (One might also ask what we mean by classes, which is something we can answer quickly: "Class" is another name for "set" that we use to avoid talking about dividing a set into sets.) We are going to use the word "relationship" casually; that is, without definition. When we want to be more precise, we will use "relation" (which we will soon define) rather than "relationship."

Exercise A.1-1	Consider the functions defined on the set $\{1, 2, 3, 4, 5\}$ by the rules $f(x) = x^5 - 15x^4 + 85x^3 - 224x^2 + 268x - 111$ and $g(x) = x^2 - 6x + 9$. Are they the same function or different functions?

Exercise A.1-2	For the two functions f and g in Exercise A.1-1, write down the set of ordered pairs $\left\{ (x, f(x)) \mid x \in \{1, 2, 3, 4, 5\} \right\}$ and the set of ordered pairs $\left\{ (x, g(x)) \mid x \in \{1, 2, 3, 4, 5\} \right\}$. How does this relate to your answer to Exercise A.1-1?

At first, Exercise A.1-1 looks silly; the two rules are different, so aren't the functions different? The point to Exercise A.1-2 is that, in fact, f and g represent the same function on the set $\{1, 2, 3, 4, 5\}$. In particular $f(i) = g(i)$ for each $i \in \{1, 2, 3, 4, 5\}$.

For a function h defined on a set X, we define the **relation** of h to be the set

$$\{(x, h(x)) \mid x \in X\}.$$

Thus, the relation of f is

$$\{(1, 4), (2, 1), (3, 0), (4, 1), (5, 4)\},$$

and the relation of g is

$$\{(1, 4), (2, 1), (3, 0), (4, 1), (5, 4)\}.$$

Two functions defined on a set X are considered to be the same function if they have the same relation. Intuitively, to specify a relationship, we specify what is related to what. We do so by putting the ordered pair (x, y) into a set of ordered pairs if and only if x and y are related. More precisely, a relation is nothing more or less than a set of ordered pairs. Here is yet another example of abstraction; we abstract the essence of the concept of a relationship as consisting of an exact specification of what is related to what. A *relation from a set X to a set Y* is a set of ordered pairs (x, y) with $x \in X$ and $y \in Y$.

There are many examples of relations that do not arise from functions. Perhaps one of the most important relations we have studied is that of congruence modulo n. We define

$$a \equiv b \pmod{n}$$

if and only if $a \bmod n = b \bmod n$, using the terminology of Section 2.1. Notice that with three bars instead of an equal sign, it is traditional to put parentheses around mod n. We say that a is congruent to b modulo n when $a \equiv b \pmod{n}$. This relation relates the set of integers to the set of integers. We say that this is a relation defined on the set Z of integers. More generally, a *relation on a set X* is a set of ordered pairs (x_1, x_2) that have both x_1 and x_2 in X.

Exercise A.1-3 | Show that

$$a \equiv b \pmod{n}$$

if and only if $a - b$ is a multiple of n.

First, if $a \bmod n = b \bmod n$, then we have $a = q_1 n + r$ and $b = q_2 n + r$ with $0 \le r < n$. Therefore, $a - b = (q_1 - q_2)n$, so $a - b$ is a multiple of n. On the other hand, if $a - b = qn$, then $a = b + qn$ and, by Lemma 2.2, $a \bmod n = b \bmod n$.

We will investigate interesting mathematical properties of this relation later in this appendix. For now, notice that the set of ordered pairs we have described is

$$\{(a, b) \mid a \bmod n = b \bmod n\} = \{(a, b) \mid a - b = kn \text{ for some } k \in Z\}.$$

This set is an infinite set because it contains, for example, the pairs (i, i) for each i in Z.

Recall that when we derived the formula $\binom{n}{k} = \frac{n!}{k!(n-k)!}$, we saw that there were $k!$ different permutations of a k-element subset of an n-element set S. Any of these permutations is equivalent for the purposes of specifying the subset. We can define two permutations to be **set-equivalent** if they are permutations of the same subset of S. This is a relation on the set of k-element permutations of X.

Yet another example of a relation on the set of integers is the **"neighbor"** relation: i is a neighbor of j if the absolute value of the difference between i and j is 1. Some pairs in this relation are $(-1, 0)$, $(0, -1)$, $(0, 1)$, $(1, 0)$, $(1, 2)$, $(2, 1)$, and $(2, 3)$. This is another example of an infinite relation.

A final example of a relation on the integers is the "less than" relation. We put the ordered pair (i, j) in the relation if $i < j$. Thus, the "less than" relation is the set

$$\{(i, j) \mid i, j \in Z \text{ and } i < j\}.$$

The "less than" relation brings up a good point: You have probably never seen anyone write $(x, y) \in <$. Instead, you would see $x < y$. If we want to refer to an arbitrary relation, we might refer to it as R, as in "Let R be a relation on the set X," and we might want to write aRb in place of $(a, b) \in R$. As with so many things in mathematics, we choose the notation that most comfortably fits the situation.

We can now begin a study that will help us understand what it means to say that a relationship divides a set into disjoint classes.

Exercise A.1-4   Write down the set of all integers n between -5 and 20 such that

a. $n \equiv 0 \pmod 5$.

b. $n \equiv 1 \pmod 5$.

c. $n \equiv 2 \pmod 5$.

d. $n \equiv 3 \pmod 5$.

e. $n \equiv 4 \pmod 5$.

Do any of the sets have any elements in common? Do you think they would if we replaced 20 with 2,000,000? Is there any number between -5 and 20 that is not in one of the sets?

Exercise A.1-5   Write down the set of all three-element permutations of $\{1, 2, 3, 4\}$ that are set-equivalent to

a. 243.

b. 123.

c. 142.

d. 134.

Do any of these sets have any permutations in common? Is every three-element permutation of $\{1, 2, 3, 4\}$ in one of these sets?

Exercise A.1-6 Write down the set of all neighbors in Z of

   a. 0.

   b. 1.

   c. 2.

   d. 3.

Do any of these sets have any elements in common?

Exercise A.1-7 Write down the set of all positive integers less than

   a. 2.

   b. 3.

   c. 4.

Do any of these sets have any elements in common?

Exercise A.1-8 Consider the relation on the positive integers given by $\{(s, s^2) \mid s$ is an integer$\}$. Write down the set of all integers related to the following.

   a. 1.

   b. 2.

   c. 3.

Is every positive integer going to be in one of the sets $\{s \mid s$ is related to $n\}$ for some integer n?

In Exercise A.1-4, the sets are $[0] = \{-5, 0, 5, 10, 15, 20\}$, $[1] = \{-4, 1, 6, 11, 16\}$, $[2] = \{-3, 2, 7, 12, 17\}$, $[3] = \{-2, 3, 8, 13, 18\}$, and $[4] = \{-1, 4, 9, 14, 19\}$. No two of these sets have any elements in common, and each integer between -5 and 20 is in one of them, so they give us a partition of $\{-5, -4, -3, -2, -1, 0, 1, 2, \ldots, 20\}$. Thus, our relation divides the integers between -5 and 20 into disjoint classes. We expect the same thing to happen if we use integers between -5 and 2,000,000, or $-5,000,000$ and 2,000,000.

In Exercise A.1-5, the set of all lists equivalent to 243 is the set $\{243, 234, 423, 432, 342, 324\}$. The set of lists equivalent to 123 is $\{123, 132, 213, 231, 312, 321\}$. Those equivalent to 142 are $\{142, 124, 214, 241, 412, 421\}$. Those equivalent to 134 are $\{134, 143, 314, 341, 413, 431\}$. No two of these sets of lists have any elements in common, and every three-element permutation of $\{1, 2, 3, 4\}$ is in one of the sets. Therefore, these sets form a partition of the set of all three-element permutations of $\{1, 2, 3, 4\}$. Thus, the "set-equivalence" relation divides the three-element permutations of $\{1, 2, 3, 4\}$ into disjoint classes.

In Exercise A.1-6, the set of neighbors of 0 is {−1, 1}, of 1 is {0, 2}, of 2 is {1, 3}, and of 3 is {2, 4}. Each of 1 and 2 is in two of these sets. For example, 1 is in {−1, 1} and {1, 3}. Therefore, the "neighbor" relation does not divide the integers into disjoint classes.

In Exercise A.1-7, the set of positive integers less than 2 is {1}, the set of positive integers less than 3 is {1, 2}, and the set of integers less than 4 is {1, 2, 3}. The number 1 is in each of these sets, so the "less than" relation does not divide the positive integers into disjoint classes.

In Exercise A.1-8, the set of elements related to 1 is {1}, to 2 is {4}, and to 3 is {9}. Because the only elements related to something are squares, the relation does not divide *all* the integers, or *all* the positive integers, into disjoint classes.

We see that a relation on a set S gives us classes consisting of the set of all elements related to x for each $x \in S$. These classes might or might not contain every element of S, and they might or might not be disjoint. What makes a relation divide a set into disjoint classes? Several properties that relations can have help us explain when a relation divides a set into disjoint classes.

Exercise A.1-9

We say that a relation R defined on a set X is **reflexive** if for every x in X, we have $(x, x) \in R$, or, in other notation, xRx. For example the "less than or equal to" relation is a reflexive relation defined on the set of integers. Determine whether each of the following relations is reflexive.

a. The "congruence modulo n" relation on the integers

b. The "set-equivalence" relation on the k-element permutations of an n-element set

c. The "neighbor" relation on the integers

d. The "less than" relation on the integers

Exercise A.1-10

We say that a relation R, defined on a set X, is **symmetric** when $(a, b) \in R$ if and only if $(b, a) \in R$ (or aRb if and only if bRa). For example, the relation of being a sibling is a symmetric relation on the set of people, but the relation of being a sister is not. Determine whether each of the following relations is symmetric.

a. The "congruence modulo n" relation on the integers

b. The "set-equivalence" relation on the k-element permutations of an n-element set

c. The "neighbor" relation on the integers

d. The "less than" relation on the integers

Exercise A.1-11	To say that a relation R on a set X is **transitive** means that whenever $(x, y) \in R$ and $(y, z) \in R$, then $(x, z) \in R$. In our other notation, to say R is transitive means that whenever xRy and yRz, then xRz. The "greater than" relation on the real numbers is an example of a transitive relation. The "is a father of" relation on the set of people is not a transitive relation. Determine whether each of the following relations is transitive.

     a. The "congruence modulo n" relation on the integers

     b. The "set-equivalence" relation on the k-element permutations of an n-element set

     c. The "neighbor" relation on the integers

     d. The "less than" relation on the integers

In Exercise A.1-9, we see that congruence modulo n and set-equivalence are reflexive relations, but the "neighbor" relation and the "less than" relation are not. To see why congruence modulo n is reflexive, notice that $a - a = 0$ and $0 = 0 \cdot n$, so the difference between a and a is a multiple of n. It is perhaps even easier to note that $a \bmod n = a \bmod n$, so that $a \equiv a \pmod n$. On the other hand, the "neighbor" relation is not reflexive, because $|a - a| \neq 1$.

In Exercise A.1-10, the "congruence modulo n," "set-equivalence," and "neighbor" relations are symmetric, while the "less than" relation is not. Again, for one example, if a list L_1 of distinct elements lists the elements of a subset K of S and the list L_2 of distinct elements lists the elements of the same subset, then L_1 lists the same set as L_2 and L_2 lists the same elements as L_1. Thus, the two lists are set-equivalent in either order, so the relation is symmetric.

Finally, in Exercise A.1-11, the "equivalence modulo n," "set-equivalence," and "less than" relations are transitive, but the "neighbor" relation is not. To see that equivalence modulo n is transitive, note that if $a - b = kn$ and $b - c = hn$, then addition gives $a - c = (k + h)n$. To see that the "neighbor" relation is not transitive, note that 1 is a neighbor of 2, and 2 is a neighbor of 3, but 1 is not a neighbor of 3.

You'll notice that the two relations that divide their sets into disjoint classes are reflexive, symmetric, and transitive. The relation of congruence mod n gives one set, or class, of integers for each possible remainder. Two integers are related if they have the same remainder. The relation of set-equivalence makes two lists equivalent if they are lists of the same set. In both of these cases, there is an idea of sameness, and two objects are related if they are the same in this sense. Think about how we use the word "same." Anything should be the same as itself. If a is the same as b, then we would expect b to be the same as a. If a is the same as b and b is the same as c, then we expect that a is the same as c as well. Thus, these are three properties that any relation we use to capture an idea of sameness must satisfy. In fact, as we shall see, there is a sense in which these three properties abstractly capture the idea of sameness. We redefine an **equivalence relation** on a set S to be a relation on S that is reflexive, symmetric, and transitive.

Theorem A.1	Let R be an equivalence relation on S. Then for each pair of elements x and y, the sets $S_x = \{z \mid (x, z) \in R\}$ and $S_y = \{z \mid (y, z) \in R\}$ are either equal or disjoint. Further, with S_x as defined here, the set

$$\{S_x \mid x \in S\}$$

is a partition of the set S—that is, it is a set of disjoint sets whose union is S.

Proof: Suppose that S_x and S_y are as defined in the theorem, and suppose that $z \in S_x \cap S_y$. Then (x, z) and (y, z) are in R. By symmetry, $(z, y) \in R$. By transitivity, $(x, y) \in R$. By transitivity once again, for all z with $(y, z) \in R$, we have that $(x, z) \in R$ as well, so that $z \in S_x$. But then, by the definitions of S_x and S_y, we have that $S_y \subseteq S_x$. Exactly the same kind of argument (starting out with S_y and S_x in that order at the beginning of the first sentence of the proof) shows that $S_x \subseteq S_y$. Therefore, if $S_x \cap S_y \neq \emptyset$, then $S_x = S_y$. Thus, S_x and S_y are either equal or disjoint.

Because x is in the set S_x by the reflexive law, the union of the sets of $\{S_x \mid x \in S\}$ is S. According to the previous paragraph, the sets are mutually disjoint, so $\{S_x \mid x \in S\}$ is a partition of S. ∎

The sets S_x in Theorem A.1 are called **equivalence classes**. Theorem A.1 tells us that when we have an equivalence relation, two elements are equivalent if and only if they are in the same equivalence class. So in this sense, the two elements are the same. Our next theorem tells us that if we accept that the idea of sameness (for elements of a set) can be modeled by partitioning our set into classes and saying that two things are the same if and only if they are in the same class, then the defining properties of equivalence relations capture exactly what we mean by sameness. (We use the terminology for partitions introduced in Section 1.1.)

Theorem A.2	Let P be a partition of a set S. If we define the relation R by

$$R = \{(x, y) \mid x \text{ and } y \text{ are in the same block of } P\},$$

then the relation R is an equivalence relation whose equivalence classes are the blocks of P.

Proof: See Problem 5. ∎

Congruence Relations

There are some special properties of the relation of congruence mod n that we have frequently used in our computations mod n.

Exercise A.1-12 Suppose that $x \equiv y \pmod 5$. What can you say about $x + 3$ and $y + 3$ that is interesting? What about $3x$ and $3y$?

Exercise A.1-13 Suppose $x_1 \equiv y_1 \pmod 5$ and $x_2 \equiv y_2 \pmod 5$. What can you say about $x_1 + x_2$ and $y_1 + y_2$? What about $x_1 x_2$ and $y_1 y_2$?

In Exercise A.1-12, we can write

$$(x + 3) \bmod 5 = (x \bmod 5 + 3 \bmod 5) \bmod 5$$
$$= (y \bmod 5 + 3 \bmod 5) \bmod 5 = (y + 3) \bmod 5 \,,$$

by successive uses of Lemma 2.3. Therefore, $x + 3$ is congruent to $y + 3$. More succinctly, we can observe that $x - y = k \cdot 5$ for some integer k, and thus, $x + 3 - (y+3) = x - y = k \cdot 5$. Again we can conclude that $x + 3 \equiv y + 3 \pmod 5$. Similarly, if $x - y = k \cdot 5$, then $3x - 3y = 3k \cdot 5$, so $3x \equiv 3y \pmod 5$. The same kinds of computations apply in Exercise A.1-13 to show that $x_1 + x_2 \equiv y_1 + y_2 \pmod 5$ and $x_1 x_2 \equiv y_1 y_2 \pmod 5$. In summary, we have the following theorem.

Theorem A.3 If $x_1 \equiv y_1 \pmod n$ and $x_2 \equiv y_2 \pmod n$, then

$$x_1 + x_2 \equiv y_1 + y_2 \pmod n \tag{A.1}$$

and

$$x_1 x_2 \equiv y_1 y_2 \pmod n \,. \tag{A.2}$$

Because the proof of this theorem may be modeled on the computations preceding it, we save it as a problem. Equation A.1 is often described by saying that congruence mod n is a **congruence relation for addition**. Equation A.2 is often described by saying that congruence mod n is a **congruence relation for multiplication**.

Exercise A.1-14 Using $[x]_5$ to stand for the equivalence class containing x for congruence mod 5, does the rule $f([x]_5, [y]_5) = [x + y]_5$ define a function from pairs of equivalence classes to equivalence classes? What about the rule $g([x]_5, [y]_5) = [x \cdot y]_5$?

Exercise A.1-15 Suppose that we define two integers to be equivalent if they have the same number of digits. Explain briefly why this is an equivalence relation. Let $\langle x \rangle_{\mathrm{dig}}$ stand for the equivalence class containing x. Does the rule $h(\langle x \rangle_{\mathrm{dig}}, \langle y \rangle_{\mathrm{dig}}) = \langle x + y \rangle_{\mathrm{dig}}$ define a function from ordered pairs of equivalence classes to equivalence classes?

In Exercise A.1-14, the fact that congruence mod 5 is a congruence relation for addition tells us that if we take $x' \in [x]_5$ and $y' \in [y]_5$, then $[x + y]_5 = [x' + y']_5$. Because congruence mod 5 is a congruence relation for multiplication, $[xy]_5 = [x'y']_5$. Thus, the relations f and g are indeed functions. Exercise A.1-15 shows us that we need the idea of a congruence relation to define f and g, because while

$$h\big(\langle 1\rangle_{\mathrm{dig}}, \langle 2\rangle_{\mathrm{dig}}\big) = \langle 3\rangle_{\mathrm{dig}} = \langle 1\rangle_{\mathrm{dig}} \,,$$

we also have

$$h\big(\langle 9\rangle_{\mathrm{dig}}, \langle 2\rangle_{\mathrm{dig}}\big) = \langle 11\rangle_{\mathrm{dig}} \neq \langle 1\rangle_{\mathrm{dig}} \,,$$

even though $\langle 9\rangle_{\mathrm{dig}} = \langle 1\rangle_{\mathrm{dig}}$. Here is a way to make these observations more intuitive. Notice that the answer to Exercise A.1-14 can be re-expressed by saying that the idea that we add two equivalence classes by the rule $[x] + [y] = [x + y]$ makes sense when $[x]$ and $[y]$ are equivalence classes for congruence mod 5. However, it doesn't make sense when $[x]$ and $[y]$ are equivalence classes for an arbitrary equivalence relation on the integers. The same holds for multiplication. The reason it made sense to define addition and multiplication of congruence classes mod n was that congruence mod n is a congruence relation for addition and multiplication. The reason it didn't make sense for the "number of digits" relation is that this equivalence relation is not a congruence relation.

We define (or perhaps redefine) Z_n to mean the set of equivalence classes of the "congruence mod n" relation with the operations $[x]_n + [y]_n = [x + y]_n$ and $[x]_n \cdot [y]_n = [x \cdot y]_n$. It is now straightforward to demonstrate that all the rules of algebra that we are used to (with the exception of those using division or multiplicative inverses) hold for these new operations. For example, to prove the commutative law for addition in Z_n, we would write

$$[x]_n + [y]_n = [x + y]_n = [y + x]_n = [y]_n + [x]_n \,.$$

This definition of Z_n gives us the same structure as the definition of Z_n that we gave in Section 2.1. The difference between the two constructions is similar to the difference between saying that rational numbers are fractions in lowest terms and that rational numbers are equivalence classes of fractions where a/b is equivalent to ac/bc.

Important Concepts, Formulas, and Theorems

1. *Relation of a function.* For a function h defined on a set X, we define the *relation* of h to be the set

$$\{(x, h(x)) \mid x \in X\} \,.$$

2. *Relation.* A *relation* from a set X to a set Y is a set of ordered pairs whose first elements are in X and whose second elements are in Y.

3. *Relation on a set.* A *relation on a set* X is a set of ordered pairs (x_1, x_2) that have both x_1 and x_2 in X.

4. *Congruence modulo n.* We define

$$a \equiv b \pmod{n}$$

if and only if $a \bmod n = b \bmod n$. We read this as "a is congruent to b modulo n." Equivalently, $a \equiv b \pmod{n}$ if and only if $a - b$ is an integer (positive, negative, or zero) multiple of n.

5. *Reflexive.* We say that a relation R, defined on a set X, is *reflexive* if for every x in X, we have $(x, x) \in R$, or, in other notation, xRx.

6. *Symmetric.* We say that a relation R, defined on a set X, is *symmetric* when, for every a and b in X, $(a, b) \in R$ if and only if $(b, a) \in R$, or, in other notation, aRb if and only if bRa.

7. *Transitive.* To say that a relation R, defined on a set X, is *transitive* means that if $(x, y) \in R$ and $(y, z) \in R$, then $(x, z) \in R$ for all x, y, and z in X. In other notation, R is transitive means that if xRy and yRz, then xRz for all x, y, and z in X.

8. *Equivalence relation.* We define an *equivalence relation* on a set S to be a relation on S that is reflexive, symmetric, and transitive.

9. *Equivalence class.* An equivalence relation on the set X gives a partition of X into blocks given by $S_x = \{z \mid (x, z) \in R\}$. The blocks of this partition are called *equivalence classes*. Further, given a partition of a set S into blocks $B_1, B_2, \ldots, B_n$, the relation defined by "x is related to y if and only if they are both in the same block of the partition" is an equivalence relation defined on S, and its equivalence classes are the blocks of the partition.

Problems

1. Write the ordered pairs of the congruence mod 4 relation on the set

$$\{-6, -5, -4, -3, -2, -1, 0, 1, 2, 3, 4\} .$$ a/h

2. Determine whether the following relations are equivalence relations.

   a. "Is a brother of or is," on the set of people a/h

   b. "Is a sibling of or is," on the set of people a/h

   c. "x is related to y if $|x - y| \leq 2$," on the set of integers a/h

3. Explain why the relation given by x is related to y if $x^2 = y^2$ is an equivalence relation on the integers, and describe the equivalence classes.

4. Let S be the set of permutations of $\{0, 1, 2, 3, 4, 5, 6, 7, 8\}$, thought of as lists. Define $a_0, a_1, \ldots, a_8$ to be related to $b_0, b_1, \ldots, b_8$ if there is an integer k so that $b_i = a_{(i+k) \bmod 9}$. Is this relation an equivalence relation? What does it have to do with seating people around a round table? a/h

5. Prove Theorem A.2.

6. Prove Theorem A.3.

7. Determine whether the rule $g([x], [y]) = [x \cdot y]$ defines a function when $[x]$ stands for the equivalence class containing x for congruence mod n. Determine whether the rule $h([x], [y]) = [x \cdot y]$ defines a function when $[x]$ stands for the equivalence class containing x for the "has the same number of digits" equivalence relation. a/h

8. Prove the commutative law for multiplication of equivalence classes of integers mod n.

9. Prove the associative law for multiplication of equivalence classes of integers mod n. a/h

10. Show that adding the equivalence class $[0]_n$ to $[x]_n$ gives $[x]_n$. What is the corresponding statement about multiplication?

Section 1.1 (pages 7–8)

1. $n(n-1)/2$. You get this many if the original ordering is the reverse of the sorted ordering.

3. $52 \cdot 51 = 2652$

4. $52 \cdot 51/2 = 1326$

5. $52 \cdot 51 \cdot 50 = 132{,}600$

6. $10 \cdot 9 = 90$

7. $\binom{10}{2} = 45$

8. $10 \cdot \binom{9}{2}$ or $\binom{10}{2} \cdot 8$

9. _Hint:_ Think about a club that needs to choose a president and a two-person committee to advise the president.

12. $10 \cdot 9 = 90$

14. $5 \cdot 3 \cdot 3 \cdot 3 = 135$

Section 1.2 (pages 17–19)

2. $f_1(1) = a$, $f_1(2) = a$, $f_1(3) = a$; $f_2(1) = a$, $f_2(2) = a$, $f_2(3) = b$; $f_3(1) = a$, $f_3(2) = b$, $f_3(3) = a$; $f_4(1) = a$, $f_4(2) = b$, $f_4(3) = b$; $f_5(1) = b$, $f_5(2) = a$, $f_5(3) = a$; $f_6(1) = b$, $f_6(2) = a$, $f_6(3) = b$; $f_7(1) = b$, $f_7(2) = b$, $f_7(3) = a$; $f_8(1) = b$, $f_8(2) = b$, $f_8(3) = b$. None are one-to-one; all but f_1 and f_8 are onto.

4. t^s

6. $\binom{n}{k}$. If $k > n$, the answer is zero.

8. $2 \cdot 4! \cdot 4! = 1152$

10. $\binom{20}{3} = 1140$

12. $2\binom{10}{4}\binom{20}{4}4!4! = 2 \cdot 10^{\underline{4}}20^{\underline{4}} = 1{,}172{,}102{,}400$

14. $\left(\binom{10}{3} + \binom{10}{2} + \binom{10}{1}\right) \cdot 3 \cdot 8 = 5280$

16. $\binom{12}{5}$; $\binom{5}{2}\binom{4}{2}\binom{3}{1} = 180$; $\binom{5}{2}\binom{4}{2}\binom{3}{1} + \binom{5}{2}\binom{5}{2}\binom{2}{1} = 380$

18a. _Hint:_ You want to define $g(y)$ to be a certain x. In terms of f, what is this x and how do you know it exists?

18b. _Hint:_ Suppose g and h both satisfy the definition of being inverses to f. What can you say about $g(y)$ and $h(y)$ for any y equal to $f(x)$ for some x?

Section 1.3 (pages 26–28)

1. 220; 220; $\binom{n}{k}$ equals $\binom{n}{n-k}$.

3a. $x^5 + 5x^4 + 10x^3 + 10x^2 + 5x + 1$

3d. $x^5 - 5x^4 + 10x^3 - 10x^2 + 5x - 1$

5. $\frac{10!}{3!3!4!} = 4200$; _Hint:_ Label three of the chairs green.

7. Let $N - K$ stand for the set of all elements of N that are not in K. Then $f(K) = N - K$.

8. $\binom{m+n}{n}$ or $\binom{m+n}{m}$

10. *Hint:* You can think of one of the two things that the first sentence asks you to count as a(n) (ordered) list of three-element sets.

11. $20 \cdot 19 \cdot 18 \cdot 17 \cdot \binom{16}{3} = 65,116,800$;

$20 \cdot 19 \cdot 18 \cdot 17 \cdot \binom{20}{3} = 132,559,200$

13. *Hint:* Does the order in which k and $n - k$ appear in the denominators matter? For a second proof, in how many ways could you choose the elements that you *don't* want in a subset?

15. *Hint:* The ugly proof uses the formulas. The pretty proof explains why both sides count the same collection of sets.

17. *Hint:* What is $1 - 1$?

19. Partial answer: False

Section 1.4 (pages 38–41)

1. $(n - 1)!$

3. $\binom{5}{2}$; $\frac{5!}{2 \cdot 3!}$

5. $n!(n - 1)!$

7. $\binom{k}{n} n! n^{\overline{k-n}} = \frac{k!(k-1)!}{(k-n)!(n-1)!}$

9. $\binom{n+k-1}{k}$

11. $\frac{1}{n+1} \binom{2n}{n}$

13. *Hint:* What can you say about the sizes of the equivalence classes?

16a. n^k

16c. $\binom{n+k-1}{k}$

16e. $n^{\underline{k}}$

16g. $\binom{n}{k}$

16i. $n^{\underline{k}}$

16k. $n^{\underline{k}}$

Section 2.1 (pages 54–56)

1. 14 mod 9 = 5; −1 mod 9 = 8; −11 mod 9 = 7

3. EBOB FP X JBPPXDB

5. 11; 12

7. $(x \cdot 4)$ mod 9 = 1; because $7 \cdot 4 = 28$, you have that $(1/4)$ mod 9 = 7; $(1/3)$ mod 9 does not exist.

9. Partial answer:

+	0	1	2	3	4	5	6
0	0	1	2	3	4	5	6
1	1	2	3	4	5	6	0
2	2	3	4	5	6	0	1
3	3	4	5	6	0	1	2
4	4	5	6	0	1	2	3
5	5	6	0	1	2	3	4
6	6	0	1	2	3	4	5

11. Yes; yes; no; yes.

13. *(Big) Hint:* What possible values can $(x + a)$ mod n take on as functions of x and a? Note, we assume $0 \le x$ and $a < n$.

16. The associative law says $x \cdot_n (y \cdot_n z) = (x \cdot_n y) \cdot_n z$. As a hint for the rest of the problem, think about Lemma 2.3.

Section 2.2 (pages 70–71)

1. Yes; 133 mod m.

3. No for 10; yes for 11.

5. It is either zero or one.

7. 42

9. The first suggestion is not safe. She computes q^{-1} mod p using the extended GCD algorithm. The second is safe with a large p, so far as we know. The wiretapper could try all powers of q until she finds one such that $q^i \equiv q^a$. Then, she computes $(q^b)^i$. If p is large, this is impractical. Also, if the wiretapper knew how to take logarithms to the base q in Z_p, she could compute the $\log_q$ of q^a. But nobody knows a fast way to compute logarithms in Z_p.

11. GCD is 18; $x = 11$; $y = -13$.

13. $x = 85$

15. Yes, gcd(j, k) is a divisor of gcd(r, k), and if gcd$(r, k) = 1$, then gcd$(j, k) = $ gcd(r, k).

17. gcd$(F_i, F_{i+1}) = 1$; $x = (-1)^{i-1} F_i$; $y = (-1)^i F_{i-1}$

19. lcm$(x, y) = xy / $ gcd(x, y)

21. $4 \cdot_6 x = 4$ has a solution in Z_6.

23. The recursive description of Euclid's extended GCD algorithm gives a basis for a recursive proof of the theorem.

Section 2.3 (pages 81–82)

1. 4, 2, 1, 4, 2, 1, ...; 4, 6, 4, 6, 4, ...
3. They are all 1.
5. 1176; 1; 18; 19; 105. y^d mod p need not determine x.
7a. 1
7b. 1
7c. 67
9. 0, p, $2p$, $3p$, ..., $(p-1)p$ have no multiplicative inverses; 1; no; it is 0.
11a. $mx + nz = 1$
11b. *Hint:* The substitutions give $k = kmx + knz = cnmx + bmnz$.
14. *Hint:* x^{n-1} mod $n = 1$ tells you that x has a multiplicative inverse in Z_n.

Section 2.4 (pages 89–90)

1. 4
3. About 12 billion; about 12 trillion; insignificant in comparison.
5. 10 and 23
7. 10^{120}; a lot closer; no.
9. It doesn't make sense because you would need $a^{e_1 e_2}$ mod $n = a^{e_1 e_2 \bmod n}$ mod n. Try simple examples to see if this rule holds.
11. It would make sense if a has a multiplicative inverse, but not otherwise.
13. 103; 100 encrypts to 111, and 111^{103} mod $209 = 100$.
16. *Hint:* The word "signature" is being used in a very broad sense. Bob needs to do something to convince the world that he and only he is the person who gave them a certain piece of information, which we refer to as his signature of the document.

Section 3.1 (pages 102–103)

1a.

s	t	$(s \vee t) \wedge (\neg s \vee t) \wedge (s \vee \neg t)$
T	T	T
T	F	F
F	T	F
F	F	F

1b.

s	t	u	$(s \Rightarrow t) \wedge (t \Rightarrow u)$
T	T	T	T
T	T	F	F
T	F	T	F
T	F	F	F
F	T	T	T
F	T	F	F
F	F	T	T
F	F	F	T

1c.

s	t	u	$(s \vee t \vee u) \wedge (s \vee \neg t \vee u)$
T	T	T	T
T	T	F	T
T	F	T	T
T	F	F	T
F	T	T	T
F	T	F	F
F	F	T	T
F	F	F	F

4. *Hint:* Give the truth table for $s \Rightarrow t$ and $\neg s \vee t$, and compare them or construct a double truth table.
5. *Hint:* Construct truth tables for both statements or construct a double truth table.
7a. s **7b.** s **7c.** T **7d.** F
9. *Hints:* One way to use the distributive law is "backward"—that is, start with $(s \vee t) \wedge (u \vee t)$ and change it to $(s \wedge u) \vee t$. Another way to use it is to write $(s \wedge t) \vee (u \wedge v) = ((s \wedge t) \vee u) \wedge ((s \wedge t) \vee v)$.
12. $(\neg s \vee t) \wedge (s \vee \neg t)$ or $(\neg s \wedge \neg t) \vee (s \wedge t)$
14. No. *Hint for second question:* What do you get when you apply DeMorgan's laws to $\neg(\neg s \vee \neg t)$?
16. *Hint:* Why were we allowed to say "$q \neq q^*$ or $r \neq r^*$" in our proof?

Section 3.2 (pages 115–117)

1. 1, 2, and 3; 1, 2, and 3; all real numbers between 1 and 3; no.
3. $\forall x \in R \ (x^2 > 0)$

7a. False

7b. True

7c. False

7d. True

8. Partial answer: Yes, there are two universal quantifiers.

10. For all positive integers n there is an integer m larger than n such that there is a polynomial equation $p(x) = 0$ of degree m that has a real solution.

11a. Partial answer: False

11b. Partial answer: False

13a. Partial answer: False

13b. Partial answer: False

13c. Partial answer: False

13d. Partial answer: True

15. Partial answer: "For all" and "there exist" do not commute.

Section 3.3 (pages 125–126)

1a. Converse: If the hose reaches the tomatoes, then the hose is 60 ft long. Contrapositive: If the hose doesn't reach the tomatoes, then the hose is not 60 ft long.

1b. Converse: Mary goes for a walk only if George goes for a walk. Contrapositive: Mary doesn't go for a walk only if George doesn't go for a walk.

1c. Converse: If Pamela recites a poem, then Andre asked for a poem. Contrapositive: If Pamela doesn't recite a poem, then Andre didn't ask for a poem.

4. Partial answer: This means that for all m and n in the integers, if m and n are odd, then $m + n$ is even.

6. Partial answer: No

7. *Hint:* To get started, assume the negation of $x \neq 1$—that is, assume that $x = 1$. Try to use this to show the negation of $x^2 - 2x \neq -1$, and use contrapositive inference.

9. *Hint:* Try either contraposition or contradiction.

11. *Hint:* Contraposition and contradiction are two possible methods.

12. *Hint:* Experiment with some small values for n to help you decide whether the statement is true.

14. *Hint:* Try contradiction. If there is a biggest prime n, what do you know about prime factors of $n! + 1$?

Section 4.1 (pages 141–143)

1a. i. No ii. It is $1 - (1/3)^i$. iii. Yes iv. It is $1 - (1/3)^{n-1}$ v. $2/3 + 2/9 + \cdots + 2/3^{n-1} + 2/3^n = 1 - (1/3)^{n-1} + 2/3^n = 1 - (1/3)^n$ vi. The assumption is wrong. vii. The formula is true. viii. $p(k-1) \Rightarrow p(k)$

1b. i. The base case is $2/3 = 1 - 1/3$. ii. The inductive hypothesis is $2/3 + 2/9 + \cdots + 2/3^{k-1} = 1 - (1/3)^{k-1}$. iii. Denoting the formula you have to prove by $p(n)$, you would prove $p(k)$ based on assuming $p(k-1)$, thereby showing that $p(k-1) \Rightarrow p(k)$. You may have written your answers so that they involve the variable n rather than the variable k. iv. $2/3 + 2/9 + \cdots + 2/3^{n-1} + 2/3^n = 1 - (1/3)^{n-1} + 2/3^n = 1 - (1/3)^n$. v. $2/3 + 2/9 + \cdots + 2/3^k = 1 - (1/3)^k$ for all positive integers k. vi. $p(k) \Rightarrow p(k+1)$, where $p(k)$ is the previous formula.

3. Abbreviated answer: Base case: $1 \cdot 2 = 6/3$ when $n = 1$; *inductive hypothesis:* $1 \cdot 2 + 2 \cdot 3 + \cdots + (n-1)n = (n-1)n(n+1)/3$. Add $n(n+1)$ to both sides and simplify to get $n(n+1)(n+2)/3$ in the inductive step.

5. Abbreviated answer: Base cases: $m \leq n$ (You could also do $m = 1$, but the multiple base cases make the proof flow more smoothly.); *inductive hypothesis:* When $0 \leq k < m$, $\exists$ unique integers q and r with $k = qn + r$ and $0 \leq r < n$; *inductive step:* Go from $k = m - n$ to $k = m$. (You can either make uniqueness part of the inductive proof or prove it separately.)

7. Abbreviated answer: Base cases: $n = 8, 9, 10$; *inductive hypothesis:* When $8 \leq k < n$, you can make k cents in postage; *inductive step:* Go from $n - 3$ to n.

9. Abbreviated answer: Base case: $n = 2$ is given; *inductive hypothesis:* The size of a union of n disjoint sets is the sum of their sizes; *inductive step:* The union of $n + 1$ sets is the union of the first n unioned with the last set. The size of the union of the first n is given by the inductive hypothesis. The size of the union of that set with the last set is given by the sum principle for two sets.

12. *Hint:* Look for similarities between this problem and the proof of Theorem 1.3.

14. *Hint:* Suppose you know that the weak principle of mathematical induction holds. Suppose further that you know $p(b)$ holds and the implication $p(b) \wedge p(b+1) \wedge \cdots \wedge p(n-1) \Rightarrow p(n)$ holds for all $n > b$. Let $q(n)$ be the statement $p(b) \wedge p(b+1) \wedge \cdots \wedge p(n)$. See what you can do with weak induction applied to $q(n)$.

Section 4.2 (pages 155–156)

3. *Hint:* Try iterating the recurrence, or use the formula for first-order linear recurrences, or guess the formula and prove by induction that you are right. Partial answer: The difference is that for this recurrence the coefficient of 2^n is larger.

4. *Hint:* Try iterating the recurrence, or use the formula for first-order linear recurrences, or guess the formula and prove by induction that you are right. Partial answer: The difference is that for this recurrence you have 3^n instead of 2^n.

5. *Hint:* Try iterating the recurrence, or use the formula for first-order linear recurrences, or guess the formula and prove by induction that you are right. Partial answer: The difference is that here the solution grows as a linear function of n rather than as an exponential function of n.

6. m^2; m^3; m^n

8. $M(n+1) = 2M(n) + 2000$;
$M(n) = 2^{n-1}M(1) + 2000(2^{n-1} - 1)$

10. $T(n) = \Theta(n)$

12. $T(n) = 2^{n+1} + 2^n(n^2(n+1)^2/4)$

14. $T(n) = (n+1)r^n$

16. $T(n) = s(r^{n+1} - s^{n+1})/r(s-r)$

Section 4.3 (pages 168–170)

2. $2n \log n + 2n$

3. $T(n) = \Theta(n^2)$

5. $T(n) = \Theta(n)$

7. $(5/2)n - 1/2$

9a. $T(n) = \Theta(n^3)$

9b. $T(n) = \Theta(n^3 \log n)$

9d. $T(n) = \Theta(\log n)$

10a. $T(n) = n((4n^2 - 1)/3)$

10b. $T(n) = n^3(\log_2 n + 1)$

10d. $T(n) = \log_4 n + 1$

12. *Hint:* Try substituting $b = 2^{\log b}$ into b^n; then see what you get if you take logs to the base b.

14a. $T(n) = O(n^2)$

14c. $T(n) = \Theta(\log \log n)$

14d. $T(n) = O(n \log^2 n)$

15a. Yes **15b.** Yes **15d.** Yes

17. $S(n) = \Theta(c^n)$

Section 4.4 (pages 185–186)

1a. $T(n) = \Theta(n^3)$

1b. $T(n) = \Theta(n^3 \log n)$

1d. $T(n) = \Theta(\log n)$

3. $T(n) = \Theta(n^{\log_2 3})$

5. $T(n) = \Theta(n \log n)$

7. *Hint:* Use the fact that $x = y^{\log_y x}$.

9. *Hint:* Prove by induction that if $1 \leq x < 2^n$, then $t(x)$ is uniquely determined.

Section 4.5 (pages 199–200)

1. *Hints:* You want to find two constants n_0 and k such that $T(n) \leq kn$ whenever $n > n_0$. It helps to write $kn/4 = kn - 3kn/4$. This should lead you to decide that you want $k \geq (4/3)c$.

3. *Hints:* You want to find $n_0 > 0$ and $k > 0$ so that $T(n) \leq kn \log_3 n$ for $n \geq n_0$. One thing to notice is that n_0 can't be 1, so it must be at least 3. If you replace the 2 with a 3, you get the same result with a bit more careful work. Changing the base for the logarithm doesn't change the big O bound.

5. No

7a. $T(n) = O(n^2)$

7b. *Hint:* You'll find you need a stronger inductive hypothesis than the natural choice. Try proving that $T(n) \leq k_1 n^2 - k_2 n \log n$.

8. Yes

11. Infinitely many answers are possible. Here is one for each question: $f(x) = 2x$; $f(x) = 2^x$.

13. $T(n) = O(n \log n)$

Section 4.6 (pages 211–211)

1. $T(n) = O(n)$

3. Partial answer: $T(n) = O(n)$, $S(n) = O(n)$; To compare the solutions, compare the recursion trees level by level.

7. $\Theta(n \log n)$

8. $T(n) = O(n^2)$

Section 5.1 (pages 223–224)

1. 5/16; 1/2
3. .72; n has to be 5; n still has to be 5.
5. You would get 3/11, which doesn't make sense.
7. *Hint:* The number of ways to get five heads in ten flips is $\binom{10}{5}$.
9. 33/16660, which is approximately .00198.
11. 7/128, which is about .0546875; 121/128, which is about .9453125.
13. No
15a. Drawing an ace and a king from the spades is more likely.
15b. Drawing an ace and a king from the spades is more likely.

Section 5.2 (pages 234–236)

1. 3/4
3. 11/36
5. 10/13
7. 50
9. $\sum_{i=0}^{n}(-1)^i\binom{n}{i}(n-i)! = \sum_{i=0}^{n}(-1)^i\frac{n!}{i!} = \sum_{i=2}^{n}(-1)^i\frac{n!}{i!}$; $\sum_{i=0}^{n}\frac{(-1)^i}{i!} = \sum_{i=2}^{n}\frac{(-1)^i}{i!}$
11. $\sum_{k=0}^{m}(-1)^k\binom{m}{k}\frac{(m-k)^n}{m!}$
13. $\sum_{i=0}^{n}(-1)^i\binom{n}{i}\binom{n+k-i(m+1)-1}{n-1}$
14. $\sum_{k=0}^{n}(-1)^k 2^k\binom{n}{k}\frac{(2n-k-1)!}{(2n-1)!}$
16. $\sum_{k=0}^{n}(-1)^k n!\binom{2n-k-1}{k}(n-k-1)!$
17. $\sum_{k=0}^{j}(-1)^k\binom{j}{k}\binom{n}{mk}(mk)!\frac{(n-mk+j-1)!}{(j-1)!}$

Section 5.3 (pages 248–249)

1. 1/2
3. Yes
5. Each pair of events is independent, but we wouldn't want to say they are mutually independent.
7. Partial answer: 1/5; 1/4
9. Each of the three probabilities is $20/120 = 1/6$.
11. If E and F are independent, one of the events must have probability 0.
13. Partial answer: You should switch.

Section 5.4 (pages 263–265)

2. First three questions: $p^3(1-p)^3$; last question: $\binom{6}{3}p^3(1-p)^3$
3. $\binom{10}{8}.5^{10} = .0439453125$; $\binom{10}{8}.5^{10} + \binom{10}{9}.5^{10} + \binom{10}{10}.5^{10} = .0546875$
5. \$3.50
7. 4
9. 6
11. Subtract the number of wrong answers from the number of right answers.
13. $\binom{10}{5}\binom{5}{3}p^3q^2r^5$; $\frac{n!}{i!j!k!}p^iq^jr^k$
14. *Hint:* There are (at least) four different solution techniques available: Induction, a "story" about choosing subsets, taking derivatives of both sides of a formula you know to give something related to the formula you want to know, and substituting the "quotient of factorials" formula into the left side of the formula you want and converting the result to the right side of the formula you want. We leave it to you to decide which of these strategies is most helpful. Several strategies might be equally helpful.
16. The expected amount of money on any of the three draws is 40/3 cents. Thus, the expected total amount of money you draw is 40 cents. No, it doesn't change.
18. We give one example to give you the idea; you should give another one. Roll one die. Let X be the number of dots on top, and let Y be the total number of dots on the other five sides. Then $E(X) = 7/2$, $E(Y) = 21 - 7/2$, and $E(XY) = 175/3$, while $E(X)E(Y) = 245/4$. There are simpler examples, but we wanted to leave them for you!
20. *Hint:* Take the derivative of $\sum_{j=1}^{\infty} x^j = 1/(1-x)$ and multiply both sides by x.
21. One possible answer is $X = (1/(1-p))^i$, where i is the number of the trial with the first success. Another is $X = i^i$.

Section 5.5 (pages 275–278)

1a. $1/d$ **1b.** $1/d$ **1c.** c/d **1d.** c/d **1e.** yes
3. d
5. At least six.

7. For $n = 2$, you get $1 \leq 2^2/2^2$, $2 \leq 2^2/1$, $1 \leq 2^2/2^2$. For $n = 3$, you get $1 \leq 3^3/3^3$, $3 \leq 3^3/2^2$, $3 \leq 3^3/2^2$, $1 \leq 3^3/3^3$.

9. *Hint:* If X_i is the number of occupied locations and Y_i is the number of empty ones, then $X_i + Y_i = k$.

11a. The expected time for unsuccessful search is $1 + n/k$.

11b. The expected running time for a successful search for is $1 + n/2k - 1/2k$.

13. *Hint:* $\lim\limits_{n \to -\infty} (1 + 1/n)^n = \lim\limits_{h \to 0}(1 + h)^{1/h} = e$. Also, $\lim\limits_{n \to -\infty} (1 + 1/n)^n = \lim\limits_{h \to 0^-} (1 + h)^{1/h} = e$.

15. *Hint:* Try substituting $\log n / \log \log n$ for x into the equation $x^x = n$ to see how close it comes to being a solution. Then experiment with multiples of $\log n / \log \log n$. This should help you figure out upper and lower bounds on solutions.

18. *Hints:* Draw rectangles of width 1 above and below the curve, starting at $x = 1$ and going to the right to $x = n+1$. You can show that the area above the lower rectangles and included in the upper ones is $1 - 1/(n + 1)$. Convert the upper rectangles to trapezoids above the curve, and you will reduce the difference in areas by a factor of one half. Now use the fact that the integral of $1/x$ is $\ln x$ to approximate the harmonic numbers

Section 5.6 (pages 292–294)

1. $X_1 + X_2 + \cdots + X_n$ is the number of times you assign a value to L. The expected number of times you assign a value is H_n, the nth harmonic number.

3. *Hint:* It should be clear why the sum is $O(n \log n)$. Think what you can say about the largest half of the terms of the sum.

5. By the master theorem.

7. *Hint:* Try analyzing a recursion tree. You could also try induction.

11. $1/2$; $11/16$; while your upper bound will be smaller, it is not clear that the potential time savings is worth the extra complexity.

13. *Hint:* The first key has probability $1/2$ of being in the middle half of the sorted list of keys.

15. Final answer: $T(n) = O(\log n)$

Section 5.7 (pages 305–307)

1. 0; 1.2; 1.2

3. 30 cents; 400 (assumes expected value is in cents); 60 cents; 0 (We leave the final question for you to answer.)

5. Approximately .95.

7. *Hint:* Use Theorem 5.28.

9. Expected amount of money on each draw is $40/3$ cents. For each draw, the variance is $72\frac{2}{9}$. For the sum of the two draws, the expected amount of money is $80/3$ cents, and the variance is $1853/27 \approx 68.63$.

11. $35/12$; $\sqrt{35/12}$; $35n/12$; $\sqrt{35n/12}$

13. Inconsistent; consistent

15. Partial answer: $1/4$

17. *Hints:* The variance for the number of successes is no more than $n/4$. You might also use Problem 16.

19. 80%; multiply the number of questions by $\sqrt{.18}/.4$, or about 1.125.

Section 6.1 (pages 321–322)

1. 1, 11, 7, 5

3. Vertex 2 has degree 7.

5. The cycles with vertex sets $\{9, 15\}$ and $\{10, 11, 12, 13, 14\}$ but not the cycles with vertex set $\{1\}$.

7. One example is $\{2, 9, 11, 12\}$. There are others. All the largest induced K_n's have size 4.

9. *Hint:* What do you know about the number of vertices and edges in each connected component?

11.

13. No

15. G is a tree.

17.

19. 2, 3, 4, 5, 7, 11, 2

Section 6.2 (pages 334–335)

1. $\{e_1, e_2, e_3\}$, $\{e_1, e_2, e_5\}$, $\{e_1, e_3, e_4\}$, $\{e_1, e_3, e_5\}$, $\{e_1, e_4, e_5\}$, $\{e_2, e_3, e_4\}$, $\{e_2, e_3, e_5\}$, $\{e_2, e_4, e_5\}$

3. We give the edge sets rather than drawing them. The root vertex is always vertex 1. 1. $\{1, 2\}$, $\{2, 3\}$, $\{3, 4\}$, $\{4, 5\}$; 2. $\{1, 2\}$, $\{2, 3\}$, $\{3, 4\}$, $\{3, 5\}$; 3. $\{1, 2\}$, $\{2, 3\}$, $\{2, 4\}$, $\{3, 5\}$; 4. $\{1, 2\}$, $\{2, 3\}$, $\{2, 4\}$, $\{2, 5\}$; 5. $\{1, 2\}$, $\{1, 5\}$, $\{2, 3\}$, $\{3, 4\}$; 6. $\{1, 2\}$, $\{1, 5\}$, $\{2, 3\}$, $\{2, 4\}$; 7. $\{1, 2\}$, $\{1, 3\}$, $\{1, 4\}$, $\{2, 5\}$; 8. $\{1, 2\}$, $\{1, 3\}$, $\{1, 4\}$, $\{1, 5\}$

5. Many examples are possible. We give the edge set of one: $\{\{1, 2\}, \{1, 3\}, \{3, 4\}, \{1, 5\}, \{5, 6\}, \{6, 7\}\}$.

7.

9. $d = \lfloor \log_2(n) \rfloor$.

11. *Hint:* What is the best way to start an inductive proof of a statement about binary trees?

13. *Hint:* What is the best way to use induction to prove a statement about rooted trees?

Section 6.3 (pages 351–353)

1a. 1, 2, 3, 4, 5, 1, 4, 2, 5, 3, 1
1b. No Eulerian circuit.
1c. No Eulerian circuit.
1d. 1, 5, 2, 6, 3, 7, 4, 8, 1, 6, 4, 5, 3, 8, 2, 7, 1
3. 2
5. For odd n.
7. $n > 1$
9a. If m and n are nonzero and both even or if one is 0 and the other is 1, the graph is Eulerian.
9b. If m and n are greater than 1 and are equal, the graph is Hamiltonian.
11. No such circumstances.
13. *Hint:* Look carefully at the proof of Dirac's theorem.
15. *Hint:* You have to prove two implications, and one is easy. Would it help to prove that if G_i is Hamiltonian, then G_{i-1} is Hamiltonian?

Section 6.4 (pages 369–370)

1. Possible answer: $\{\{a, 5\}, \{b, 2\}, \{c, 3\}, \{d, 6\}, \{e, 4\}\}$

3. $S = \{a, c, d, f\}$; $N(S) = \{2, 3, 4\}$
5. Partial answer: $N(S)$ was a subset of the minimum vertex cover.
7. Yes
9. No; no
11. Possible answer: The complete graph K_7.
13. Partial answer: True.
15. The sum is v.

Section 6.5 (pages 384–385)

1. 2
3. Partial answer: Yes
5. 3; 3
7. *Hint:* This is a situation where greed is good!
9. 3
11. $t(t - 1)^{n-1}$
13. *Hint:* If a graph has no triangles, what can you say about how the number of edge-face pairs compares with the number of faces?
15. *Hint:* If all vertices have degree 4 or more, how does the sum of the degrees of the vertices relate to the number of vertices? Is this consistent with Problem 13?
17. 5; no
19. 7, 9, and 8, respectively; In fact, 7 is the chromatic number.

Appendix A (pages 396–397)

1. $\{(-6, -6), (-6, -2), (-6, 2), (-5, -5), (-5, -1), (-5, 3), (-4, -4), (-4, 0), (-4, 4), (-3, -3), (-3, 1), (-2, -2), (-2, 2), (-2, -6), (-1, -1), (-1, 3), (-1, -5), (0, 0), (0, -4), (0, 4), (1, 1), (1, -3), (2, 2), (2, -2), (2, -6), (3, 3), (3, -1), (3, -5), (4, 4), (4, 0), (4, -4)\}$
2a. No
2b. Yes
2c. No
4. Partial answer: Two lists are equivalent if they represent seating people around the table in the same order.
7. The first rule defines a function. The second one doesn't.
9. $([x]_n \cdot [y]_n) \cdot [z]_n = [(x \cdot y) \cdot z]_n = [x \cdot (y \cdot z)]_n = [x]_n \cdot ([y]_n \cdot [z]_n)$

Bibliography

[1] Manindra Agrawal, Neeraj Kayal, and Nitin Saxena. PRIMES is in P. http://www.cse.iitk.ac.in/news/primality.html, 2002. For updated information, see http://crypto.cs.mcgill.ca/~stiglic/PRIMES_P_FAQ.html.

[2] W. R. Alford, A. Granville, and C. Pomerance. There are infinitely many Carmichael numbers. *Ann. of Math,* 140: 703–722, 1994.

[3] Kenneth Appel and Wolfgang Haken. Every planar map is four colorable. *Bull. Amer. Math. Soc.,* 82: 711–712, 1976.

[4] Jon L. Bentley, Dorthea Haken, and James B. Saxe. A general method for solving divide-and-conquer recurrences. *SIGACT News,* 12(3): 36–44, 1980.

[5] Claude Berge. Two theorems in graph theory. *Proceedings of the National Academy of Sciences, USA,* 43: 842–844, 1957.

[6] Claude Berge. *Graphs and Hypergraphs.* 1st ed. Amsterdam: North Holland, 1973.

[7] Norman L. Biggs, E. Keith Lloyd, and Robin J. Wilson. *Graph Theory 1736–1936.* 1st ed. Oxford: Clarendon Press, 1976.

[8] Manuel Blum, Robert W. Floyd, Vaughan Pratt, Ronald Rivest, and Robert E. Tarjan. Time bounds for selection. *Journal of Computer and System Sciences,* 7(4): 448–461, 1973.

[9] Kenneth P. Bogart. *Discrete Mathematics.* 1st ed. Boston: Houghton Mifflin, 1988.

[10] Kenneth P. Bogart. *Introductory Combinatorics.* 3rd ed. Boston: Harcourt-Academic Press, 2000.

[11] Alan Cobham. The intrinsic computational difficulty of functions. In *Proceedings of the 1964 Congress for Logic, Methodology, and the Philosophy of Science,* 24–30. Amsterdam: North Holland, 1964.

[12] Stephen Cook. The complexity of theorem proving procedures. In *Proceedings of the Third Annnual ACM Symposium on Theory of Computing,* Association for Computing Machinery, 151–158, 1971.

[13] Thomas H. Cormen, Charles E. Leiserson, Ronald L. Rivest, and Clifford Stein. *Introduction to Algorithms.* 2nd ed. Cambridge, MA: McGraw-Hill, 2001.

[14] Richard Crandall and Carl Pomerance. *Prime Numbers: A Computational Perspective.* 2nd ed. New York: Springer-Verlag, 2005.

[15] Whitfield Diffie and Martin Hellman. New directions in cryptography. *IEEE Transactions on Information Theory,* IT-22(6): 644–654, 1976.

[16] Jack Edmonds. Paths, trees, and flowers. *Canadian Journal of Mathematics,* 17: 449–467, 1965.

[17] Michael R. Garey and David S. Johnson. *Computers and Intractability: A Guide to the Theory of NP-Completeness.* 1st ed. New York: W. H. Freeman, 1979.

[18] Martin C. Golumbic. *Algorithmic Graph Theory and the Perfect Graph Conjecture.* 2nd ed. Amsterdam: Elsevier, 2004. First published 1980 by Academic Press.

[19] Jonathan L. Gross and Jay Yellen, eds. *Handbook of Graph Theory.* Vol. 25, *Discrete Mathematics and Its Applications.* 1st ed. Boca Raton, FL: CRC Press, 2003.

[20] C.A.R. Hoare. Algorithm 63 (PARTITION) and algorithm 65 (FIND). *Communications of the ACM,* 4(7): 321–322, 1961.

[21] Richard M. Karp. *Reducibility among Combinatorial Problems,* 85–103. New York: Plenum Press, 1972.

[22] Richard M. Karp. An introduction to randomized algorithms. *Discrete Applied Mathematics,* 34: 165–201, 1991.

[23] John G. Kemeny, J. Laurie Snell, and Gerald L. Thompson. *Finite Mathematics.* 3rd ed. Englewood Cliffs, NJ: Prentice-Hall, 1974.

[24] Donald E. Knuth. Big omicron and big omega and big theta. *ACM SIGACT News,* 8(2): 18–23, 1976.

[25] L. A. Levin. Universal sorting problems. *Problemy Peredachi Informatsii,* 9(3): 265–266, 1973.

[26] G. L. Miller. Riemann's hypothesis and tests for primality. *Journal of Computer and Systems Science,* 13: 300–317, 1976.

[27] Michael O. Rabin. Probabilistic algorithm for testing primality. *Journal of Number Theory,* 12(1): 128–138, 1980.

[28] R. L. Rivest, A. Shamir, and L. Adleman. A method for obtaining digital signatures and public-key cryptosystems. *CACM,* 21: 120–126, February 1978.

[29] Neil Robertson, Daniel P. Sanders, Paul D. Seymour, and Robin Thomas. The four color theorem. *Journal of Combinatorial Theory,* 70: 2–44, 1997.

[30] Kenneth Rosen. *Discrete Mathematics and Its Applications.* 4th ed. New York: McGraw-Hill, 1999.

[31] Kenneth Rosen. *Elementary Number Theory and Its Applications.* 4th ed. Boston: Addison-Wesley, 2000.

[32] Robin Thomas. An update on the four color theorem. *Notices of the American Mathematical Society,* 45(7): 848–859, 1998.

[33] Douglas B. West. *Introduction to Graph Theory.* 2nd ed. Upper Saddle River, NJ: Prentice Hall, 2001.

[34] Robin J. Wilson. *Four Colors Suffice: How the Map Problem Was Solved.* 1st ed. Princeton, NJ: Princeton University Press, 2002.

Index

inference, 117–126
  contradiction 121–124
  direct, 117–119
  indirect, 120–124
  rules of, 119–122, 124–125
initial condition for a recurrence, 144, 154
injection, 11, 17
insertion sort, 279–281, 291
instance of a problem, 349n
integers
  mod n, 44–53, 54
  summing consecutive, 3
  Z as notation for, 106
internal vertex, 330, 333
intersection graphs, 371–372, 383
interval graphs, 373–375, 383, 384
  coloring of, 374–375
interval representation, 373, 383
Introduction to Algorithms
      (Cormen et al.), 174
inverse function, 19
inverses, multiplicative mod n, 56–59, 68
  computing, 68
  every nonzero element has inverse if n is
      prime, 68, 69
  Fermat's Little Theorem and, 87, 89
  greatest common divisors and, 61–62, 68, 69
  necessary and sufficient condition, 61, 69
  solutions to equations and, 57, 69
  uniqueness of, 59, 68
isolated vertices, 315
iterating a recurrence, 146–147, 155, 162

K

k-cycle problem, 348–349
k-element multisets, 36, 38
k-element permutations, 13–14, 17
  falling factorials and, 14, 17
  k-element subsets and ,14–16, 17
k-element subsets, 14–16, 17
  binomial coefficients and, 16, 17,
      19–22, 28–30
keys (cryptographic)
  private, 44, 53
  public, 47, 53
  secret, 47, 53
keys (as data identifiers), 214–216
Knuth, Donald E., 14
König-Egerváry theorem, 366, 369
Königsberg Bridge problem, 336
kth falling factorial power of n, 14, 17

L

labeling, 24–25, 26, 29–30, 31
large numbers, law of, 306–307

laws
  associative, 50, 53
  Chebyshev's, 306
  commutative, 50, 53, 395
  DeMorgan's, 96–97, 101
      and negation of quantified statements, 112
  distributive, 50, 53, 93, 96, 101
  large numbers, of, 306–307
  transitive, 30n, 40, 82, 119, 302, 396
leaf vertex, 330, 333
length
  of a path, 312, 321
  of a walk, 312
"less than" relation, 389
Levin, Leonid, 348
lexicographic ordering, 14, 95n
linearity of expectation, 257, 259,
      263, 297
linear-time algorithms, 202n, 210
lists, 10, 16
  adjacency, 326
  functions and, 10–12
  lexicographic order of, 14
  ordered, binary search in, 157–158
  product principle and counting, 10, 16
  sets vs., 16
  selection of the ith largest element of a list,
      200–211, 284–286, 291
  sorting
      by mergesort, 158, 167, 186–187
      by insertion sort, 7, 279–281, 291
      by quicksort, 286–289, 291
      by selection sort, 1
logarithms
  bases for, 168
      standard notation for base 2, 161n
      relative growth rates, 164
  fundamental facts about, 167, 168, 172, 183
  properties of, 168, 183
logic, 91–126
  equivalence and implication in, 91–103
      conditional and contrapositive,
          120–121, 125
      DeMorgan's laws and, 96–97, 101
      equivalent statements, 91–93, 96, 101
      implication, 97–98, 99, 100, 102, 118–119
      quantified statements, of, 105, 108,
          110–111, 114–115
      truth tables and, 93–97, 99
  inference and proof in, 117–126
      contradiction, 121–124
      direct, 117–119
      indirect, 120–124
      rules of, 119–122, 124–125
  variables and quantifiers in, 104–117
logical connectives, 94, 99, 101

probability, hashing and (*continued*)

  independent trials process and,
      241–242, 248

  keys all hashing to different
      locations, 217–219

  independence, 238, 239–240, 247

    independent events, 238, 247

    independent random variables,
        299–301, 304

    product principle for independent
        probabilities, 239

    symmetry of, 240, 247

  independent trials process, 241–242, 248

    Bernoulli trials, 250–251, 263

  primality testing and, 246–247

    Fermat's Little Theorem and, 86–87

    Miller-Rabin algorithm for, 87–88, 89

    probability and, 246–247

    randomized algorithms and, 88

  principle of inclusion and exclusion for,
      227–232, 234

  random variables, 249–265, 294–307

  sample space, 214–215

  tree diagrams illustrating, 242–246, 248

  uniform distribution, 219–221, 222

  union of events and, 224–226, 233

probability measure, 215, 222

probability weight, 215

problem

  NP-complete, 85, 347–350,
      351, 373

  graph decision, 346, 350

  instance of, 349n

problems (named)

  bookcase, 35–36

  clique, 349

  coupon-collector, 269–270

  derangement, 230, 234

  factoring, 85

  Hamiltonian cycle, 348

  Hamiltonian path, 347

  hatcheck, 230, 234

  independent set, 349

  k-cycle, 348–349

  Königsberg Bridge, 336

  ménage problem, 236

  Monty Hall problem, 249

  register assignment, 373

  satisfiability, 348

  selection, 200

  Tower of Hanoi, 143–144

product notation, 10, 16

product principle, 4–6, 7, 9–10, 16

  for independent probabilities, 239–240, 247

proof of quantified statements, 109, 110,
    113, 114

proof techniques with names, examples

  conditional, 59, 63, 118, 318–319

  indirect. *See* contradiction and contraposition

  by contradiction, 58, 62–63, 87,
      123–124, 344

  by contraposition, 120–121, 123, 318,

  direct. *See* conditional, sum principle,
      product principle, induction, universal
      generalization

  by induction, 144, 148, 151, 189–193, 313,
      319, 323, 331–332, 338–339

  by product principle, 5, 9–10, 14

  by smallest counterexample, 62–63

  by sum principle, 9, 21–23

  universal generalization, 118, 323

proper coloring, 371–372, 373, 374, 383

  chromatic number and, 372–373, 374

  five color theorem and, 380–383

  four color theorem and, 376, 383

  greedy algorithm for, 375, 384

  interval graphs and, 373–375

  NP-completeness and, 373

  planarity and, 375–376

pseudocode, xv

pseudoprimes, 88

  Miller-Rabin algorithm for, 87–88, 89

  probability and, 246–247

  randomized algorithms and, 88

public-key cryptography/cryptosystems,
    46–48, 53

  private key (secret key), 76

  public key, 47, 76

  secret key, 47

Q

quantified statements, 105–114

  equivalence of, 108, 110–111

  meaning of, 109–110

  negation of, 110–112, 115

    DeMorgan's laws and, 112

    "switching quantifiers and pushing
        negation inside," 111–112

  proof of, 109–110, 113–114, 115, 118–119

    universal generalization, 118

  rewriting for larger universes, 108–109,
      114–115

  standard notation for, 106–107

  truth of, 109–110, 115

  variables and, 107–108, 114

quantifiers, 105–106, 114

  existential, 105, 114

  implicit quantification and, 112–113

  short notation for, 106–107

  universal, 105, 114

queue, 325, 332